T0233735

Lecture Notes in Bioinformatics 9271

Subseries of Lecture Notes in Computer Science

LNBI Series Editors

Sorin Istrail
Brown University, Providence, RI, USA
Pavel Pevzner
University of California, San Diego, CA, USA
Michael Waterman
University of Southern California, Los Angeles, CA, USA

LNBI Editorial Board

Alberto Apostolico
Georgia Institute of Technology, Atlanta, GA, USA
Søren Brunak
Technical University of Denmark Kongens Lyngby, Denmark
Mikhail S. Gelfand
IITP, Research and Training Center on Bioinformatics, Moscow, Russia
Thomas Lengauer
Max Planck Institute for Informatics, Saarbrücken, Germany
Satoru Miyano
University of Tokyo, Japan
Eugene Myers
Max Planck Institute of Molecular Cell Biology and Genetics Dresden, Germany
Marie-France Sagot
Université Lyon 1, Villeurbanne, France
David Sankoff
University of Ottawa, Canada
Ron Shamir
Tel Aviv University, Ramat Aviv, Tel Aviv, Israel
Terry Speed
Walter and Eliza Hall Institute of Medical Research Melbourne, VIC, Australia
Martin Vingron
Max Planck Institute for Molecular Genetics, Berlin, Germany
W. Eric Wong
University of Texas at Dallas, Richardson, TX, USA

More information about this series at http://www.springer.com/series/5381

Alessandro Abate · David Šafránek (Eds.)

Hybrid Systems Biology

Fourth International Workshop, HSB 2015
Madrid, Spain, September 4–5, 2015
Revised Selected Papers

 Springer

Editors
Alessandro Abate
Department of Computer Science
University of Oxford
Oxford
UK

David Šafránek
Faculty of Informatics
Masaryk University
Brno
Czech Republic

ISSN 0302-9743   ISSN 1611-3349   (electronic)
Lecture Notes in Computer Science
ISBN 978-3-319-26915-3   ISBN 978-3-319-26916-0   (eBook)
DOI 10.1007/978-3-319-26916-0

Library of Congress Control Number: 2015958329

© Springer International Publishing Switzerland 2015
This work is subject to copyright. All rights are reserved by the Publisher, whether the whole or part of the material is concerned, specifically the rights of translation, reprinting, reuse of illustrations, recitation, broadcasting, reproduction on microfilms or in any other physical way, and transmission or information storage and retrieval, electronic adaptation, computer software, or by similar or dissimilar methodology now known or hereafter developed.
The use of general descriptive names, registered names, trademarks, service marks, etc. in this publication does not imply, even in the absence of a specific statement, that such names are exempt from the relevant protective laws and regulations and therefore free for general use.
The publisher, the authors and the editors are safe to assume that the advice and information in this book are believed to be true and accurate at the date of publication. Neither the publisher nor the authors or the editors give a warranty, express or implied, with respect to the material contained herein or for any errors or omissions that may have been made.

Printed on acid-free paper

This Springer imprint is published by SpringerNature
The registered company is Springer International Publishing AG Switzerland

Preface

The 4th International Workshop on Hybrid Systems Biology (HSB 2015) was held during September 4–5, 2015, at the Facultad de Matemáticas, Universidad Complutense de Madrid.

HSB 2015 was co-located with the week-long Madrid Meet 2015, which also hosted CONCUR 2015, QEST 2015, and FORMATS 2015, among other scientific events. Previous editions of the HSB Workshops were held in Newcastle upon Tyne (UK, with CONCUR 2012), Taormina (IT, within ECAL 2013), and Vienna (AT, within VSL 2014).

The scope of the HSB 2015 workshop has broadened since the earlier editions, and now covers the general area of dynamical models in biology. HSB 2015 retained the emphasis on *hybrid* approaches – by no means restricted to a narrow class of mathematical models – and in particular stressed the importance of taking advantage of techniques developed separately in different areas. Topics featured at the workshop included models of metabolic, signaling, and genetic regulatory networks; models of tissues; biological applications of quantitative and formal analysis techniques; parametric and non-parametric system identification techniques; efficient techniques for combined and heterogeneous simulations of biological models; modeling languages for biological systems; models coping with incomplete and uncertain biological information; stochastic and hybrid models in biology; hierarchical approaches for multi-scale, multi-domain analysis; abstraction, approximation, discretization, and model reduction techniques; control architectures in biological systems; and modeling and synthesis for synthetic biology.

HSB 2015 was a packed two-day event, featuring invited talks, single-track regular podium sessions, and an interactive session with posters and software tool demos. We hosted about 40 registered participants, plus two invited speakers, and a constant inflow of attendees from other co-located events at Madrid Meet 2015. The 30-minute contributed talks were of high quality and the participation lively, interactive, and stimulating.

In all, 46 Program Committee (PC) members helped to provide at least four reviews (with in some cases up to six) of the submitted contributions, out of which 13 high-quality articles were accepted to be presented during the single-track sessions, and appear (possibly after further feedback from a shepherding process by the PC members) as full papers in these proceedings. The articles were bundled in four thematic sessions, which is reflected in the organization of these proceedings: statistical analysis; analysis and verification of continuous and hybrid models; quantitative analysis of biological models; and application of advanced models on case studies. In the afternoon of the first day we also hosted a poster/demo session, with 11 presentations, of which five with interactive tool demonstrations.

A highlight of HSB 2015 was the presence of two high-profile invited speakers, whom we selected also in view of the breadth of the event: computer sciences and control and dynamical systems; theoretical work and laboratory experiments. Luca Cardelli, principal researcher at Microsoft Research Cambridge (UK) and Royal Society research professor at the Department of Computer Science, University of Oxford (UK), gave a seminar titled "Morphisms of Reaction Networks." Mustafa Khammash, professor of Control Theory and Systems Biology at the Department of Biosystems Science and Engineering at ETH Zürich (CH), gave a talk titled "Cyber-genetics: Synthetic Circuits and Systems for the Precise Control of Living Cells."

Further details on HSB 2015 are featured on the website: http://hsb2015.fi.muni.cz

Finally, a few words of acknowledgment are due. Thanks to David de Frutos Escrig and to Fernando Rosa Velardo (UCM, Spain) for the supportive and can-do attitude, as well as the local seamless organization of the Madrid Meet 2015. Thanks to Springer for hosting the HSB proceedings in its *Lecture Notes in Bioinformatics, a sub-series of Lecture Notes in Computer Science.* Thanks to Luca Bortolussi and Ezio Bartocci from the Steering Committee of HSB for support and encouragement, to all the PC members and additional reviewers for their work in ensuring the quality of the contributions to HSB 2015, and to all the participants for contributing to this memorable event.

September 2015   Alessandro Abate
                                                                         David Šafránek

Organization

Program Committee Chairs

Alessandro Abate	University of Oxford, UK
David Šafránek	Masaryk University, Czech Republic

Program Committee

Alessandro Abate	University of Oxford, UK
Ezio Bartocci	Vienna University of Technology, Austria
Gregory Batt	Inria Paris-Rocquencourt, France
Sergiy Bogomolov	IST Austria, Austria
Luca Bortolussi	University of Trieste, Italy
Kevin Burrage	University of Oxford, UK
Luca Cardelli	Microsoft Research, Cambridge, UK
Pieter Collins	Maastricht University, The Netherlands
Milan Češka	University of Oxford, UK
Neil Dalchau	Microsoft Research, Cambridge, UK
Thao Dang	VERIMAG/CNRS, Grenoble, France
Hidde De Jong	Inria Grenoble - Rhône-Alpes, France
Alexandre Donzé	University of California, Berkeley, USA
François Fages	Inria Rocquencourt, France
Eric Fanchon	TIMC-IMAG Laboratory, Grenoble, France
Giancarlo Ferrari-Trecate	University of Pavia, Italy
Elisa Franco	University of California at Riverside/Caltech, USA
Sicun Gao	MIT CSAIL, USA
Radu Grosu	Vienna University of Technology, Austria
Joao Hespanha	University of California, Santa Barbara, USA
Jane Hillston	University of Edinburgh, UK
Agung Julius	Rensselaer Polytechnic Institute, USA
Heinz Koeppl	Technical University Darmstadt, Germany
Hillel Kugler	Microsoft Research, Cambridge, UK
Marta Kwiatkowska	University of Oxford, UK
Pietro Liò	University of Cambridge, UK
Oded Maler	VERIMAG/CNRS, Grenoble, France
Andrzej Mizera	University of Luxembourg, Luxembourg
Stefan Müller	Austrian Academy of Sciences, Vienna, Austria
Chris Myers	University of Utah, USA
Nicola Paoletti	University of Oxford, UK
Ion Petre	Åbo Akademi University, Finland

Tatjana Petrov	IST Austria, Austria
Carla Piazza	University of Udine, Italy
Nir Piterman	University of Leicester, UK
Alberto Policriti	University of Udine, Italy
David Šafránek	Masaryk University, Czech Republic
Guido Sanguinetti	University of Edinburgh, UK
Abhyudai Singh	University of Delaware, USA
Kateřina Staňková	Maastricht University, The Netherlands
P.S. Thiagarajan	National University of Singapore, Singapore
Jana Tůmová	Royal Institute of Technology, Sweden
Aljoscha Wahl	Delft University of Technology, The Netherlands
Verena Wolf	Saarland University, Germany
Boyan Yordanov	Microsoft Research, Cambridge, UK
Paolo Zuliani	Newcastle University, UK

Additional Reviewers

Adimoolam, Arvind	Kyriakopoulos,	Shmarov, Fedor
Daca, Przemyslaw	Charalampos	Traynard, Pauline
Islam, Md. Ariful	Milios, Dimitrios	Wang, Qinsi
Krüger, Thilo	Panchal, Charmi	Woodhouse, Steven
	Rogojin, Vladimir	

Keynote Abstracts

Cybergenetics: Synthetic Circuits and Systems for the Precise Control of Living Cells

Mustafa Khammash

Department of Biosystems Science and Engineering, ETH Zürich

In his 1948 book, Cybernetics, Norbert Wiener put forth a vision in which the study of control and communication in the animal and the machine are unified. The field of Cybernetics-translated from the Greek as "the art of steering" – was thus born. Predating the discovery of the structure of DNA and the ensuing molecular biology revolution, cybernetic applications in the life sciences at the time were limited. More than 65 years later, the convergence of powerful genetic manipulation techniques, novel measurement technologies for measuring cellular constituents, and actuation tools for affecting cellular events is enabling a new area of research in which control theoretic ideas are used for regulating cellular processes at the gene level. To highlight the genetic aspects of this research and in keeping with Wiener's vision, we refer to this area as Cybergenetics.

This presentation describes several cybergenetic applications in which the analysis, design, and implementation of control systems in living cells is achieved. Using computer control coupled with optogenetic and flow cytometry technologies, we show that automated feedback loops can achieve precise and extremely robust control of gene expression in living cells. We then demonstrate that biomolecular controllers can be realized entirely inside living cells and then used to achieve autonomous regulation of gene expression.

Finally, we present a new control theory for the integral control of gene expression in a stochastic setting. We show that such stochastic integral control utilizes just a few molecules to achieve robust steady-state tracking and perfect adaptation and, remarkably, that it leads to closed-loop systems that are more robust than their deterministic counterparts.

Morphisms of Reaction Networks

Luca Cardelli[1,2]

[1] Microsoft Research, Cambridge
[2] Department of Computer Science, University of Oxford

The mechanisms underlying complex biological systems are routinely represented as networks. Network kinetics is widely studied, and so is the connection between network structure and behavior. But it is the relationships between network structures that can reveal similarity of mechanism.

We define morphisms (mappings) between reaction networks that establish structural connections between them. Some morphisms imply kinetic similarity, and yet their properties can be checked statically on the structure of the networks. In particular we can determine statically that a complex network will emulate a simpler network: it will reproduce its kinetics for all corresponding choices of reaction rates and initial conditions. We use this property to relate the kinetics of many common biological networks of different sizes, also relating them to a fundamental population algorithm. Thus, structural similarity between reaction networks can be revealed by network morphisms, elucidating mechanistic and functional aspects of complex networks in terms of simpler networks.

In recent joint work, we established a correspondence between network emulation and a notion of backward bisimulation for continuous systems. An emulation morphism establishes a bisimulation relation over the union of two networks, and a bisimulation relation over a network can be seen as an emulation morphism from the full network to the reduced network of its equivalence classes. Along this correspondence, we obtain minimization algorithms for chemical reaction networks, which are of interest for model execution, and algorithms to discover morphisms between networks, which are of interest for model understanding.

References

1. Cardelli, L.: Morphisms of reaction networks that couple structure to function. BMC Syst. Biol. **8**(1), 84 (2014)
2. Cardelli, L., Tribastone, M., Tschaikowski, M., Vandin, A.: Forward and backward bisimulations for chemical reaction networks. In: Aceto, L., de Frutos Escrig, D. (eds.) 26th International Conference on Concurrency Theory (CONCUR 2015), volume 42 of Leibniz International Proceedings in Informatics (LIPIcs), pp. 226–239, Dagstuhl, Germany (2015). Schloss Dagstuhl–Leibniz-Zentrum fuer Informatik

Contents

Statistical Analysis

Reconstructing Statistics of Promoter Switching from Reporter Protein
Population Snapshot Data.   3
  Eugenio Cinquemani

Comparative Statistical Analysis of Qualitative Parametrization Sets   20
  Adam Streck, Kirsten Thobe, and Heike Siebert

Analysis and Verification of Continuous and Hybrid Models

Parallelized Parameter Estimation of Biological Pathway Models   37
  R. Ramanathan, Yan Zhang, Jun Zhou, Benjamin M. Gyori,
  Weng-Fai Wong, and P.S. Thiagarajan

High-Performance Discrete Bifurcation Analysis for Piecewise-Affine
Dynamical Systems.   58
  Luboš Brim, Martin Demko, Samuel Pastva, and David Šafránek

Integrating Time-Series Data in Large-Scale Discrete Cell-Based Models. . . .   75
  Louis Fippo Fitime, Christian Schuster, Peter Angel, Olivier Roux,
  and Carito Guziolowski

Approximate Probabilistic Verification of Hybrid Systems   96
  Benjamin M. Gyori, Bing Liu, Soumya Paul, R. Ramanathan,
  and P.S. Thiagarajan

Quantitative Analysis of Biological Models

Synthesising Robust and Optimal Parameters for Cardiac Pacemakers
Using Symbolic and Evolutionary Computation Techniques   119
  Marta Kwiatkowska, Alexandru Mereacre, Nicola Paoletti,
  and Andrea Patanè

Model-Based Whole-Genome Analysis of DNA Methylation Fidelity   141
  Christoph Bock, Luca Bortolussi, Thilo Krüger, Linar Mikeev,
  and Verena Wolf

Studying Emergent Behaviours in Morphogenesis Using Signal
Spatio-Temporal Logic   156
  Ezio Bartocci, Luca Bortolussi, Dimitrios Milios, Laura Nenzi,
  and Guido Sanguinetti

Efficient Reduction of Kappa Models by Static Inspection of the Rule-Set . . .   173
   Andreea Beica, Calin C. Guet, and Tatjana Petrov

Application of Advanced Models on Case Studies

Model Checking Tap Withdrawal in C. Elegans 195
   Md. Ariful Islam, Richard De Francisco, Chuchu Fan, Radu Grosu,
   Sayan Mitra, and Scott A. Smolka

Solving General Auxin Transport Models with a Numerical Continuation
Toolbox in Python: PyNCT 211
   Delphine Draelants, Przemysław Kłosiewicz, Jan Broeckhove,
   and Wim Vanroose

Analysis of Cellular Proliferation and Survival Signaling by Using Two
Ligand/Receptor Systems Modeled by Pathway Logic 226
   Gustavo Santos-García, Carolyn Talcott, and Javier De Las Rivas

Posters and Tool Demos. 247

Author Index 249

Statistical Analysis

Reconstructing Statistics of Promoter Switching from Reporter Protein Population Snapshot Data

Eugenio Cinquemani[✉]

INRIA Grenoble – Rhône-Alpes, 655 Avenue de L'Europe, Montbonnot,
38334 Saint-Ismier CEDEX, France
eugenio.cinquemani@inria.fr

Abstract. The use of fluorescent reporter proteins is an established experimental approach for dynamic quantification of gene expression over time. Yet, the observed fluorescence levels are only indirect measurements of the relevant promoter activity. At the level of population averages, reconstruction of mean activity profiles from mean fluorescence profiles has been addressed with satisfactory results. At the single cell level, however, promoter activity is generally different from cell to cell. Making sense of this variability is at the core of single-cell modelling, but complicates the reconstruction task. Here we discuss reconstruction of promoter activity statistics from time-lapse population snapshots of fluorescent reporter statistics, as obtained e.g. by flow-cytometric measurements of a dynamical gene expression experiment. After discussing the problem in the framework of stochastic modelling, we provide an estimation method based on convex optimization. We then instantiate it in the fundamental case of a single promoter switch, reflecting a typical random promoter activation or deactivation, and discuss estimation results from *in silico* experiments.

Keywords: Identification · Gene regulatory networks · Doubly stochastic process

1 Introduction

Gene expression dynamics and regulatory interactions have been the object of intensive study in the last decades. Mathematical modelling of gene expression dynamics profits from a variety of experimental monitoring techniques, allowing one to quantify the activity of one or several genes of interest over time. Among these, fluorescent or luminescent reporter protein techniques have proven to be an extremely valuable approach [11]. The principle of these techniques is to place the coding sequence of light-emitting reporter proteins under the control of the promoter of the gene of interest, so that, when the gene is expressed, new reporter molecules are synthesized and can be quantified via light detection techniques.

In a deterministic setting, reporter proteins have been used with success for the inference of regulatory interactions in bacteria and simple eukaryotes

© Springer International Publishing Switzerland 2015
A. Abate and D. Šafránek (Eds.): HSB 2015, LNBI 9271, pp. 3–19, 2015.
DOI: 10.1007/978-3-319-26916-0_1

(see [1, 23, 28] and references). However, fluorescence (or luminescence) levels provide only an indirect quantification of promoter activity. In dynamical conditions, it is being recognized that equating fluorescence levels with promoter activity may lead to serious errors in the inference of regulatory interactions [28], and efforts have been dedicated to the problem of reconstructing promoter activity from the observed population-average fluorescence levels (see e.g. [36]).

In a single cell framework, where stochastic fluctuations of gene expression are of crucial importance [7, 22, 31], estimation of stochastic models of gene regulation is enabled by the most recent single-cell monitoring techniques, providing either time correlation [30, 32, 35] or population statistics [9, 17, 34] of gene expression in single cells. However, the intrinsic complexity of the problem has so far limited results to the estimation of unknown model parameters [9, 14, 29, 34] or the selection of a best model among small pools of alternative model structures [18].

Nonetheless, it is reasonable to expect that accounting for stochasticity is not only necessary to make sense of single-cell data [12], but it may also provide a boost for the inference of unknown regulatory interactions [4], as much as random effects are known to help parameter estimation [17].

Toward this goal, similar to the deterministic setting, a key problem is the reconstruction of the promoter activity statistics from population snaphot data, i.e. statistics of reporter abundance in biological samples from the cell population collected at different points in time. Different from the deterministic setup, where linear ODEs trivially relate reporter mean profile with mean promoter activity, recovery of promoter statistics turns out to be hardly solved by linear inversion (i.e. deconvolution [36]) methods unless the laws governing promoter activity are deterministic and known to a certain extent.

In a companion paper [3], we have considered a random telegraph model [22, 29] of reporter gene expression, whereby synthesis of new reporter molecules is turned on and off in accordance with a switching process describing the promoter state (active or inactive). Under the assumption of fixed switching rates, we have addressed the identifiability of the model parameters, and started looking at the reconstruction of promoter statistics via deconvolution methods.

Here, in the same modelling framework, we instead focus on the case where the laws governing the switching of the promoter are completely unknown, which is by definition the case in the context of interaction network inference, and address the following problem.

Objective: Reconstruct promoter activity statistics from reporter protein population snapshot data, without assumptions on the laws governing promoter switching.

Different from e.g. [6, 13, 20], where gene expression profiles of individual cells are used to reconstruct single-cell promoter activities via nonparametric noise approximations, the problem addressed here is therefore the estimation of promoter activity statistics over a cell population from the distribution of reporter abundance in different samples of the same population collected at different times. This data is rather straightforward to produce, e.g., via flow cytometry experiments. Nonparametric modelling is used in our approach for the unknown statistics of promoter activation and deactivation, whereas the stochastic transcription and translation kinetics triggered by promoter activation are modelled in detail.

Developed in the paper are methods to retrieve from experimental data the necessary biological information to address the problem of stochastic network inference. Because variability is fundamental in single-cell regulation, we are especially interest in second-order moments of promoter activity. In particular, time-correlation (autocovariance function) naturally embeds infomation on the underlying process dynamics [15] and is therefore sought. Network inference itself is instead not addressed here, and will be the object of later work (see [19] for a recent work on the topic).

In the next section we present the modelling framework for random reporter gene expression and population snapshot data. In Sect. 3 we present the inference method proposed and simulation results showing its performance in a fundamental case study. Conclusions and perspectives of the work are drawn in Sect. 4.

Notation. For a generic matrix M, M_c denotes its cth column, $M_{r,c}$ its element of row r and column c, M^T its transpose. For generic random variable X and event A, $\mathscr{E}[X]$ denotes expectation of X and $\mathscr{E}[X|A]$ its conditional expectation given A.

2   Modelling Reporter Expression Dynamics

Synthesis of reporter protein molecules, discussed in Sect. 2.1 below, can be described by standard gene expression models in terms of transcription of mRNA molecules and their subsequent translation into protein molecules. This is often completed by a maturation step, that takes newly synthesized proteins into their visible form. For the sake of this work, we will not distinguish between immature and mature state. This constitutes no loss of generality for reporters taking a fixed time to mature. Otherwise, generalization of our work is pretty straightforward. Mathematical modelling of the monitoring of reporter gene expression is discussed in the subsequent Sect. 2.2. We consider population snapshot data, e.g. reporter fluorescence measured at different times of a dynamical experiment in different cell samples from a common population. This data can be easily obtained e.g. via flow cytometry (as considered e.g. in [17, 34]), though our results are of wider applicability.

2.1   Stochastic Modelling of Gene Expression

Let M and P denote the (reporter) mRNA and protein species, respectively. Gene expression is often described in terms of the reaction system

$$\emptyset \xrightarrow{k_m \cdot f} M \qquad\qquad M \xrightarrow{d_m} \emptyset \qquad\qquad (1)$$

$$M \xrightarrow{k_p} M + P \qquad\qquad P \xrightarrow{d_p} \emptyset \qquad\qquad (2)$$

Here f is a deterministic, possibly continuous profile common to all cells of a population (see e.g. [16,17,32,34]) that captures the overall activity of the relevant promoter. For the specified profile f, at the single-cell level, stochastic reaction kinetics are considered [22]. Let $X^f(t) = [X_1^f(t)\ X_2^f(t)]^T$ denote the bivariate stochastic process describing the abundance (molecule count) of M and P at time t in the generic cell. Writing the stochastic chemical kinetics in terms of these variables leads to the so-called Chemical Master Equation [8,26]. Let $\mu^f(t)$ be the mean vector and $\Sigma^f(t)$ the covariance matrix of $X^f(t)$. For later use, also define the matrix of uncentered second-order moments of X^f, $\mathcal{M}^f = \Sigma^f + (\mu^f) \cdot (\mu^f)^T$. It can be shown (see e.g. [10]) that these moments obey

$$\dot{\mu}^f = SW\mu^f + Sw_0, \tag{3}$$

$$\dot{\Sigma}^f = SW\Sigma^f + \Sigma^f W^T S^T + S\mathrm{diag}(W\mu^f + w_0^f)S^T, \tag{4}$$

where

$$S = \begin{bmatrix} 1 & -1 & 0 & 0 \\ 0 & 0 & 1 & -1 \end{bmatrix}, \qquad W = \begin{bmatrix} 0 & 0 \\ d_m & 0 \\ k_p & 0 \\ 0 & d_p \end{bmatrix}, \qquad w_0^f = \begin{bmatrix} k_m \cdot f \\ 0 \\ 0 \\ 0 \end{bmatrix}$$

are, in the order, the stoichiometry matrix for reactions (1)–(2) and the coefficients of the corresponding reaction rates $W \cdot X^f + w_0^f$. Letting

$$z^f = \begin{bmatrix} \mu_1^f & \mu_2^f & \Sigma_{1,1}^f & \Sigma_{2,2}^f & \Sigma_{1,2}^f \end{bmatrix}^T, \tag{5}$$

Equations (3)–(4) can also be written in the vector form

$$\dot{z}^f = \begin{bmatrix} -d_m & 0 & 0 & 0 & 0 \\ k_p & -d_p & 0 & 0 & 0 \\ d_m & 0 & -2\,d_m & 0 & 0 \\ k_p & d_p & 0 & -2\,d_p & 2\,k_p \\ 0 & 0 & k_p & 0 & -d_m - d_p \end{bmatrix} \cdot z^f + \begin{bmatrix} k_m \\ 0 \\ k_m \\ 0 \\ 0 \end{bmatrix} \cdot f. \tag{6}$$

(A reduction of this system to four equations is possible since one may show that $\mu_1^f(t) = \Sigma_{1,1}^f(t)$ at all t, see e.g. [21].) Thus, for a fixed promoter activation profile $f(\cdot)$, gene expression reporter statistics are governed by a system of linear differential equations, which is often used to describe statistics collected from many cells of a same population [25,34].

However, in individual cells, promoter activity rather follows a random pattern, switching between an "off" state, where mRNA synthesis is disabled, and an "on" state, where mRNA synthesis is enabled. This is captured by the reaction system

$$P_{\mathrm{off}} \xrightarrow{\lambda_+} P_{\mathrm{on}} \qquad\qquad P_{\mathrm{on}} \xrightarrow{\lambda_-} P_{\mathrm{off}} \tag{7}$$

$$P_{\mathrm{on}} \xrightarrow{k_m} P_{\mathrm{on}} + M \qquad\qquad M \xrightarrow{d_m} \emptyset \tag{8}$$

$$M \xrightarrow{k_p} M + P \qquad\qquad P \xrightarrow{d_p} \emptyset \tag{9}$$

where P_{on} and P_{off} stand for active and inactive promoter species, respectively, and the switching rates λ_+ and λ_- generally depend on promoter regulation. For this system, denote with $[X(t)^T \; F(t)]^T$, where $X(t)^T = [X_1(t) \; X_2(t)]$, the process where $X_1(t)$ and $X_2(t)$ are the count of mRNA and reporter protein molecules at time t, in the order, and $F(t)$ is the state of the promoter at the same time (0 in presence of P_{off}, 1 in presence of P_{on}). For simple dynamical models of F, e.g. for fixed switching rates λ_+ and λ_-, an augmented linear ODE system can be obtained for the joint moments of $[X^T \; F]^T$, which in principle allows for the inference of the statistics of F via linear inversion methods [3]. Unfortunately, switching rates are typically unknown and may themselves vary across cells and time due to stochastic promoter regulation mechanisms. In this case, F is a so-called doubly stochastic process [5]. In order to comply with the objective of Sect. 1, our first goal is to establish a relationship between the statistics of F and X_2 that is valid regardless of the switching laws of F. In the sequel, we only require that the probability laws of F (in fact, the laws of the joint process $[X \; F]^T$) are the same in every cell, and denote with dP_F the corresponding probability measure.

We make one standing assumption as follows.

Assumption 1 (Granger Causality [15]). *There is no feedback from X to F, i.e., at any time t, the future of F is conditionally independent on the past of X given the past of F.*

In practice, this means that species M and P do not influence the regulation of the promoter. This is a legitimate assumption for most reporter systems, e.g. all systems where reporter proteins and regulatory proteins are physically different molecules, and may still be acceptable in many more cases, where the stochastic effects of feedback are sufficiently mild. (In a relevant context, the notion of causality is also considered in [2].)

Now consider the unconditional moments of process X, $\mu(t) = \mathscr{E}[X(t)]$ and $\mathscr{M}(t) = \mathscr{E}[X(t)X(t)^T]$. These may be written as $\mu(t) = \mathscr{E}[\mathscr{E}[X(t)|F]]$ and $\mathscr{M}(t) = \mathscr{E}[\mathscr{E}[X(t)X(t)^T|F]]$, where conditioning is intended to be on the whole history of process F. In the light of Assumption 1 it holds that

$$\mu^f(t) = \mathscr{E}[X(t)|F = f], \tag{10}$$

$$\mathscr{M}^f(t) = \mathscr{E}[X(t)X(t)^T|F = f], \tag{11}$$

with $\mu^f(t)$ and $\mathscr{M}^f(t)$ as above. Therefore,

$$\mu(t) = \mathscr{E}[\mu^F(t)] = \int \mu^f(t)dP_F(f) \tag{12}$$

and

$$\mathscr{M}(t) = \mathscr{E}[\mathscr{M}^F(t)] = \int \mathscr{M}^f(t)dP_F(f). \tag{13}$$

(Notice that an equality in the form (13) cannot be obtained for the variance matrix $\Sigma(t) = \mathscr{M}(t) - \mu(t) \cdot \mu(t)^T$.) For the validity of these equations (notably (10) and (11)) absence of feedback is crucial. Otherwise, conditioning on F would implicitly introduce new constraints on the stochastic dynamics of X, i.e. the evolution of $\mathscr{E}[X(t)|F = f]$ and $\mathscr{E}[X(t)X(t)^T|F = f]$ would not correspond to the differential equation system (6) [15]. In sums, together with (6), Equations (10)–(13) compute the ensemble population moments of the reporter as a weighted average of different hybrid dynamics, each composed of a possible switching promoter profile driving the linear differential evolution of the conditional moments of X. These formulas provide the basis for the inference methods developed in Sect. 3.

2.2   Population Snapshot Data

Given M measurement times $\mathscr{T} = \{t_1, \ldots, t_M\}$, we consider data $\mathscr{Y} = \{\tilde{y}(t) : t \in \mathscr{T}\}$, where $\tilde{y}(t)$ is a noisy measurement of

$$y(t) = \mathscr{E}[h(X_2(t))]. \tag{14}$$

In agreement with the previous section, X_2 also reflects stochastic fluctuations of promoter activity across different cells. Function h represents any vector of measurable functions of X_2. In particular, we will focus from now on to

$$h(x) = \begin{bmatrix} x \\ x^2 \end{bmatrix} \tag{15}$$

i.e. $y(t)$ is composed of the mean and statistical power (second-order uncentered moment) of $X_2(t)$. In practice, noisy measurements of $y(t)$ at all times $\mathscr{T}$ are easily obtained e.g. by flow cytometry experiments, where, at every $t \in \mathscr{T}$, a sample of cells is taken from the observed population and fluorescence is automatically quantified in every cell. For the typically large sample sizes of these experiments (in the order of several thousands of cells), measurements can be modelled as [24, 34]

$$\tilde{y}(t) = y(t) + e(t), \tag{16}$$

where random error $e(t)$ follows a bivariate zero-mean Gaussian distribution with covariance matrix $R(t)$ given by

$$R = \begin{bmatrix} \operatorname{var}(\tilde{\mu}_2) & \operatorname{cov}(\tilde{\mu}_2, \tilde{\mathscr{M}}_{2,2}) \\ \operatorname{cov}(\tilde{\mu}_2, \tilde{\mathscr{M}}_{2,2}) & \operatorname{var}(\tilde{\mathscr{M}}_{2,2}) \end{bmatrix} = \frac{1}{N} \begin{bmatrix} m(2) - m(1)^2 & m(3) - m(1)m(2) \\ m(3) - m(1)m(2) & m(4) - m(2)^2 \end{bmatrix}$$

(all quantities evaluated at time t, omitted from notation for brevity). Here, $\tilde{\mu}_2$ and $\tilde{\mathscr{M}}_{2,2}$ are the empirical estimators of mean and statistical power of X_2 from a sample of size S (number of cells at time t), while for any p, $m(p) = \mathscr{E}[X_2^p]$. In practice, in order to fix R, it suffices to estimate the $m(p)$ from the data [34]. In the light of the fact that different cell samples are observed at different times, we further assume that measurements $\tilde{y}(t)$ at different times are mutually independent.

3   Inference Methods and Results

In this section we outline a general approach to the reconstruction of promoter activity statistics from population snapshot data, and develop from it a practical method that addresses some among the most interesting special cases at a well affordable computational cost. For simplicity, we assume that the kinetic parameters for the synthesis and degradation of M and P are known along with the statistics of the latter at time zero, but relaxation of these assumptions is possible [3,17]. Numerical demonstration of the method's performance is given further below.

3.1   Reconstruction of Promoter Switching Statistics

The objective is to reconstruct promoter statistics of the type $\mathscr{E}[g(F)]$ from data $\mathscr{Y}$, for some measurable vector function g. We consider the measurement model (14)–(16), with output function as in (15) (reporter mean and statistical power), and wish to reconstruct the first and second-order moments of F. We address this problem in a more general fashion. Assume that the unknown dP_F belongs to some class of measures $\mathscr{P}$. Ideally, one would like to reconstruct the true dP_F by seeking the element of $\mathscr{P}$ that best explains the data $\mathscr{Y}$. Stated in this form, the problem is formidable, but one can reformulate it in an approximate fashion according to the following rationale. Let $\mathscr{F} = \{f_i : i = 1, \dots, N\}$, be a family of N piecewise continuous functions $f_i : t \mapsto \{0,1\}$. For every i let μ^{f_i} and $\mathscr{M}^{f_i}$ be defined as in (10)–(11) with f_i in place of f. Assume that, for every element $dP_F \in \mathscr{P}$, we can uniquely choose weights $p_i \in [0,1]$, with $i = 1, \dots, N$ and $\sum_i p_i = 1$, so as to minimize, for a suitable norm $\| \cdot \|$,

$$\varepsilon = \left\| \int \left[\mu^f(\cdot) \, \mathscr{M}^f(\cdot) \right] dP_F(f) - \sum_{i=1}^{N} \left[\mu^{f_i}(\cdot) \, \mathscr{M}^{f_i}(\cdot) \right] p_i \right\| \tag{17}$$

and that ε is sufficiently small for all elements of $\mathscr{P}$. We may then call $\mathscr{F}$ an approximating family for $\mathscr{P}$. For the sake of inference, we aim at reconstructing the probability weights p_i that best explain the data $\mathscr{Y}$, i.e., in the light of (17), an approximate estimate of the true laws dP_F, from which estimates of first- and second-order moments of F readily follow. For complex classes of measures $\mathscr{P}$, the choice of $\mathscr{F}$ and the solution of this problem remain a hard task. However, interesting classes of problems where N can be kept within reasonable bounds may be solved very effectively via a maximum likelihood approach.

Fix $\mathscr{F}$, and denote $p = [p_1, \dots, p_N]^T$. Let $\hat{y}(t|p) = [\hat{\mu}_2(t|p) \, \hat{\mathscr{M}}_{2,2}(t|p)]^T$ be the moments of $X_2(t)$ for a given p. In the light of the above approximation, these can be written as

$$\hat{y}(t|p) = \sum_{i=1}^{N} p_i \begin{bmatrix} z_2^{f_i}(t) \\ z_4^{f_i}(t) + \left(z_2^{f_i}(t) \right)^2 \end{bmatrix} = \sum_{i=1}^{N} p_i y^{f_i}(t), \tag{18}$$

where z^{f_i}, defined as in (5) but with f_i in place of f, is the solution of (6) under $f = f_i$. In particular, $z_2^{f_i}$ and $z_4^{f_i}$ are the predicted mean and variance of X_2

under f_i, and the definition of the y^{f_i} follows. Under measurement model (16) and assumptions thereof, the negative log-likelihood of p given the data $\mathscr{Y}$ is thus (neglecting additive constants)

$$Q(p) = \frac{1}{2} \sum_{t \in \mathscr{T}} (\tilde{y}(t) - \hat{y}(t|p))^T R(t)^{-1} (\tilde{y}(t) - \hat{y}(t|p)). \tag{19}$$

Let $R^{1/2}$ be a square symmetric factor of R, i.e. $R^{1/2} R^{1/2} = R$. Defining the matrix

$$\mathbb{M} = \begin{bmatrix} R^{-1/2}(t_1) & & \\ & \ddots & \\ & & R^{-1/2}(t_M) \end{bmatrix} \cdot \begin{bmatrix} y^{f_1}(t_1) & y^{f_2}(t_1) & \cdots & y^{f_N}(t_1) \\ \vdots & \vdots & & \vdots \\ y^{f_1}(t_M) & y^{f_2}(t_M) & \cdots & y^{f_N}(t_M) \end{bmatrix}$$

and the measurement vector $\mathbb{Y} = [\tilde{y}(t_1)^T R^{-1/2}(t_1) \cdots \tilde{y}(t_M)^T R^{-1/2}(t_M)]^T$, one may rewrite (19) as

$$Q(p) = (\mathbb{Y} - \mathbb{M} \cdot p)^T \cdot (\mathbb{Y} - \mathbb{M} \cdot p),$$

where factor $1/2$ has been neglected. Clearly $Q(p)$ is a quadratic form in p. The maximum likelihood estimation of p can then be obtained as solution of the linearly constrained, quadratic optimization problem

$$\text{Find } \hat{p} \text{ such that } Q(\hat{p}) = \min\{Q(p) : \ p \geq 0, \ \bar{1}^T p = 1\}. \tag{20}$$

where $\bar{1}$ is the length-N vector $[1 \cdots 1]^T$, and inequality is intended componentwise. Estimates of $\mathscr{E}[g(F)]$ can then be constructed as

$$\hat{\mathscr{E}}[g(F)] = \sum_{i=1}^{N} \hat{p}_i g(f_i). \tag{21}$$

In practice, solutions of (20) do not guarantee the presumable regularity over time of F, notably of

$$\hat{\mathscr{E}}[F(t)] = \sum_{i=1}^{N} \hat{p}_i f_i(t). \tag{22}$$

In the spirit of Tikhonov regularization [33], we then modify (20) by penalizing the norm of the second-order derivative of (a discrete-time version of) (22) (because we consider piecewise constant f_i, the continuous-time second-order derivative does not exist). Given a uniform grid of time points $\tau_0, \ldots, \tau_{L-1}$ (not necessarily equal to the measurement times) and a generic function f, let $\nabla_\ell f$ be second-order difference of the sequence $f(\tau_0), \ldots f(\tau_{L-1})$ at τ_ℓ, i.e. $\nabla_\ell f = f(\tau_{\ell-1}) - 2f(\tau_\ell) + f(\tau_{\ell+1})$, with $\ell = 1, \ldots, L-2$. Since

$$\nabla_\ell \left(\sum_{i=1}^{N} p_i f_i \right) = \sum_{i=1}^{N} (\nabla_\ell f_i) p_i = \mathbb{D}_\ell^T \cdot p,$$

for a tunable parameter $\alpha \geq 0$, small norm of the second derivative of (22) is attained by adding the cost term $\alpha \cdot (\mathbb{D} \cdot p)^T (\mathbb{D} \cdot p)$ to $Q(p)$, where

$$\mathbb{D} = \begin{bmatrix} \mathbb{D}_1^T \\ \vdots \\ \mathbb{D}_{L-2}^T \end{bmatrix} = \begin{bmatrix} \nabla_1 f_1 & \nabla_1 f_2 & \cdots & \nabla_1 f_N \\ \vdots & \vdots & & \vdots \\ \nabla_{L-2} f_1 & \nabla_{L-2} f_2 & \cdots & \nabla_{L-2} f_N \end{bmatrix}.$$

Thus, the regularized version of Problem (20) becomes

$$\text{Find } \hat{p}_\alpha \text{ such that } \bar{Q}_\alpha(\hat{p}_\alpha) = \min\{\bar{Q}_\alpha(p) : \ p \geq 0, \ \bar{1}^T p = 1\} \qquad (23)$$

where

$$\bar{Q}_\alpha(p) = Q(p) + \alpha \cdot (\mathbb{D} \cdot p)^T (\mathbb{D} \cdot p) = (\bar{\mathbb{Y}}\bar{\mathbb{Y}} - \bar{\mathbb{M}}\bar{\mathbb{M}}_\alpha \cdot p)^T \cdot (\bar{\mathbb{Y}}\bar{\mathbb{Y}} - \bar{\mathbb{M}}\bar{\mathbb{M}}_\alpha \cdot p)$$

and in turn

$$\bar{\mathbb{Y}} = \begin{bmatrix} \mathbb{Y} \\ \bar{0} \end{bmatrix}, \quad \bar{\mathbb{M}}_\alpha = \begin{bmatrix} \mathbb{M} \\ \alpha\mathbb{D} \end{bmatrix}$$

($\bar{0}$ denotes the zero matrix of appropriate dimension). For $\alpha = 0$, formulation (20) is recovered, whereas the larger the α, the more regular the solution. Problem (23) is convex (quadratic cost, linear constraints), and can be solved by means of fast algorithms. In practice, matrix $\mathbb{M}$ follows from the integration of (6) while $\mathbb{D}$ can be computed by standard matrix multiplications given the values $f_i(\tau_\ell)$.

An obvious choice of $\mathscr{F}$ is to partition the time interval $[t_1, t_M]$ into K uniform subintervals, and to define functions f_i taking constant value 0 or 1 over each subinterval. Of course, the scalability of this approach is generally a challenge, since the size N of the optimization problem rapidly increases with the complexity (e.g. maximum number of switches) of the possible outcomes of F. Still, in many cases of interest [12] and for a reasonable duration of the experiment, it suffices to consider a small number of switches, which allows one to approximate the laws of F with a relatively small N. In the next section, we demonstrate the approach for the prototypical, yet fundamental case study of promoter switch-on, where a single switch from state off to on occurs randomly in different cells.

3.2 Numerical Simulations

We now apply the method of Sect. 3.1 for the reconstruction of the statistics of F in the fundamental case study where the promoter becomes active in the course of the experiment by switching once from off to on. This is obtained by fixing $\lambda_- = 0$, with F starting from the zero state. Simulations of stochastic reactions are performed using STOCHKIT [27]. Estimation algorithms are implemented in MATLAB R2014a (The MathWorks Inc., Natick, Massachusetts). On a laptop equipped with quadcore Intel CPU i7, 4 Gb RAM and Fedora 21 operating system, for the case studies below, data simulation in STOCHKIT takes within seconds while setting up and solving inference takes about half a minute in a basic, non-optimized MATLAB implementation.

We first consider the case of randomly regulated switching rate λ_+, so as to simulate stochastic regulation operated by unknown transcription factors. Inspired by [12], we take realistic values $k_m = 0.5$, $d_m = 0.1$, $k_p = 0.2$, $d_p = 0.01$ (time unit is minute), while $\lambda_+ = 0.05 \cdot U$. U is itself a stochastic process (not affected by feedback) following the stochastic dynamics of the additional reaction $\emptyset \to U$ with rate 0.05. At time zero, no molecules of any species are present. We simulate $S = 10^5$ cells, i.e. generate independent random trajectories of the system over the time span $[0, 99]$, and at every time $t_k = k - 1$, with $k = 1, \ldots, M$ and $M = 100$, we randomly sample 10^4 cells from the pool of 10^5 cells, thus getting statistics of X and F that are (roughly) independent across time. Next, we define the approximating family $\mathscr{F}$ by functions each representing a possible switch from 0 to 1 at a different multiple of $T = 0.2 \min$. Formally, for $i = 0, \ldots, 5 \cdot (M - 1)$,

$$f_i(t) = \begin{cases} 0, & t < i \cdot T; \\ 1, & t \geq i \cdot T. \end{cases} \tag{24}$$

(The analytical study of the approximation (17) obtained by (24) is deferred to later work, but its viability will be apparent from our inference results.) The regularization time grid is instead fixed to $\tau_\ell = \ell$, with $\ell = 0, \ldots, L - 1$ and $L = 100$. We then solve Problem (23) using MATLAB function lsqlin with regularization parameter (empirically tuned to) $\alpha = 10^4$ (automatic tuning of this parameter is left for future investigation).

Figure 1 shows the observed fluorescence (i.e. X_2) mean and statistical power profiles as computed empirically from the simulated cells, and the fit of these quantities from the solution of Problem (23). Figure 2 shows the corresponding estimates of the mean and standard deviation of F computed via (21). An optimal fit to the data is apparent together with an accurate estimation of the moments of F, compatibly with the resolution of $\mathscr{F}$, with the exception of some deviation in the estimation of the standard deviation in the long run, where the true mean has almost reached 1 (its asymptotic value). This is likely explained by the imposed regularity, which favors the f_i with a constant tail and thus excludes late switching, therefore implying zero estimated variance (and estimated mean equal to 1) in that time period.

For comparison, fitting was repeated by changing the regularization parameter to $\alpha = 0$ (no regularity enforced). Results are also reported in Fig. 2 and are self-explanatory. Interestingly, it can be noticed that the resolution of $\mathscr{F}$ is largely underexploited in this case, i.e. probability mass is effectively distributed only over a few switching times. Zero values of several probability coefficients are naturally explained by the solution of the non-regularized optimization problem lying at the boundaries of the constraint set, as a consequence of the approximations made and of measurement noise.

Coming back to the regularized solution, note that F is pointwise a Bernoulli random variable. Therefore looking at the estimated standard deviation profile does not add much relative to inspecting the estimated mean, since the former

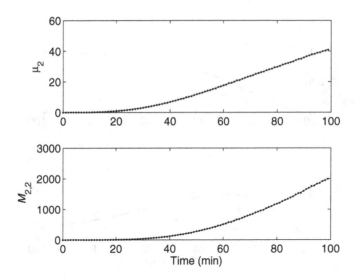

Fig. 1. Fit of measured mean and statistical power reporter profiles from the solution of Problem (23) with regularization ($\alpha = 10^4$), using $M = 100$ measurements collected every minute. Black dots: Simulated data. Blue lines: Model fits (Colour figure online).

is basically determined by the latter. What is of real interest is the ability to reconstruct the autocovariance function of F, since this embeds information about the time dynamics of the generating laws of F, i.e. the basic first step for the discovery of its regulatory laws (here simulated by means of regulator U). Figure 3 illustrates the estimation of the autocovariance function of F at different lags and times (F is a nonstationary process), as obtained from the solution of (23) above again by means of (21). The accuracy of the estimate is quite apparent.

One may argue that, due to the simplicity of the single-switch process F, estimation of the autocovariance function of F is quite trivial and uniquely determined from an estimate of the mean profile. While this statement is certainly not true for multiple-switch processes, for the present case we investigated this question leveraging on the fact that the mean of F can be reconstructed from the sole mean of X_2. We solved again Problem (23) in a modified form where the measurements of the second-order moment of X_2 are ignored, and then looked at the corresponding estimate of the autocovariance function as obtained via (21). Results are quantitatively compared with those previously obtained in Table 1. It is apparent that a significant loss in the reconstruction accuracy of all statistics of F is encountered, a sign of the additional information provided by the second moment of X_2 relative to its mean. Note that the different accuracy in the estimation of the mean of F is explained by the fact that, contrary to the intuition one would typically get from linear Gaussian processes, the mean of F explicitly enters the equations for the variance of X_2, and is hence reflected in the measured statistical power of X_2 in a nontrivial manner.

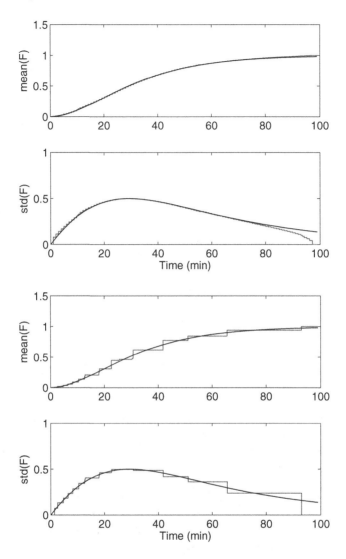

Fig. 2. Estimation of mean and standard deviation profiles of promoter state process F from the solution of Problem (23) with regularization ($\alpha = 10^4$, upper plot) and without regularization ($\alpha = 0$, lower plot), using $M = 100$ measurements collected every minute. Black lines: Statistics from the simulated cells. Blue lines: Estimates (Colour figure online).

For further comparison, we also simulated the case of a nonregulated promoter, where everything is identical to the previous case except that data are simulated with λ_+ fixed to 0.05 (so that F is a simple (reducible) Markov chain). Quantitative results are again reported in Table 1, and reconfirm all the observations from the previous example.

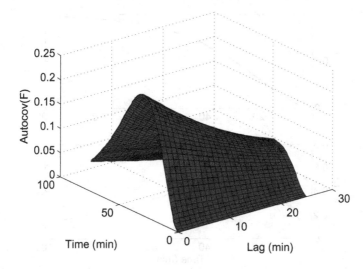

Fig. 3. Estimation of the autocovariance function of promoter state process F from the solution of Problem (23) with regularization ($\alpha = 10^4$), using $M = 100$ measurements collected every minute. Blue surface: Statistics from the simulated cells. Red surface: Estimate (Colour figure online).

Table 1. Mean squared error of the estimates of the mean, variance, and autocovariance function of the promoter state process F relative to the empirical statistics from the simulated cells, from the solutions of Problem (23) with regularization (third column), with regularization but using mean data only (fourth column), and without regularization (fifth column), based on $M = 100$ measurements collected every minute. Results are reported for a single-switch promoter with random switching rate (upper block) and fixed rate (lower block). Mean and variance estimation errors are evaluted at all measurement times, autocovariance estimation errors are evaluated at lags up to 25 min and times up to 74 min (with 1 min sampling; compare Fig. 3). For ease of reading, all table entries have been multiplied by 10^4.

Switch rates	Error on	$\alpha = 10^4$ Mean & Power	$\alpha = 10^4$ Mean only	$\alpha = 0$ Mean & Power
Random	Mean	0.41	0.85	19.05
	Variance	0.24	0.43	8.27
	Autocovariance	0.076	0.21	10.92
Fixed	Mean	0.92	1.57	10.74
	Variance	0.31	0.55	3.55
	Autocovariance	0.24	0.46	4.56

Finally, in order to evaluate the role of measurement sampling time, we repeated the whole investigation above with $t_k = 5 \cdot (k - 1)$ minutes, with $k = 1, \ldots, M$, where now $M = 20$, i.e. with sparser measurements taken over the same time span. Results are reported in Figs. 4 and 5 (random promoter

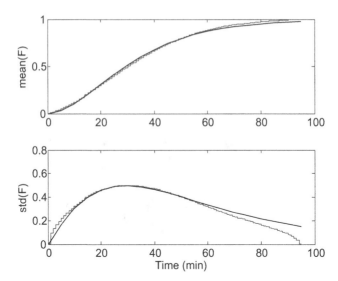

Fig. 4. Estimation of mean and standard deviation profiles of promoter state process F from the solution of Problem (23) with regularization ($\alpha = 10^4$), using $M = 20$ measurements collected every 5 min. Black lines: Statistics from the simulated cells. Blue lines: Estimates (Colour figure online).

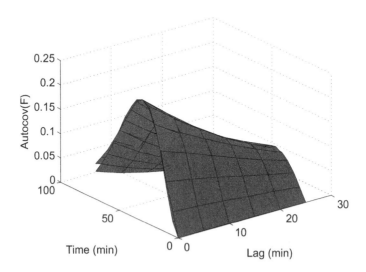

Fig. 5. Estimation of the autocovariance function of promoter state process F from the solution of Problem (23) with regularization ($\alpha = 10^4$), using $M = 20$ measurements collected every 5 min. Blue surface: Statistics from the simulated cells. Red surface: Estimate (Colour figure online).

Table 2. Mean squared error of the estimates of the mean, variance, and autocovariance function of the promoter state process F relative to the empirical statistics from the simulated cells, from the solutions of Problem (23) with regularization (third column), with regularization but using mean data only (fourth column), and without regularization (fifth column), based on $M = 20$ measurements collected every 5 min. Results are reported for a single-switch promoter with random switching rate (upper block) and fixed rate (lower block). Mean and variance estimation errors are evaluated at all measurement times, autocovariance estimation errors are evaluated at lags up to 25 min and times up to 70 min (with 5 min sampling; compare Fig. 5). For ease of reading, all table entries have been multiplied by 10^4.

Switch rates	Error on	$\alpha = 10^4$	$\alpha = 10^4$	$\alpha = 0$
		Mean & Power	Mean only	Mean & Power
Random	Mean	3.20	4.34	18.41
	Variance	1.73	2.25	4.65
	Autocovariance	1.25	1.73	6.38
Fixed	Mean	7.52	10.68	29.93
	Variance	4.75	6.55	12.16
	Autocovariance	4.01	5.62	14.38

regulation, compare Figs. 2 and 3), and in Table 2 (to be compared with Table 1). Relative to the previous results, a graceful degradation of estimation performance is observed, whereas the same considerations about the use of variance data and regularization apply.

4   Discussion and Perspectives

We have addressed the problem of reconstructing the promoter activation statistics from population snapshots of reporter fluorescence in single cells. Similar to what has been proven necessary and hence addressed in a population average context, our results enable one to recover the biological information of actual interest in a stochastic single-cell context from indirect measurements. The absence of assumptions on the promoter switch laws allows us to apply the method to unknown systems for the subsequent study of the regulatory mechanisms. We believe that approaches like ours open the way toward the exploitation of noise for learning regulatory networks that cannot be resolved solely based on average data, in a way much similar to what has been achieved in parameter estimation. Of course, applicability (scalability) of the method depends on the variety of switching profiles that a promoter is deemed capable of. In this regard, we have shown successful application of our approach to a simple case study that is however of fundamental importance, where we have proven the utility of using fluorescence moments higher than the mean not only for estimating promoter time-correlations, but also for estimating the mean activity profile. Other realistic and more general scenarios, such as a handful of on-off switches within an experimental time period, are well within reach and shall be pursued along with automated data-driven tuning of the regularization parameter, analytical study of approximations and statistical properties of the inference methods, and application to real data.

References

1. Bansal, M., Belcastro, V., Ambesi-Impiombato, A., di Bernardo, D.: How to infer gene networks from expression profiles. Mol. Syst. Biol. **3**, 78 (2007)
2. Bowsher, C.G., Voliotis, M., Swain, P.S.: The fidelity of dynamic signaling by noisy biomolecular networks. PLoS Comput. Biol. **9**(3), e1002965 (2013)
3. Cinquemani, E.: Reconstruction of promoter activity statistics from reporter protein population snapshot data. In: Accepted for the 54th IEEE Conference on Decision and Control (2015)
4. Cinquemani, E., Milias-Argeitis, A., Summers, S., Lygeros, J.: Local identification of piecewise deterministic models of genetic networks. In: Majumdar, R., Tabuada, P. (eds.) HSCC 2009. LNCS, vol. 5469, pp. 105–119. Springer, Heidelberg (2009)
5. Cox, D., Isham, V.: Point Processes. Chapman & Hall/CRC Monographs on Statistics & Applied Probability, Taylor & Francis (1980)
6. Finkenstädt, B., Heron, E.A., Komorowski, M., Edwards, K., Tang, S., Harper, C.V., Davis, J.R.E., White, M.R.H., Millar, A.J., Rand, D.A.: Reconstruction of transcriptional dynamics from gene reporter data using differential equations. Bioinformatics **24**(24), 2901–2907 (2008)
7. Friedman, N., Cai, L., Xie, X.S.: Linking stochastic dynamics to population distribution: An analytical framework of gene expression. Phys. Rev. Lett. **97**, 168302 (2006)
8. Gillespie, D.T.: The chemical Langevin equation. J. Chem. Phys. **113**, 297–306 (2000)
9. Hasenauer, J., Waldherr, S., Doszczak, M., Radde, N., Scheurich, P., Allgower, F.: Identification of models of heterogeneous cell populations from population snapshot data. BMC Bioinf. **12**(1), 125 (2011)
10. Hespanha, J.: Modelling and analysis of stochastic hybrid systems. IEE Proc. Control Theory Appl. **153**(5), 520–535 (2006)
11. de Jong, H., Ranquet, C., Ropers, D., Pinel, C., Geiselmann, J.: Experimental and computational validation of models of fluorescent and luminescent reporter genes in bacteria. BMC Syst. Biol. **4**(1), 55 (2010)
12. Kaern, M., Elston, T.C., Blake, W.J., Collins, J.J.: Stochasticity in gene expression: From theories to phenotypes. Nat. Rev. Gen. **6**, 451–464 (2005)
13. Komorowski, M., Finkenstädt, B., Harper, C., Rand, D.: Bayesian inference of biochemical kinetic parameters using the linear noise approximation. BMC Bioinf. **10**(1), 343 (2009)
14. Lillacci, G., Khammash, M.: The signal within the noise: efficient inference of stochastic gene regulation models using fluorescence histograms and stochastic simulations. Bioinformatics **29**(18), 2311–2319 (2013)
15. Lindquist, A., Picci, G.: Linear Stochastic Systems - A Geometric Approach to Modeling, Estimation and Identification. Springer, Heidelberg (2015)
16. Milias-Argeitis, A., Stewart-Ornstein, S.S.J., Zuleta, I., Pincus, D., El-Samad, H., Khammash, M., Lygeros, J.: In silico feedback for in vivo regulation of a gene expression circuit. Nat. Biotechnol. **29**, 1114–1116 (2011)
17. Munsky, B., Trinh, B., Khammash, M.: Listening to the noise: random fluctuations reveal gene network parameters. Mol. Syst. Biol. **5**, 318 (2009)

18. Neuert, G., Munsky, B., Tan, R., Teytelman, L., Khammash, M., van Oudenaarden, A.: Systematic identification of signal-activated stochastic gene regulation. Science **339**(6119), 584–587 (2013)
19. Ocone, A., Haghverdi, L., Mueller, N.S., Theis, F.J.: Reconstructing gene regulatory dynamics from high-dimensional single-cell snapshot data. Bioinformatics **31**(12), i89–i96 (2015)
20. Lafferty, J., Williams, C., Shawe-Taylor, J., Zemel, R., Culotta, A. (eds.) Advances in Neural Information Processing Systems 23, pp. 1831–1839. Curran Associates, Inc., (2010)
21. Parise, F., Ruess, J., Lygeros, J.: Grey-box techniques for the identification of a controlled gene expression model. In: Proceedings of the ECC (2014)
22. Paulsson, J.: Models of stochastic gene expression. Phys. Life Rev. **2**(2), 157–175 (2005)
23. Porreca, R., Cinquemani, E., Lygeros, J., Ferrari-Trecate, G.: Identification of genetic network dynamics with unate structure. Bioinformatics **26**(9), 1239–1245 (2010)
24. Ruess, J., Lygeros, J.: Moment-based methods for parameter inference and experiment design for stochastic biochemical reaction networks. ACM Trans. Model. Comput. Simul. **25**(2), 8 (2015)
25. Ruess, J., Milias-Argeitis, A., Summers, S., Lygeros, J.: Moment estimation for chemically reacting systems by extended Kalman filtering. J. Chem. Phys. **135**(16), 165102 (2011)
26. Samad, H.E., Khammash, M., Petzold, L., Gillespie, D.: Stochastic modelling of gene regulatory networks. Int. J. Robust Nonlin. Contr. **15**, 691–711 (2005)
27. Sanft, K.R., Wu, S., Roh, M., Fu, J., Lim, R.K., Petzold, L.R.: Stochkit2: software for discrete stochastic simulation of biochemical systems with events. Bioinformatics **27**(17), 2457–2458 (2011)
28. Stefan, D., Pinel, C., Pinhal, S., Cinquemani, E., Geiselmann, J., de Jong, H.: Inference of quantitative models of bacterial promoters from time-series reporter gene data. PLoS Comput. Biol. **11**(1), e1004028 (2015)
29. Suter, D.M., Molina, N., Gatfield, D., Schneider, K., Schibler, U., Naef, F.: Mammalian genes are transcribed with widely different bursting kinetics. Science **332**, 472–474 (2011)
30. Taniguchi, Y., Choi, P.J., Li, G.W., Chen, H., Babu, M., Hearn, J., Emili, A., Xie, X.S.: Quantifying *E. coli* proteome and transcriptome with single-molecule sensitivity in single cells. Science **329**, 533–538 (2010)
31. Thattai, M., van Oudenaarden, A.: Intrinsic noise in gene regulatory networks. PNAS **98**(15), 8614–8619 (2001)
32. Uhlendorf, J., Miermont, A., Delaveau, T., Charvin, G., Fages, F., Bottani, S., Batt, G., Hersen, P.: Long-term model predictive control of gene expression at the population and single-cell levels. PNAS **109**(35), 14271–14276 (2012)
33. Wahba, G.: Spline models for observational data. In: SIAM (1990)
34. Zechner, C., Ruess, J., Krenn, P., Pelet, S., Peter, M., Lygeros, J., Koeppl, H.: Moment-based inference predicts bimodality in transient gene expression. PNAS **21**(109), 8340–8345 (2012)
35. Zechner, C., Unger, M., Pelet, S., Peter, M., Koeppl, H.: Scalable inference of heterogeneous reaction kinetics from pooled single-cell recordings. Nat. Methods **11**, 197–202 (2014)
36. Zulkower, V., Page, M., Ropers, D., Geiselmann, J., de Jong, H.: Robust reconstruction of gene expression profiles from reporter gene data using linear inversion. Bioinformatics **31**(12), i71–i79 (2015)

Comparative Statistical Analysis of Qualitative Parametrization Sets

Adam Streck[✉], Kirsten Thobe, and Heike Siebert

Freie Universität Berlin, Berlin, Germany
adam.streck@fu-berlin.de

Abstract. The problem of model parametrization is a core issue for all varieties of mathematical modelling in biology. This problem becomes more tractable when qualitative modelling is used, since the range of parameter values is finite and consequently it is possible to enumerate and evaluate all possible parametrizations of a model. If such an approach is undertaken, one usually obtains a vast set of parametrizations that are scored for various properties, e.g. fitness. The usual next step is to take the best scoring parametrization. However, as noted in recent works [1,4], there is knowledge to be gained from examining sets of parametrizations based on their scoring. In this article we extend this line of thought and introduce a comprehensive workflow for comparing such sets and obtaining knowledge from the comparison.

Keywords: Qualitative modelling · Statistical inference · Big data · Parameter identification · Data mining

1 Introduction

One of the key tasks in the field of systems biology is reverse engineering of regulatory and signalling networks [6]. A researcher is usually presented with sets of experimental data and observations and tries to design a model of the mechanics of the system. As the model can be constructed using various modelling frameworks, there is a zoo of methods for tasks like network inference, parameter identification etc., each having its particular set of pros and cons. In our work we are employing the so-called Thomas Networks [17] framework, whose main purpose is to provide insights into qualitative, high-level behaviour. A particular feature of this framework is that the values governing the behaviour of the model—its parameters—have a finite domain and thus it is possible to enumerate and evaluate all the options. At this point, two additional problems arise. Firstly, when evaluating a *parametrization* (a particular set of parameter values) of the network, one is usually focusing again only on some abstract, qualitative feature, e.g. whether the system is stable or not. While this is useful information, its binary nature means that the set of all possible parametrizations is simply split in two, one part having the feature, the other not. This poses a problem if one is aiming to pick an optimal parametrization, as all the members in each of

© Springer International Publishing Switzerland 2015
A. Abate and D. Šafránek (Eds.): HSB 2015, LNBI 9271, pp. 20–34, 2015.
DOI: 10.1007/978-3-319-26916-0_2

the two sets are between themselves indistinguishable. Secondly, considering all the options leads to a rapid combinatorial explosion and while quite huge sets can still be easily manipulated and stored by the computer, it becomes swiftly infeasible for the researcher to keep a mental insight into the structure of the parametrization pool.

Fig. 1. Our workflow starts with the enumeration (1) of all possible parametrizations that fit the expectations of the modeller about the structure. Each of the parametrizations is evaluated for certain properties (2) like dynamical behaviour, and the result of the evaluation is stored with the parametrization as its *label*. After parametrizations are labelled the user can select (3) a subset of these that seem of special interest. This selection can then be analysed using various tools (4) and compared to other selections (5). The selection or analysis can then be refined based on the newly gained knowledge.

Similarly to recent works of other authors in the field [1,4] we propose to shift the focus from individual parametrizations to sets thereof. Following on our previous work [5,16], we introduce a unified workflow for parameter identification, illustrated in Fig. 1. This workflow combines formal and statistical methods with the aim of maximizing the amount of qualitative knowledge obtained from data. All the methods presented in the article have been implemented in the tool TREMPPI, whose preliminary version is available at [13]. All of the methods are illustrated on a toy running example and later the functionality of our workflow is demonstrated on a case study of Hepatocyte Growth Factor (HGF) signalling, based on data of [2].

2 Background

In this section we define the notions necessary for our workflow. Most of the terms are illustrated in Fig. 2 on a toy example.

The topology of a biological system is encoded as a directed *regulatory graph* (RG) $G = (V, \rho, E)$ where V is a set of named *components*, $\rho : V \to \mathbb{N}$ is the *maximum activity* label s.t. each component can adapt an integer from $[0, \rho(v)]$, denoting its current *activity level*, and $E \subseteq V \times \mathbb{N} \times V$ is a set of *regulations* s.t. for each $(u, t, v) \in E$ it holds that $t \leq \rho(u)$. For $(u, t, v) \in E$ the value t denotes a *threshold* i.e. the lowest *activity level* of u at which the regulation can affect v. Additionally, we introduce a threshold function $\theta : V \times V \to 2^{\mathbb{N}}$ s.t. $\theta(u, v) = \{t \mid (u, t, v) \in E\}$ and its extended version Θ s.t. $\Theta(u, v) = \theta(u, v) \cup \{0, \rho(u) + 1\}$ for any pair $u, v \in V$. Moreover, if $(u, t, v) \in E$ then $t_-, t_+ \in \Theta(u, v)$ denote the closest lower and higher element of t, i.e. have $\uparrow^{\Theta}$ the ordinal successor function in Θ, then $\uparrow^{\Theta}(t_-) = t$ and $\uparrow^{\Theta}(t) = t_+$. A regulation becomes effective when the

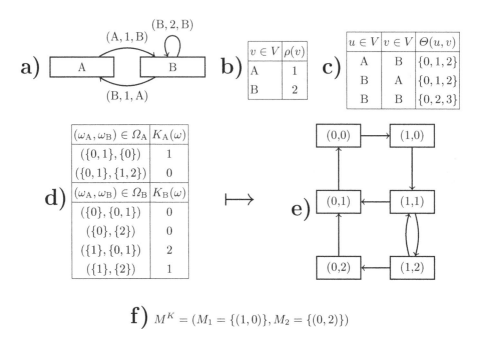

Fig. 2. A simple Thomas Network. **a)** The regulatory graph, **b)** its components, and **c)** regulators. **d)** One of the 324 possible parametrizations of the network. **e)** The asynchronous dynamics encoded by the parametrization. **f)** An example property. Note that the property is satisfied by K, i.e. $((1,0),(1,1),(1,2),(0,2)) \models M^K$.

activity of the respective regulator reaches the threshold value. Consequently, the thresholds divide the range of activity levels of a component into so-called activity intervals $I_v^u = \{[t, \uparrow^{\ominus} (t)) \mid t \in \theta(u,v) \cup \{0\}\}$. Note that if $\theta(u,v) = \emptyset$ then $I_v^u = [0, \rho(u)+1)$. For each component we can then create a set of configurations of components of the system, called *regulatory contexts*, where the behaviour of a component $v \in V$ can qualitatively differ from the other contexts, denoted and defined $\Omega_v = \prod_{u \in V} I_v^u$. Note that $\omega \in \Omega_v$ is a vector of length $|V|$ and consequently we use the notation ω_u for its u-th element. The qualitative behaviour of a component is then fully described through a *partial parametrization* $K_v : \Omega_v \to [0, \rho(v)]$, as explained in the following paragraph. Note that for each $v \in V$ the set Ω_v is sufficient to obtain the set of regulators of v. The *parametrization* $K = (K_v)_{v \in V}$ therefore fully suffices to derive both the behaviour and the topology of a network. We will further use K as an identifier of a single model and $\mathcal{K}^G$ to denote the set of all possible parametrizations of a regulatory graph G. If G can be arbitrary (but fixed) we use simply $\mathcal{K}$.

The asynchronous behaviour of a parametrized regulatory graph $G = (V, \rho, E)$ is then captured in a so-called *transition system* (TS), which is a pair $(S^K, \to^K)$ where $S^K = \prod_{v \in V} \rho(v)$ is a set of states and $\to^K \subseteq S^K \times S^K$ is a transition relation obtained from a parametrization $K \in \mathcal{K}^G$ s.t. $s \to^K s'$ if and only if one of the following two, mutually exclusive conditions holds:

$$\forall v \in V : s'_v = s_v \wedge K_v(s) = s_v,$$

$$\exists u \in V, \forall v \in V \setminus \{u\} : s'_v = s_v \wedge s_u \neq s'_u = K_u(s).$$

To examine a behaviour of a TS we use the model checking procedure, for details please refer to [14]. This method allows to query whether the system satisfies a certain formally described property. In case of our system we verify whether a path of a certain kind exists in a TS, more specifically, whether a path that matches a sequence of *measurements* exists. Formally a sequence of measurements is described via a vector $M^K = (M_1, \ldots, M_m)$ for some $m \in \mathbb{N}$ where for any $i \in [1, m]$ it holds that $M_i \subseteq S^K$. Understandably, a state $s \in S^K$ matches a measurement M_i *iff* also $s \in M_i$. Then a path $w = (s_1, \ldots, s_n) \in (\to^K)^{n-1}$ satisfies M^K *iff* there is a vector $I = (i_1, \ldots, i_m)$ of indices such that for any $k \in [1, m]$ it holds that $s_{i_k} \in M_k$ (measurements are matched) and for each pair $k, l \in [1, m]$ we have that if $k < l$ then also $i_k < i_l$ (ordering is preserved). The path w is then called the *witness* of satisfaction of M^K by $(S^K, \to^K)$, written $w \models M^K$.

3   Labels

Having a model, one is usually interested in what its properties are, e.g. which of the regulations are effective, how it behaves dynamically etc. We call functions that provide such information *labels*. We recall some previously introduced labels [5,16] one can assign to a parametrization, now updated to fit the workflow, and some new ones. As the domain of a label usually depends on the respective regulatory graph, we use the symbol l for a label in general, and l^K to denote that the label depends on the graph encoded by K and is evaluated under K.

All the labels are illustrated in Fig. 3 on the toy example in Fig. 2.

3.1   Sign

This label is based on the usual interpretation of an effect of a regulation:

$$(u, t, v) \in E \text{ is activating} \iff \exists \omega \in \Omega_v : K_v(\omega_{u \leftarrow [t, t_+]}) > K_v(\omega_{u \leftarrow [t_-, t]}),$$

$$(u, t, v) \in E \text{ is inhibiting} \iff \exists \omega \in \Omega_v : K_v(\omega_{u \leftarrow [t, t_+]}) < K_v(\omega_{u \leftarrow [t_-, t]}),$$

where $\omega_{u \leftarrow [t_-, t]}$ denotes that the regulatory interval ω_u is substituted by $[t_-, t]$. From this definition we derive the SIGN label $Sign^K : E^K \to \{0, +, -, 1\}$ where:

$$Sign^K(e) = \begin{cases} 0 & \text{iff } e \text{ is not } activating \text{ and not } inhibiting, \\ + & \text{iff } e \text{ is } activating \text{ and not } inhibiting, \\ - & \text{iff } e \text{ is not } activating \text{ and } inhibiting, \\ 1 & \text{iff } e \text{ is } activating \text{ and } inhibiting. \end{cases}$$

Structural Labels	
$Sign^K(A,1,B)$	+
$Sign^K(B,1,A)$	-
$Sign^K(B,2,B)$	-
$Indegree^K(A)$	1
$Indegree^K(B)$	2
$Indegree^K(SUM)$	3
$Bias^K(A)$	1
$Bias^K(B)$	2
$Impact^K(A,1,B)$	0.905
$Impact^K(B,1,A)$	-1
$Impact^K(B,2,B)$	-0.302

a)

b)

Regulatory Functions	
F_A^K	1&B{0}
F_B^K	1&A{1}&B{2}+2&A{1}&B{0,1}

c)

Property Labels	
$Cost^K(M^K)$	4
$Robustness^K(M^K)$	0.25

Fig. 3. Illustrative labels for the toy example from Fig. 2. **a)** All the possible structural labels. For clarity we use the symbol SUM to denote the sum of the INDEGREE values. **b)** The REGULATORY FUNCTION labels corresponding to the given parametrization. **c)** The dynamic labels for the measurement series M^K from Fig. 2f.

Note that the 0 value means that the regulation has no effect on its target and could be removed without affecting the dynamics, which is utilized in the following label INDEGREE. The value 1 describes the situation where a regulation has ambiguous semantics, not meeting the so-called *Snoussi condition* [10], which is usually contrary to the expectation of the modeller about the system.

3.2   Indegree

This self-explanatory label counts the number of effective incoming regulations. Formally we denote $Indegree^K : V^K \rightarrow \mathbb{N}$ the number of non-zero incoming regulations, defined as:

$$Indegree^K(v) = |\{(u,t,v) \in E \mid u \in V, Sign(u,t,v) \neq 0\}|.$$

Additionally, the function is extended to capture the sum of the INDEGREE values of all the components, such that $Indegree^K(V) = \sum_{v \in V^K} Indegree^K(v)$. The sum of INDEGREE values is of a special interest, as quite often one is interested in structures that are minimal w.r.t. number of regulations.

3.3   Cost

The COST [14] of a measurement series is equal to the number of states of its shortest witness. The value is of interest under the assumption that a shorter witness in general means a lower number of qualitative changes and in turn a slower energy consumption by the system. Even in the cases where the energy assumption is not realistic (e.g. due to different time scales) the COST value still reflects on how functionally complex the system is. Thus, one is usually interested in minimizing it.

Denote $\mathcal{M}^K$ the set of possible measurement series for S, then the label has the form $Cost^K : \mathcal{M}^K \to \mathbb{N}_0$ s.t. if there is no witness for M^K then $Cost^K(M^K) = 0$, otherwise $Cost^K(M^K) = min\{m \mid \exists w \in (\to^K)^{m-1} : w \models M^K\}$.

3.4   Robustness

The ROBUSTNESS [14] label $Robustness^K : \mathcal{P}^K \to [0,1]$ is closely related to the COST label. In general terms it denotes the probability that M such that $Cost^K(M) = m$ will be satisfied by a random walk of length m that starts from M_1. Formally:

$$Robustness^K(M) = \frac{|\{w \in (\to^K)^{m-1} \mid w_1 \in M_1, w \models M\}|}{|\{w \in (\to^K)^{m-1} \mid w_1 \in M_1\}|},$$

where w_1 denotes the first state on the path w. Understandably, if a measurement is not satisfiable, then there are no witnesses and the dividend and therefore also the ROBUSTNESS is equal to 0.

This particular notion of ROBUSTNESS reflects on the ability of the model to keep the requested behaviour even though uncertainty is introduced to the model through the modelling framework. The non-determinism of the simulation arises in states where the qualitative behaviour in reality depends on quantitative nuances indistinguishable by our abstraction. The higher the ROBUSTNESS of the model w.r.t. a measurement series, the less sensitive the model is to these quantitative nuances, respectively to perturbation in these.

3.5   Impact

The IMPACT label represents the relation between a regulator and its target via the function $Impact^K : E^K \to [-1,1]$. We have introduced this value in [16] and here we present a definition that uses regulatory contexts as its domain. For a regulation $(u,t,v) \in E$ we obtain the IMPACT of u on v by computing the correlation of the activity level of the regulator and the respective parameter value. As we are interested only in parameters that are directly affected by this regulation, we take a subset of regulatory contexts on the border of the threshold value t. These we list as an arbitrarily ordered vector $\Omega_v^t = (\omega \in \Omega_v \mid \omega_u \in \{[t_-,t),[t,t_+)\})$. To indicate presence or absence of the said regulation, we use the function $Pres_u : \Omega_u \to [0,\rho(u)]$ that projects the activity interval of u on its lower boundary, i.e. if $\omega_u = [t_-,t)$ then $Pres_u(\omega) = t_-$. The IMPACT of (u,t,v) is then equal to the Pearson correlation coefficient between the image of Ω_v^t under $Pres_u$ and K_v:

$$Impact^K(u,t,v) = \frac{cov(Pres_u(\omega)_{\omega \in \Omega_v^t}, K_v(\omega)_{\omega \in \Omega_v^t})}{std(Pres_u(\omega)_{\omega \in \Omega_v^t}) \cdot std(K_v(\omega)_{\omega \in \Omega_v^t})},$$

where cov is the covariance and std is the standard deviation. This value is quite helpful when one is searching for the key regulators of a certain component. The further the value is from 0, the more prominent the regulation is.

3.6 Bias

By the term BIAS we here mean the general tendency of a parametrization to push a component towards higher or lower activity levels. The BIAS label $Bias^K :$ $V \rightarrow [0,1]$ is obtained simply as $Bias^K(v) = \sum_{\omega \in \Omega_v} K_v(\omega) \cdot |\Omega_v|^{-1} \cdot \rho(v)^{-1}$. For a Boolean component this coincides with the notion as defined by other authors, e.g. [9].

As a component has in general more effect on the other components at higher activity levels, the BIAS label allows to distinguish the components whose presence seems to be crucial for the activity of the network.

3.7 Regulatory Function

While not being a label *per se* we also assign a logical REGULATORY FUNCTION, providing a more human-readable description of a parametrization. In particular, we describe each partial parametrization as a Post Algebra [7] expression in a disjunctive normal form (DNF) of cardinality $max\{\rho(v) \mid v \in V^K\}$.

A Post Algebra expression P in DNF of cardinality n is in our case described using the grammar:

$$P \rightarrow M|M \mid M$$
$$M \rightarrow V\&A \mid V$$
$$V \rightarrow 0 \mid \cdots \mid n$$
$$A \rightarrow A\&A \mid v\{L\}$$
$$L \rightarrow LL \mid V$$

where $v \in V, |, \&, \{, \}, 0, \ldots, n$ are terminals, and P, M, V, A, L are non-terminals. The semantics are such that an atom, i.e. an expression of the form $v\{L\}$, evaluates to n if the variable v is at a level listed in L and to 0 otherwise. The binary operator & evaluates to the smaller of its operands and the binary operator $|$ evaluates to the bigger of its operands. E.g. consider the function in Fig. 3b and an interpretation $A = 1, B = 1$. Then we can do the following valuation:

$$1\&A\{1\}|2\&B\{0,2\} \mapsto 1\&2|2\&B\{0,2\} \mapsto 1\&2|2\&0 \mapsto 1|2\&0 \mapsto 1|0 \mapsto 1.$$

Note that in the Boolean case the operator & corresponds to the logical conjunction, $|$ to the disjunction, $v\{1\}$ to the simple v, and $v\{0\}$ to $\neg v$.

We obtain the REGULATORY FUNCTION label by enumerating all the prime implicants and joining them via a disjunction.

4 Parametrization Sets Analysis and Comparison

While the individual parametrizations can be at least partially ordered by the values of their labels, it is only seldom that a single parametrization would appear as an optimal one. Moreover, even if one aims to find a parametrization that scores the best in all the metrics, i.e. minimum COST, maximum ROBUSTNESS, minimum INDEGREE etc., usually there are multiple parametrizations with the best score or those that are pairwise incomparable. We therefore focus on so-called *selections*, i.e. sets of parametrizations that fit certain criteria on the labels and analyse the whole selection.

a)

all parametrizations

Label	#	Elements
Cost(M)	2	0:66.67, 4:33.33,
F_A	4	0:25, 1:25, B{0}:25, B{12}:25,
F_B	81	0:1.23, 1:1.23, 1&A:1.23, 1&A&B{...
Indegree(A)	2	0:50, 1:50,
Indegree(B)	3	0:3.7, 1:14.81, 2:81.48,
Indegree(SUM)	4	0:1.85, 1:9.26, 2:48.15, 3:40.74,
K_A(B{0})	2	0:50, 1:50,
K_A(B{1})	2	0:50, 1:50,
K_B(A{0},B{0,1})	3	0:33.33, 1:33.33, 2:33.33,
K_B(A{0},B{2})	3	0:33.33, 1:33.33, 2:33.33,
K_B(A{1,2},B{0,1})	3	0:33.33, 1:33.33, 2:33.33,
K_B(A{1,2},B{2})	3	0:33.33, 1:33.33, 2:33.33,
Sign(A,1,B)	4	+:33.33, -:33.33, 0:11.11, 1:22.22,
Sign(B,1,A)	3	+:25, -:25, 0:50,
Sign(B,2,B)	4	+:33.33, -:33.33, 0:11.11, 1:22.22,

–

Label	#	Elements
Cost(M)	2	0:66.67, 4:-66.67,
F_A	4	0:-35, 1:25, B{0}:-15, B{12}:25,
F_B	81	0:1.23, 1:1.23, 1&A:1.23, 1&A&B{...
Indegree(A)	2	0:-10, 1:10,
Indegree(B)	3	0:-36.3, 1:-45.19, 2:81.48,
Indegree(SUM)	4	0:-18.15, 1:-50.74, 2:28.15, 3:40.74,
K_A(B{0})	2	0:-10, 1:10,
K_A(B{1})	2	0:-50, 1:50,
K_B(A{0},B{0,1})	3	0:33.33, 1:33.33, 2:-66.67,
K_B(A{0},B{2})	3	0:33.33, 1:33.33, 2:-66.67,
K_B(A{1,2},B{0,1})	3	0:13.33, 1:-6.67, 2:-6.67,
K_B(A{1,2},B{2})	3	0:13.33, 1:-6.67, 2:-6.67,
Sign(A,1,B)	4	+:33.33, -:-26.67, 0:-28.89, 1:22.22,
Sign(B,1,A)	3	+:25, -:-15, 0:-10,
Sign(B,2,B)	4	+:33.33, -:-33.33, 0:-88.89, 1:22.22,

Cost(p) = 4, Robustness(p) = 1, Sign(B,2,B) = 0

Label	#	Elements
Cost(M)	1	4:100,
F_A	2	0:60, B{0}:40,
F_B	3	1&A|2&1A:40, 2:40, 2&1A:20,
Indegree(A)	2	0:60, 1:40,
Indegree(B)	2	0:40, 1:60,
Indegree(SUM)	3	0:20, 1:60, 2:20,
K_A(B{0})	2	0:60, 1:40,
K_A(B{1})	1	0:100,
K_B(A{0},B{0,1})	1	2:100,
K_B(A{0},B{2})	1	2:100,
K_B(A{1,2},B{0,1})	3	0:20, 1:40, 2:40,
K_B(A{1,2},B{2})	3	0:20, 1:40, 2:40,
Sign(A,1,B)	2	-:60, 0:40,
Sign(B,1,A)	2	-:40, 0:60,
Sign(B,2,B)	1	0:100,

b)

Label	Count	Min	Max	Mean
K_A(B{0})	162	0	1	0.5
K_A(B{1})	162	0	1	0.5
K_B(A{0},B{0,1})	216	0	2	1
K_B(A{0},B{2})	216	0	2	1
K_B(A{1,2},B{0,1})	216	0	2	1
K_B(A{1,2},B{2})	216	0	2	1
Indegree(A)	162	0	1	0.5
Indegree(B)	312	0	2	1.777...
Indegree(SUM)	318	0	3	2.277...
Bias(A)	243	0	1	0.5
Bias(B)	320	0	1	0.5
Impact(B,1,A)	162	-1	1	0
Impact(A,1,B)	248	-1	1	4.386...
Impact(B,2,B)	248	-1	1	1.370...
Cost(M)	108	0	4	1.333...
Robustness(M)	108	0	1	0.1875

Label	Count	Min	Max	Mean
K_A(B{0})	160	0	0	0.099...
K_A(B{1})	162	0	1	0.5
K_B(A{0},B{0,1})	211	-2	0	-1
K_B(A{0},B{2})	211	-2	0	-1
K_B(A{1,2},B{0,1})	212	0	0	-0.19...
K_B(A{1,2},B{2})	212	0	0	-0.19...
Indegree(A)	160	0	0	0.099...
Indegree(B)	309	0	1	1.177...
Indegree(SUM)	314	0	1	1.277...
Bias(A)	241	0	0.5	0.3
Bias(B)	315	-0.5	0	-0.30...
Impact(B,1,A)	160	0	1	0.4
Impact(A,1,B)	245	0	1	0.6
Impact(B,2,B)	248	-1	1	1.370...
Cost(M)	103	-4	0	-2.66...
Robustness(M)	103	-1	0	-0.8125

Label	Count	Min	Max	Mean
K_A(B{0})	2	0	1	0.4
K_A(B{1})	0	0	0	0
K_B(A{0},B{0,1})	5	2	2	2
K_B(A{0},B{2})	5	2	2	2
K_B(A{1,2},B{0,1})	4	0	2	1.2
K_B(A{1,2},B{2})	4	0	2	1.2
Indegree(A)	2	0	1	0.4
Indegree(B)	3	0	1	0.6
Indegree(SUM)	4	0	2	1
Bias(A)	2	0	0.5	0.2
Bias(B)	5	0.5	1	0.8
Impact(B,1,A)	2	-1	0	-0.4
Impact(A,1,B)	3	-1	0	-0.6
Impact(B,2,B)	0	0	0	0
Cost(M)	5	4	4	4
Robustness(M)	5	1	1	1

c) A REGULATION graph.

d) A CORRELATION graph.

Fig. 4. Reports produced by TREMPPI for the graph in Fig. 2. For the interactive version please see [12]. **Left:** Reports for $\mathcal{K}^G$. **Right:** Reports for $\mathcal{K}^{G,\Psi}$ with $\Psi \equiv Cost(p) = 4 \wedge Robustness(p) = 1 \wedge Sign(B, 2, B) = 0$. **Middle:** A comparison left - right. **a)** A QUALITATIVE report. The label F_B is not fully listed. **b)** A QUANTITATIVE report. **c)** A REGULATION graph. **d)** A CORRELATION graph.

Have a parametrization space $\mathcal{K}$ and a sequence of predicates $\Phi = \Phi_1, \ldots, \Phi_n$ where $\Phi_i : \mathcal{K} \to \mathbb{B}$ for each $i \in [1, n]$. A *selection by* Φ we call the set of parametrizations denoted $\mathcal{K}^\Phi$ s.t. for each $K \in \mathcal{K}$ we have that $K \in \mathcal{K}^\Phi$ if and only if $\bigwedge_{i=1}^n \Phi_i(K)$ holds true. As the selections may contain millions or more parametrizations in size, approaches that allow to evaluate the whole selection at once are necessary to gain understanding of the nature of the selection. We present four different methods, each used to depict some of the labels in a manner that generalizes the values of the labels from members of the selection to the whole selection. A visual representation of such data is then called a *report*. Additionally, each of the reports features an individual method of *comparison*—having two different selections $\mathcal{K}^\phi$, $\mathcal{K}^\Psi$ we create a third report which illustrates the difference between the two selections. This we denote using the minus $(-)$ symbol, illustrating the fact that it is a non-commutative difference operation. All the reports are illustrated in Fig. 4 on the example network from Fig. 2. Each report provides a comparison between the set of all 324 parametrizations, i.e. a selection by $\Phi \equiv true$ and a selection where the M^K from Fig. 3 has minimal COST and maximal ROBUSTNESS and where the self-regulation of the component B is not present, i.e. a selection by $\Psi \equiv (Cost(series) = 4 \land Robustenss(series) = 1 \land Sign(B, 2, B) = 0)$.

4.1   Explicit Qualitative Report

The first tool we employ is a QUALITATIVE summary, which describes an image of a label in the selection, i.e. all the distinct label values that appear in the selection and their frequency in percent. Have a label $l : X \to Y$, where X, Y are some sets and a selection $\mathcal{K}^\Phi$. For example in the case $l = Sign^K$, we have $X = E$ and $Y = \{0, +, -, 1\}$. For each value $x \in X$ we then set:

$$qual(\mathcal{K}^\Phi, l, x) = (size(\mathcal{K}^\Phi, l, x), elems(\mathcal{K}^\Phi, l, x)),$$
$$size(\mathcal{K}^\Phi, l, x) = |elems(\mathcal{K}^\Phi, l, x)|,$$
$$elems(\mathcal{K}^\Phi, l, x) = \{(y, q) \mid q = |\{K \in \mathcal{K}^\Phi \mid l^K(x) = y\}| \cdot 100 \cdot |\mathcal{K}^\Phi|^{-1}\}.$$

A comparison of two selections $\mathcal{K}^\Phi, \mathcal{K}^\Psi$, denoted $qual(\mathcal{K}^\Phi, l, x) - qual(\mathcal{K}^\Psi, l, x)$, is obtained by subtracting the two pairs, where $elems(\mathcal{K}^\Phi, l, x) - elems(\mathcal{K}^\Psi, l, x)$ is computed as:

$$\{(y, q^\Phi - q^\Psi) \mid (y, q^\Phi) \in elems(\mathcal{K}^\Phi, l, x), (y, q^\Psi) \in elems(\mathcal{K}^\Psi, l, x)\}.$$

Since the set of parametrizations is finite, all values have finite domain and are thus suitable to this form of presentation. However, in the case of labels that project to rational numbers, i.e. ROBUSTNESS, BIAS, and IMPACT values, the size of the image quite often threatens to be almost as big as the selection itself, therefore we chose not to include them in the QUALITATIVE report.

4.2 Explicit Quantitative Report

Similarly to the previous, we summarize the overall nature of quantitative labels, i.e. those whose image is a subset of rational numbers, using the quadruple:

$$quan(\mathcal{K}^{\Phi}, l, x) = (count(\mathcal{K}^{\Phi}, l, x), min(\mathcal{K}^{\Phi}, l, x), max(\mathcal{K}^{\Phi}, l, x), mean(\mathcal{K}^{\Phi}, l, x)),$$
$$count(\mathcal{K}^{\Phi}, l, x) = |\{K \in \mathcal{K}^{\Phi} \mid l^{K}(x) \neq 0\}|,$$
$$min(\mathcal{K}^{\Phi}, l, x) = min\{l^{K}(x) \mid K \in \mathcal{K}^{\Phi}\},$$
$$max(\mathcal{K}^{\Phi}, l, x) = max\{l^{K}(x) \mid K \in \mathcal{K}^{\Phi}\},$$
$$mean(\mathcal{K}^{\Phi}, l, x) = \sum_{K \in \mathcal{K}^{\Phi}} l^{K}(x) \cdot |\mathcal{K}^{\Phi}|^{-1}.$$

The difference between the QUANTITATIVE reports of two selections $\mathcal{K}^{\Phi}$, $\mathcal{K}^{\Psi}$ is then set simply as the subtraction of the two quadruples.

Note that the *count* has a somewhat special meaning, as the 0 value is of particular interest for some of the labels. In the case of COST for example, it denotes that the respective measurement series is not satisfiable or for SIGN it states that the edge is absent.

4.3 Inferred Regulation Graph

Based on IMPACT and SIGN, we can summarize the average effect of regulations of a sample. The IMPACT can be easily extended from a parametrization to a sample as $Impact^{\mathcal{K}^{\Phi}}(e) = \sum_{K \in \mathcal{K}^{\Phi}} Impact^{K}(e) \cdot |\mathcal{K}^{\Phi}|^{-1}$ for each $e \in E$. For the SIGN we take a supremum under the partial ordering $0 < - < 1, 0 < + < 1$, i.e. for any $e \in E$ we set $Sign^{\mathcal{K}^{\Phi}}(e) = sup\{Sign^{K}(e) \mid K \in \mathcal{K}^{\Phi}\}$. Lastly we depict the FREQUENCY of a regulation, which states how often a regulation is active at all, i.e. has a non-zero SIGN, formally $Frequency^{\mathcal{K}^{\Phi}}(e) = |\{K \in \mathcal{K}^{\Phi} \mid Sign^{K}(e) \neq 0\}|$.

Visually, the IMPACT value is mapped to a color gradient of the regulation edge with the color red representing the value -1, yellow representing 0, and green representing 1. The FREQUENCY is mapped to the width of an edge. When the FREQUENCY is equal to 0, the edge is then displayed as dotted. Lastly, the SIGN is reflected in the shape of the head of the edge. The $+$ SIGN is mapped to a pointed arrow shape, the $-$ to a rectangle shape (also known as *blunt arrow*), the 1 to a combination of both and the 0 to a circle.

To create a comparison, the IMPACT and FREQUENCY values are directly subtracted. The SIGN can not be clearly interpreted in the comparison and for simplicity it is kept from the minuend. Note that the subtraction means that the result lies behind the original boundaries of a value. The color gradient is therefore stretched to the range $[-2, 2]$ and a negative FREQUENCY value is depicted by a dashed edge.

4.4   Correlation Graph

Similarly to the REGULATION graph we also create a correlation graph, based on the BIAS label. The label extended similarly to the IMPACT label, i.e. $Bias^{\mathcal{K}^{\Phi}}(v) = \sum_{K \in \mathcal{K}^{\Phi}} Bias^{K}(v) \cdot |\mathcal{K}^{\Phi}|^{-1}$ for $v \in V$. Additionally, one is usually interested in whether there is a relation between activities of multiple components, e.g. if one component seems to be taking over if another is missing. This is obtained as the correlation between the BIAS of individual components in a sample, i.e.:

$$Correlation^{\mathcal{K}^{\Phi}}(v, u) = \frac{cov(Bias^{K}(v)_{K \in \mathcal{K}^{\Phi}}, Bias^{K}(u)_{K \in \mathcal{K}^{\Phi}})}{std(Bias^{K}(v)_{K \in \mathcal{K}^{\Phi}}) \cdot std(Bias^{K}(u)_{K \in \mathcal{K}^{\Phi}})}.$$

The CORRELATION value is mapped to a color gradient in the same manner as the IMPACT value in the REGULATIONS graph. The BIAS value is mapped to the width of the border of the respective component in a manner similar to the edge width in the case of the FREQUENCY value.

To create a comparison both values are simply subtracted.

5   Case Study

To provide a practical demonstration of our methodology, we have utilized the data provided by D'Allesandro *et al.* in their study of hepatocyte growth factor (HGF) signalling [2].

In the original article the authors constructed a core network, illustrated in Fig. 5, with a set of regulations that are with high certainty present. Afterwards, a qualitative method is used to find an optimal structure that combines the core network with a subset of possible edges. To this end the authors obtained a rich set of experimental data, which they later discretized to meet the needs of their qualitative framework. The discretized data features measurements of 6 components in 6 different experimental set-ups. In each of the experiments one or two of the components of the network are inhibited and later the HGF stimuli is added. Additionally the authors provide a control measurement where no inhibition is present.

In the study, the data are present as fold-change comparisons between some of the experiments. For each component there are 9 time-points measured, however in the discretized form these are divided at the time of 30 min into an early and late response, as it is expected that around that time feedback effects start to play a role in the behaviour of the system. As the fold-change scheme is not suitable for encoding as a time-series, we reinterpreted the data into a measurement scheme, where the fold change translates to a difference between two measurements. This means that from two fold-changes we obtain three measurements. The particular values for the measurements were determined in the following manner:

– In the experiment a Met inhibitor was used that blocks the receptor of the pathway and thereby downregulates all signalling processes even under

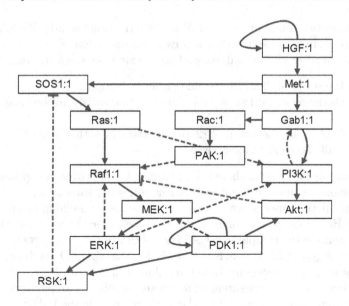

Fig. 5. The structure of the model that was identified as optimal in [2]. The regulations denoted by a full line constitute the core network, whereas those that are dashed are added from the pool of optional regulations. In the enumeration step we place requirements on the edges corresponding to the SIGN label, in particular the full edges with a pointed arrow are required to have the + SIGN, the dashed edges with a pointed to have either + or 0, and the dashed edges with a blunt arrow to have either − or 0.

stimulation. The fold-change comparison to the control shows a significant downregulation in all read-outs therefore we conclude that the control state has active read-outs.

- For other set-ups, if there is a significant decrease [2] in the fold change, the component is expected to be at the level 0 after the change.
- Likewise, if there is a significant increase, the component is expected to be at the level 1.
- If there is no significant change, no requirement is placed on the value.
- We require the full monotonicity in the behaviour, i.e. if a value of a component does not change between two timepoints, we require that it cannot change in the simulation either. If the value differs between two measurements, we require that there is exactly one change of that value. For details please refer to [15].

Altogether we have obtained 5 properties, which are detailed in the supplement [11]. Even though this interpretation of the data is quite strict, we obtained that even the core network is capable of satisfying all the experiments for each of the possible parametrizations. We have therefore focused instead on the structure identified as optimal by the authors to see whether addition of some of the optional edges can disrupt the expected function of the network. This optimal network is depicted in Fig. 5. This is a slightly simplified version of the original, which features two components for RSK in an activation cascade. We joined this cascade into a single node, which is a preserving operation [8].

For the purposes of the analysis we have created three selections:

1. *ALL* is the set of all 223776 possible parametrizations.
2. *VALID* is the set of 149184 parametrizations that satisfy all the measurement series.
3. *MINCOST* is the set of 135072 parametrizations that have the minimal COST for all the measurement series.

As can be immediately seen, there exist parametrizations over the optional edges that render one or more of the measurement series not satisfiable.

Data of all the reports are in an interactive form available in the supplement [11]. Here we provide some of the possible observations about the data. Firstly we analyse the comparison $VALID - ALL$. In the QUANTITATIVE report we see that $K_{\text{MEK}}(\text{PDK1} = 0, \text{Raf} = 1)$ is for the $VALID$ set bound to the value 1, meaning it is necessary that Raf1 alone can activate MEK. We also see that three out of the five measurement series are satisfied by all parametrizations, whereas the remaining two are satisfied exactly by those in the $VALID$ selection, meaning that both place the same requirement on the behaviour of the network. In the QUALITATIVE report we can see that the function MEK = Raf1&PDK1 is completely missing, in accordance with the quantitative observation, and the functions MEK = Raf1 and $MEK = \text{Raf1}|\text{PDK1}$ are now present with the same FREQUENCY, suggesting that the (PDK1, 1, MEK) regulation is superfluous and should probably be removed. In the REGULATIONS graph we then see that both the FREQUENCY and the IMPACT of PDK1 on MEK decreases and lastly in the CORRELATIONS report we see an increase in the BIAS of MEK.

Secondly we analyse the comparison $MINCOST - VALID$. From the QUANTITATIVE report we see that there is a slight increase in BIAS of Raf1. In the QUALITATIVE report we can see that 14 regulatory functions for Raf1 disappear completely, however as there are still 134 remaining, this does not provide too much information. A much cleaner picture can be gained from the REGULATIONS graph where we can see a decrease in the IMPACT and the FREQUENCY of ERK on Raf1 in favour of both inhibition by Akt and activation by PAK.

6   Conclusion

We have presented numerous methods for analysis and evaluation of qualitative parametrizations sets. The methods are gathered in two groups: labels, which evaluate individual parametrizations, and reports, which subsequently evaluate whole parametrizations sets. All of the methods are experimentally implemented in the tool TREMPPI [13]. The performance of the methods and the implementation is sufficient for application to realistic problems, as illustrated on a case study of HGF signalling.

While the method was mainly developed due to the specific nature of the problem of parameter identification in Thomas Networks, we believe that the general approach should be also applicable to more complicated frameworks. The approach should be readily convertible to frameworks which are, in certain ways, only an extension of the Thomas method, e.g. piece-wise affine models [3]. However even for frameworks with infinite parametrization pool, the label-report-compare approach could work if combined with an appropriate sampling method.

As the tools we were using for the purposes of this article are becoming more mature, we would like to make them more available for public use. To this end a public web-service version of TREMPPI is planned.

References

1. Alexopoulos, L.G., Saez-Rodriguez, J., Cosgrove, B.D., Lauffenburger, D.A., Sorger, P.K.: Networks inferred from biochemical data reveal profound differences in toll-like receptor and inflammatory signaling between normal and transformed hepatocytes. Mol. Cell. Proteomics **9**(9), 1849–1865 (2010)
2. DAlessandro, L.A., Samaga, R., Maiwald, T., Rho, S.-H., Bonefas, S., Raue, A., Iwamoto, N., Kienast, A., Waldow, K., Meyer, R., Schilling, M., Timmer, J., Klamt, S., Klingmller, U.: Disentangling the complexity of HGF signaling by combining qualitative and quantitative modeling. PLoS Comput. Biol. **11**(4), e1004192 (2015)
3. de Jong, H.: Modeling and simulation of genetic regulatory systems: a literature review. J. Comput. Biol. **9**(1), 67–103 (2002)
4. Guziolowski, C., Videla, S., Eduati, F., Thiele, S., Cokelaer, T., Siegel, A., Saez-Rodriguez, J.: Exhaustively characterizing feasible logic models of a signaling network using answer set programming. Bioinformatics **30**, 2320–2326 (2013)
5. Klarner, H., Siebert, H., Bockmayr, A.: Time series dependent analysis of unparametrized Thomas networks. IEEE/ACM Trans. Comput. Biol. Bioinf. **99**, 1338–1351 (2012)
6. Lee, W.-P., Tzou, W.-S.: Computational methods for discovering gene networks from expression data. Briefings Bioinf. **10**(4), 408–423 (2009)
7. Miller, D.M., Thornton, M.A.: Multiple Valued Logic: Concepts and Representations, vol. 2. Morgan & Claypool Publishers, San Rafael (2007)
8. Saadatpour, A., Albert, R., Reluga, T.C.: A reduction method for boolean network models proven to conserve attractors. SIAM J. Appl. Dyn. Syst. **12**(4), 1997–2011 (2013)
9. Shmulevich, I., Kauffman, S.A.: Activities and sensitivities in boolean network models. Phys. Rev. Lett. **93**(4), 048701 (2004)
10. Snoussi, E.H.: Qualitative dynamics of piecewise-linear differential equations: a discrete mapping approach. Dyn. Stab. Syst. **4**(3–4), 565–583 (1989)
11. Streck, A.: HGF network analysis (2015). http://dibimath.github.io/HGF_4_8_12/. Accessed 18 June 2015
12. Streck, A.: HSB 2015 example model data (2015). http://dibimath.github.io/HSB_2015/. Accessed 18 June 2015
13. Streck, A.: TREMPPI source repository (2015). https://github.com/xstreck1/TREMPPI/. Accessed 18 June 2015
14. Streck, A., Siebert, H.: Extensions for LTL model checking of Thomas networks. In: Advances is Systems and Synthetic Biology, vol. 14, pp. 101–114. EDP Sciences (2015)

15. Streck, A., Thobe, K., Siebert, H.: Analysing cell line specific EGFR signalling via optimized automata based model checking. In: Roux, O., Bourdon, J. (eds.) CMSB 2015. LNCS, vol. 9308, pp. 264–276. Springer, Heidelberg (2015)
16. Thobe, K., Streck, A., Klarner, H., Siebert, H.: Model Integration and crosstalk analysis of logical regulatory networks. In: Mendes, P., Dada, J.O., Smallbone, K. (eds.) CMSB 2014. LNCS, vol. 8859, pp. 32–44. Springer, Heidelberg (2014)
17. Thomas, R.: Regulatory networks seen as asynchronous automata: a logical description. J. Theoret. Biol. **153**(1), 1–23 (1991)

Analysis and Verification of Continuous and Hybrid Models

Parallelized Parameter Estimation
of Biological Pathway Models

R. Ramanathan[1(✉)], Yan Zhang[1], Jun Zhou[1], Benjamin M. Gyori[2],
Weng-Fai Wong[1], and P.S. Thiagarajan[2]

[1] School of Computing, National University of Singapore, Singapore, Singapore
{ramanathan,zhangyan,zhoujun,wongwf}@comp.nus.edu.sg
[2] Laboratory of Systems Pharmacology, Harvard Medical School, Boston, USA
{benjamin_gyori,thiagu}@hms.harvard.edu

Abstract. We develop a GPU based technique to analyze bio-pathway models consisting of systems of ordinary differential equations (ODEs). A key component in our technique is an online procedure for verifying whether a numerically generated trajectory of a model satisfies a property expressed in bounded linear temporal logic. Using this procedure, we construct a statistical model checking algorithm which exploits the massive parallelism offered by GPUs while respecting the severe constraints imposed by their memory hierarchy and the hardware execution model. To demonstrate the computational power of our method, we use it to solve the parameter estimation problem for bio-pathway models. With three realistic benchmarks, we show that the proposed technique is computationally efficient and scales well with the number of GPU units deployed. Since both the verification framework and the computational platform are generic, our scheme can be used to solve a variety of analysis problems for models consisting of large systems of ODEs.

Keywords: Biopathway models · Ordinary differential equations · Graphics processing units · BLTL · Statistical model checking · Parameter estimation

1 Introduction

We advocate a generic platform-aware technique to study the dynamics of large bio-pathways models. Specifically we focus on a well-established formalism, a system of ordinary differential equations (ODEs), to model the dynamics of the pathway models. For such systems we implement an analysis method on a multi-core platform consisting of a pool of general-purpose graphical processing units (GPGPUs or simply GPUs).

A system of ODEs together with a (initial) set of values for the initial concentrations and the rate constant values was formulated in [29] as a model of a biochemical network that takes into account cell-cell variability in a population.

This research was supported by the Singapore MOE grant MOE2013-T2-2-033.

© Springer International Publishing Switzerland 2015
A. Abate and D. Šafránek (Eds.): HSB 2015, LNBI 9271, pp. 37–57, 2015.
DOI: 10.1007/978-3-319-26916-0_3

These ODE systems will be high dimensional with no closed form solutions. To get around this, a probabilistic approximation technique accompanied by a statistical model checking procedure was developed by assuming a probability distribution over the set of initial values. Since the present paper is essentially an adaptation of this technique for an efficient GPU based implementation we shall begin with a brief description of theoretical underpinnings of this technique.

Variations in the initial concentrations of species and kinetic rate constants across a cell population are typical. To cater for these variations, we assume an initial probability distribution over the range of initial values and rate constants. We then suppose that the states of the system are observed at only the discrete time points $\{0, 1, 2, \ldots\}$ (with the unit of time being chosen suitably). In fact, we assume that the behavior of the system is of interest only up to a fixed maximum time point t_K. The constant K is chosen based on the application at hand. For instance, in the parameter estimation problem the last time point for which experimental data is available will be used to determine K. We choose bounded linear time temporal logic (BLTL) to specify dynamical properties and the BLTL specifications of interest are designed to respect this time bound. The atomic propositions used in the specifications will assert that the current value of a continuous variable lies in a given interval (with rational end points). This will in effect discretize the value space of the variables as well. We then impose the condition that the vector field associated with the ODE systems is C^1-continuous, which is a justifiable assumption in the context of biochemical reaction networks. As a consequence, one can construct a σ-algebra over TRJ^K, the set of trajectories of length K, and define a probability measure over this space.

Due to the deterministic dynamics, the probability measure over the space of trajectories will correspond to the probability measure over the space of measurable subsets of the set of initial states induced by the prior distribution over the set of initial values. More importantly, the set of initial states of the trajectories that satisfy a BLTL formula will be measurable in this σ-algebra and hence can be assigned a probability value. As a result, one has to merely sample the initial states and the parameter values according to the given initial distribution to carry out the sequential hypothesis testing procedure (or any statistical model checking procedure based on BLTL specifications). As an aside, this idea breaks down in the case of hybrid dynamics due to mode switchings and one needs to develop new technical machinery to induce a probability measure over the space of trajectories. A preliminary version of this extension is presented in a paper in the present volume [12].

The theory sketched above is used in [29] to carry out parameter estimation as follows. A conjunction of BLTL formulas describe the available experimental time-course data as well as known qualitative properties. One then deploys the statistical model checking procedure to evaluate the goodness of the current estimates for unknown parameter values. With the help of an evolutionary search strategy one then searches through the parameter space to obtain a good set of parameter values. The estimated values are then validated using test data that was not made available to the estimation procedure. For high dimensional systems with many unknown parameters, one will have to call the SMC procedure many times and for each such call one will have to generate sufficiently many trajectories to ensure the termination of the SMC procedure. Consequently, the computational cost induced by the repeated executions of the SMC procedure can be quite high. This is the motivation for the GPU based implementation of the above mentioned parameter estimation procedure developed in this paper.

Obviously, one can numerically generate trajectories in parallel on a GPU. Thus it is tempting to take for granted an easy parallel implementation and a corresponding increase in performance. This is, however, not the case. The memory hierarchy of a GPU and its single-instruction multiple-thread (SIMT) organization of its arithmetic units constitute severe constraints. A naïve implementation will often perform no better than (and in some cases worse than!) a sequential implementation. GPUs are, however, an attractive candidate since they are available off-the-shelf and can offer performance that is comparable to the more-expensive and less-available multi-core platforms. Furthermore, it is possible to form large pools of GPUs in a scalable and cost effective way using cloud services. Therefore, the effort required to overcome the architectural constraints of GPUs may well be worth it and this is the hypothesis we pursue here.

In simplified terms, the iterative parameter estimation procedure consists of:

(i) Encode the experimental data and known qualitative trends as a BLTL formula φ (as detailed in Sect. 3).
(ii) Fix the required confidence level and the false positives and negatives rates w.r.t. which one wishes to verify φ.
(iii) Guess a current value for each unknown parameter.
(iv) Evaluate the goodness of these estimated parameters by repeatedly generating trajectories till the statistical test associated with the SMC procedure terminates.
(v) If the outcome is yes then the current estimate is a good one. If not, guess a new set of values using the evolutionary search strategy and iterate.

Thus it is step (iv) which is ripe for parallelization. However just generating a numerical trajectory is not enough. One must evaluate if it satisfies φ which is of course easy to do. However only a small amount of memory will be available in the vicinity of a GPU core. Hence the generated trajectories need to be sent up through a number of levels in the memory hierarchy, each of which is significantly slower than the previous one. This will all but eliminate the performance gains obtained by generating the trajectories in parallel. Hence one must verify whether a generated trajectory satisfies φ on the fly without having to store the whole trajectory. Again this is not difficult to do though one must minimize the amount of intermediate data (typically Boolean combinations of the subformulas of φ that still need to be satisfied) to be kept track of. However the obvious online procedures will involve branching that is based on the current requirements and this will clash with the hardware parallelism available in GPUs. At the level of a single core, groups of parallel threads called *warps* are scheduled to run the compiled code, which at each step, execute the same machine instruction in a lock-step fashion. This is the heart of GPU's execution model. If two threads in a warp take different branches, the warp will have to be executed twice, once for each branch. This so called *branch divergence* causes severe performance degradation [22]. To avoid this, we construct a *deterministic* automaton-based online model checking technique. This is the main technical construction in the paper. It turns out that it is better to store the automaton (as a look-up table) in the intermediate storage shared by the cores and hence we also implement a standard latency hiding technique to mitigate the data transfer delays between this shared store and the global store (using which the rest of the parameter estimation steps are carried out) during model checking.

We have evaluated the performance of our GPU based parameter estimation procedure on a number of biopathway models drawn from the biomodels data base. Specifically we have applied our method to the EGF-NGF pathway, the segmentation clock pathway and the thrombin-dependent MLC-phosphorylation pathway. Our results show that one can hope to achieve significant performance gains especially by deploying a pool of GPUs. The present implementation can be further optimized and a similar strategy can be followed to solve the sensitivity analysis problem. As pointed out in the concluding section we also feel that other analysis problems concerning bio-pathways can be tackled using the approach developed in this paper.

1.1 Related Work

The problem of efficiently generating numerical trajectories on GPUs for large systems of ODEs has been studied in literature [23,27,35]. In our implementation, we adopt Liu *et al.*'s approach that encompasses a heterogeneous group of GPU threads where the memory-access threads and the trajectory computing threads are separated into different warps to achieve latency hiding and hence scalability.

Efficient methods for model checking probabilistic systems have been studied [7,8,21,31,33]. The statistical model checking (SMC) approach initiated by Younes and Simmons [34] based on the sequential probability ratio test proposed by Wald [32] has turned out to be a fruitful one and is adopted here. SMC usually involves checking whether an individual trace satisfies a given temporal specification. When the specification is a BLTL formula, this is known as *BLTL path checking*. Kuhtz and Finkbeiner show that the path checking problem can be parallelized by unrolling the BLTL formulas into Boolean circuits [18]. Barre *et al.* adopt the MapReduce framework [10] to verify a single large trace using distributed computing [3]. However, it is not clear how these methods can be implemented on a GPU-based platform.

On the other hand, Barnat *et al.* take an automata-theoretic approach to parallel model checking of a restricted class of multi-affine ODE systems [1,2]. The ODE model dynamics is first approximated as a rectangular abstraction automaton and a given LTL property is translated into a Büchi automaton that represents its negation. A parallel model checker then looks for an accepting cycle in the product automaton by symbolically exploring the state space.

But this approach tends to over-approximate the model dynamics. Oshima *et al.* present a FPGA-based framework for the checking of BLTL specifications with applications on partial differential equations [28]. Their method also involves a Büchi automaton construction but requires a large set of trajectories to be stored in the hardware before a property can be verified. In contrast our online method is based on GPUs, which we believe are more accessible and scalable. Further our focus is on ODE systems that arise in bio-pathway models.

In recent years, statistical model checking has become a building block to solve complex problems. David *et al.* apply SMC using analysis of variance (ANOVA) to find the optimal set of parameters of a network of stochastic hybrid automata [9]. Jha *et al.* show how the parameter synthesis problem for stochastic systems can be approached using statistical model checking [16]. In this paper, we focus on efficient parallelization techniques for traditional analysis tasks based on SMC, especially parameter estimation [5]. In particular, we realize the approach proposed by Palaniappan *et al.* [29] on a GPU-based platform. The key new ingredient is a novel *deterministic* online BLTL path checking procedure that fits in with the requirements of the GPU platform.

The paper is organized as follows. First Sect. 2 introduces the ODE dynamics and the syntax and semantics of BLTL. Section 3 formulates the online verification problem, and describes our automata-theoretic solution to this problem. In the subsequent Sect. 4, we develop the GPU based solution to the parameter estimation problem with the online verification procedure serving as the kernel. In the subsequent Sect. 5 we perform a number of performance case studies, and in Sect. 6 we summarize and point to future research directions.

2   Background

2.1   ODEs and Trajectories

In the present setting, a biochemical network is modeled as a system of ODEs. Assume that there are n molecular species $\{x_1, x_2, \ldots, x_n\}$ involved in the network. For each x_i, an equation of the form $\frac{dx_i}{dt} = f_i(\mathbf{x}, \Theta_i)$ describes the kinetics of the reactions that produce and consume x_i where $\mathbf{x}$ is the concentrations of the molecular species taking part in the reactions. Θ_i consists of the rate constants governing the reaction. Each x_i is a real-valued function of time $t \in \mathbb{R}$. We assume in this section that all rate constants are known. In Sect. 4, it will become clear how unknown rate constants are handled while solving the parameter estimation problem.

To capture the cell-to-cell variability regarding the initial states, we define for each variable x_i an interval $[L_i^{init}, U_i^{init}]$ with $L_i^{init} < U_i^{init}$. The actual value of the initial concentration of x_i is assumed to fall in this interval. We set $INIT = \prod_i [L_i^{init}, U_i^{init}]$. In what follows, we let $\mathbf{v}$ to range over $\mathbb{R}^n$.

We represent our system of ODEs in the vector form, $\frac{d\mathbf{x}}{dt} = F(\mathbf{x}, \Theta)$ with $F_i(\mathbf{x}, \Theta) := f_i$. In the setting of biochemical networks, the expressions in f_i will model kinetic laws such as mass-action and Michaelis-Menten's [17]. Moreover, the concentration levels of the various species will be bounded and the behavior of the system will be of interest only up to a finite time horizon. Hence we assume in this paper that f_i is Lipschitz-continuous for each i . As a result, for each $\mathbf{v} \in INIT$ the system of ODEs will have a unique solution $\mathbf{X_v}(t)$ [15]. We are also guaranteed that $\mathbf{X_v}(t)$ is a C^0-function (i.e., continuous function) [15] and hence measurable.

For convenience, we define the flow $\Phi : \mathbb{R}_+ \times \mathbf{V} \rightarrow \mathbf{V}$ for arbitrary initial vectors $\mathbf{v}$ as $\mathbf{X_v}(t)$. Intuitively, $\Phi(t, \mathbf{v})$ is the state reached under the ODE dynamics if the system starts at $\mathbf{v}$ at time 0. We work with $\Phi_t : \mathbf{V} \rightarrow \mathbf{V}$ where $\Phi_t(\mathbf{v}) = \Phi(t, \mathbf{v})$ for every t and every $\mathbf{v} \in \mathbf{V}$. Again, Φ_t is guaranteed to be a C^0-function (in fact $1 - to - 1$) and Φ_t^{-1} will also be a C^0-function.

In our applications, given the nature of the experimental data, the states of the system will be observed only at discrete time points and only within a finite time horizon. Hence by choosing a suitable unit of time we will assume that the states of the systems are observed at the time points $0, 1, \ldots$. A *trajectory* is a finite sequence $\tau = \mathbf{v}_0 \mathbf{v}_1 \ldots \mathbf{v}_k$ such that $\mathbf{v}_0 \in INIT$ and $\Phi_1(\mathbf{v}_j) = \mathbf{v}_{j+1}$ for $0 \leq j < k$. We let TRJ denote the set of finite trajectories which model the dynamics of the ODEs system.

2.2 Time-Bounded Linear Temporal Logic

In order to encode the dynamical properties of TRJ, we will use formulas in bounded time linear temporal logic (BLTL). An atomic proposition is of the form $(x_i \geq v)$ or $(x_i \leq v)$ with $v \in \mathbb{R}$. The proposition $(x_i \geq v)$ is interpreted as "the current concentration level of x_i is greater than or equal to v". A finite set of atomic propositions, AP, is assumed to be given for a bio-pathway model.

A BLTL formula is defined as follows. First, every atomic proposition, as well as the Boolean constants *true* and *false*, is a BLTL formula. If ψ_1 and ψ_2 are BLTL formulas, $\neg \psi$ and $\psi_1 \vee \psi_2$ are BLTL formulas. Also, if ψ is a BLTL formula, so is $X\psi$. Finally, if ψ_1 and ψ_2 are BLTL formulas, and t is a positive integer then $\psi_1 U^{\leq t} \psi_2$ is a BLTL formula. The derived propositional connectives $\wedge, \supset$, etc. and the temporal operators $G^{\leq t}$ and $F^{\leq t}$ are defined in the usual way.

Let $\tau = \mathbf{v}_0 \mathbf{v}_1 \ldots \mathbf{v}_k$ be a trajectory and $0 \leq j \leq k$. The semantic relation $\tau, j \models \psi$ is defined as follows.

- $\tau, j \models (x_i \geq v)$ iff $\mathbf{v}_j(i) \geq v$. The clause $\tau, j \models (x_i \leq v)$ is defined similarly.
- $\neg$ and $\vee$ are interpreted in the usual way.
- $\tau, j \models X\psi$ iff $j < k$ and $\tau, j + 1 \models \psi$.
- $\tau, j \models \psi_1 U^{\leq t} \psi_2$ iff there exists t' such that $t' \leq t$ and $j + t' \leq k$ and $\tau, j + t' \models \psi_2$. Further, $\tau, j + t'' \models \psi_1$ for every $0 \leq t'' < t'$.

We say that τ is a *model* of ψ if $\tau, 0 \models \psi$.

The rationale of choosing BLTL instead of a more sophisticated logic is twofold. First, relevant properties of bio-pathway models, especially in the context of parameter estimation and sensitivity analysis are linear time properties defined over a bounded time horizon. Second, BLTL has enough expressive power to characterize properties relating to bio-pathway models while being a very simple temporal logic to work with. Hence we choose BLTL over other commonly-used formalisms, such as continuous stochastic logic and metric temporal logic.

3   Online Model Checking Procedure

The problem of BLTL path checking involves determining whether a BLTL formula is satisfied by a trajectory. According to the BLTL semantics, it is easy to see that the truth value of a BLTL formula can be decided by trajectories with finite length. Online BLTL path checking requires only the current valuation of the atomic propositions as input. At each step, it evaluates the BLTL formula under the current valuation and generates a new formula that represents the "obligation" in the following step. The procedure terminates when the formula under consideration becomes either true or false, indicating a satisfaction or falsification of the original formula.

Such an algorithm can be easily implemented on CPUs. On the other hand, to achieve a good performance on GPUs, one must address the problem of branch divergence, which occurs when two GPU threads choose different code segments under the evaluation of a condition as illustrated in the following example.

Example 1 (Branch Divergence). Consider BLTL formula $\phi = F^{\leq 8} G^{\leq 5} p$, where p is an atomic proposition. Expanding ϕ, we get $\phi = (p \wedge XG^{\leq 4} p) \vee XF^{\leq 7} G^{\leq 5} p$. Notice that if the current valuation is $\sigma_1 = \{p \mapsto false\}$, ϕ is reduced to $\phi_1 = F^{\leq 7} G^{\leq 5} p$; if it is $\sigma_2 = \{p \mapsto true\}$, ϕ is reduced to $\phi_2 = G^{\leq 4} p \vee F^{\leq 7} G^{\leq 5} p$.

Now we initiate two GPU threads to check whether ϕ is satisfied for two different trajectories. Naively, we implement each thread as *if σ_1 then check ϕ_1 else check ϕ_2*. Branch divergence happens when the two trajectories take different valuations. Since GPU stream processors require that each GPU thread executes identical instructions, the two threads will process both ϕ_1 and ϕ_2 and simply discard the unrelated part, resulting in a 50 % loss of performance.   □

3.1   Automaton-Based BLTL Path Checking

To better utilize the parallelism of GPUs, we introduce an automaton-based BLTL path checking algorithm. Given a BLTL formula ψ, it is well-known that there exists a positive integer K that depends only on ψ such that for any trajectory τ whose length is greater than K, one needs to examine only a prefix of length K to determine whether τ is a model of ψ [4]. The online procedure we shall construct examines τ as it is being generated (through numerical simulation) in a lock-step fashion. Instead of generating a trajectory of length K at once, it incrementally simulates the ODE model and checks whether the current trajectory satisfies the formula ψ.

It is convenient to focus on the sequence of truth values of the atomic propositions induced by a trajectory. Let us call such a sequence *AP-sequence*. Given a trajectory $\tau = \mathbf{v}_0 \mathbf{v}_1 \ldots \mathbf{v}_k$, its induced AP-sequence is denoted as τ_{ap}, which is the sequence $P_0 P_1 \ldots P_k$ where for $0 \leq i \leq k$:
$$(x_j \bowtie v) \in P_i \text{ iff } \mathbf{v}_i(j) \bowtie v, \bowtie \in \{\leq, \geq\}.$$
We now wish to construct a deterministic automaton for ψ that accepts (rejects) an AP-sequence iff it is (not) a model of ψ.

As the first step, we replace the time constants mentioned in ψ by symbolic variables and manipulate these variables separately. To this end, we define the formula $sym(\psi)$ inductively as follows.

- $sym(\psi) = \psi$ if ψ is an atomic proposition;
- $sym(\neg\psi) = \neg sym(\psi)$ and $sym(\psi_1 \vee \psi_2) = sym(\psi_1) \vee sym(\psi_2)$;
- $sym(X\psi) = X sym(\psi)$;
- $sym(\psi_1 U^{\leq t}\psi_2) = sym(\psi_1)U^{\leq x_\alpha} sym(\psi_2)$ where $\alpha = \psi_1 U^{\leq t}\psi_2$.

Thus the subscript assigned to the symbolic variable is the sub-formula in which the time constant appears. Often for convenience we will index these variables by integers rather than concrete formulas. Thus $sym(F^{\leq 8}p \vee G^{\leq 3}q)$ will be typically represented as $F^{\leq x_1}p \vee G^{\leq x_2}q$. We refer to $sym(\psi)$ as a *symbolic* BLTL formula.

For a BLTL formula ψ, we now define the automaton $\mathcal{A}_\psi = \langle S_\psi, 2^{AP_\psi}, \rightarrow, s_{in}, \mathcal{F}\rangle$, where S_ψ is the set of states, AP_ψ is the set of atomic propositions that appear in ψ, $\rightarrow \subseteq S_\psi \times 2^{AP_\psi} \times S_\psi$ is the transition relation (to be defined below), $s_{in} \in S_\psi$ is the initial state and $\mathcal{F} \subseteq S_\psi$ are the final states.

Let $\phi_{in} = sym(\psi)$ and CL be the least set of formulas that contains the sub-formulas of $sym(\psi)$ and satisfies:

If $\psi_1 U^{\leq x}\psi_2$ is in CL then $X\psi_1 U^{\leq x}\psi_2$ is also in CL.

We let BC denote the Boolean combinations of formulas in CL. A state of the automaton is a triple of the form (ϕ, Y, V), where $\phi \in BC$, Y is the set of variables that appear in ϕ, and V is a valuation that assigns a *positive* integer to every variable in Y. We define $s_{in} = (\phi_{in}, Y_{in}, V_{in})$, where Y_{in} is the set of the symbolic variables that appear in ϕ_{in}, and V_{in} assigns to each variable in Y_{in} the corresponding value in ψ. More precisely, if x_α is in Y_{in} and $\alpha = \psi_1 U^{\leq t}\psi_2$ then $V_{in}(x_\alpha) = t$. $\mathcal{F} = \{(true, \emptyset, \emptyset), (false, \emptyset, \emptyset)\}$.

Next we define the transition relation $\rightarrow$ of $\mathcal{A}$. Let (ϕ, Y, V) and (ϕ', Y', V') be states and $P \subseteq AP_\psi$ be a set of atomic propositions. Then $(\phi, Y, V) \xrightarrow{P} (\phi', Y', V')$ is a transition iff the following conditions are satisfied.

- Suppose $\phi = p$ is an atomic proposition. If $p \in P$, then $\phi' = true$; otherwise, $\phi' = false$. In either case $Y' = V' = \emptyset$.
- Suppose $\phi = \neg\varphi$, and there exists a transition $(\varphi, Y, V) \xrightarrow{P} (\varphi', Y'', V'')$. Then $\phi' = \neg\varphi'$, $Y' = Y''$ and $V' = V''$.
- Suppose $\phi = \phi_1 \vee \phi_2$, and there exist transitions $(\phi_1, Y_1, V_1) \xrightarrow{P} (\phi_1', Y_1', V_1')$ and $(\phi_2, Y_2, V_2) \xrightarrow{P} (\phi_2', Y_2', V_2')$. Then $\phi' = \phi_1' \vee \phi_2'$, $Y' = Y_1' \cup Y_2'$, and $V'(x_i) = V_i'(x_i)$ for $x_i \in X_i$, $i \in \{1, 2\}$.
- Suppose $\phi = X\varphi$. Then $\phi' = \varphi$ and $Y' = Y$ and $V' = V$.
- Suppose $\phi = \phi_1 U^{\leq x_\alpha}\phi_2$, and there exist transitions $(\phi_1, Y_1, V_1) \xrightarrow{P} (\phi_1', Y_1', V_1')$ and $(\phi_2, X_2, V_2) \xrightarrow{P} (\phi_2', Y_2', V_2')$. Then $\phi' = \phi_2' \vee (\phi_1' \wedge X\varphi)$ where $\varphi = \phi_2$ if $V(x_\alpha) = 1$. Furthermore $Y' = Y_1' \cup Y_2'$ and V' restricted to Y_1' is V_1' and V' restricted to Y_2' is V_2'. If $V(x_\alpha) > 1$ then $\varphi = \phi_1 U^{\leq x_\alpha}\phi_2$. Furthermore $Y' = Y_1' \cup Y_2' \cup \{x_\alpha\}$ while V' restricted to Y_1' is V_1' and V' restricted to Y_2' is V_2'. In addition $V'(x_\alpha) = V(x_\alpha) - 1$.

The set of states S_ψ is given inductively:

$s_{in} \in S_\psi$. Suppose $s \in S_\psi$ and $s \xrightarrow{P} s'$. Then $s' \in S_\psi$.

It is easy to show that this automaton has the required properties. Moreover its number of states is bounded by $\ell + \Sigma_{x \in X_{in}} V_{in}(x)$ where ℓ is the number of appearances of the X operator in ψ.

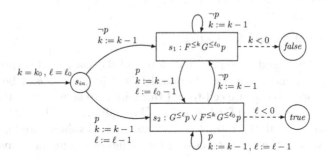

Fig. 1. Automaton for the BLTL formula $F^{\leq k_0} G^{\leq \ell_0} p$ with $k_0 = 8$ and $\ell_0 = 5$.

Example 2. Consider the BLTL formula $\psi = F^{\leq k_0} G^{\leq \ell_0} p$, where $k_0 = 8$ and $\ell_0 = 5$ are constants. Figure 1 shows a fragment of the automaton A_ψ. To avoid clutter we have not explicitly shown the symbolic variables and their valuations. The dashed arcs indicate that the input states will transit to the corresponding final states given proper valuations of atomic propositions.   □

4   Parameter Estimation

The values of many of the rate constants appearing in the ODEs and the initial concentrations of the species will often be unknown. One will have to learn them using limited experimental data. Solving this parameter estimation problem is the crucial first step towards the analysis of ODEs based bio-pathway models. Here we derive a parallel extension of the method developed by Palaniappan et al [29]. This will lead to a GPU implementation of a solution to this crucial problem.

For convenience, we shall assume–as done in the previous section–that all the initial concentrations are known but that their nominal values can vary over a cell population. The parameter estimation procedure searches through the value space of the unknown parameters to determine the "best" combination of values that can explain the given data and predict new behaviors [26]. The key step in this procedure is to determine the fit-to-data of the current set of parameter values. We use BLTL to encode both experimental time series data and known qualitative trends concerning the dynamics of the pathway. We then develop a parallel statistical model checking procedure (SMC) to determine the goodness of the given set of parameter values, while taking into account that these values can fluctuate across the population of cells that the data is based on. This procedure will numerically generate trajectories in parallel and use our online model checking method to determine if the current trajectory satisfies the given specification. Subsequently, we use a global optimization strategy known as SRES [30] to choose a new set of candidate parameter values according to the SMC based score assigned to the current set.

4.1 Statistical Model Checking

Consider an ODE-based model of a pathway and the associated notations developed in the previous section. In addition, let $\Theta = \{\theta_1, \theta_2, \ldots, \theta_m\}$ be the set of all rate constants. To capture cell-to-cell variability we assume that the range of values for each θ_j is $[L^j, U^j]$ for $1 \leq j \leq m$. We shall present the SMC procedure while assuming that all the rate constants –as interval values– are known. They are to be viewed as the current guess of the values of the unknown parameters.

An implicit assumption is that the value of a rate constant, when fixed initially, does not change during the time evolution of the dynamics, although this value can be different for different cells. To verify whether the ODE system satisfies a property $Pr_{\geq r}\psi$, where ψ is a BLTL formula and $r \in [0,1)$, we use a statistical model checking procedure based on Younes and Simmons' method [34]. Assume that we are given a distribution (usually uniform) over $INIT$ and $\Pi_j[L^j, U^j]$. Then the notion $Pr_{\geq r}\psi$ standing for "the probability of a trajectory chosen randomly according to the given distribution over $INIT$ and $\Pi_j[L^j, U^j]$ satisfying the formula ψ is $\geq r$"- can be precisely defined [29].

Accordingly, we test the alternative pair of hypotheses : $H_0 : p \geq r + \delta$ and $H_1 : p \leq r - \delta$ where p is the standard probability measure of the set of trajectories that meet the specification ψ and δ is the user defined indifference region. α and β signify the type-I error and type-II error bounds respectively. We use Wald's sequential probability ratio test (SPRT) [32] for the sequential hypothesis test. In SPRT, random samples are drawn iteratively and we update the SPRT ratio, q_m at the end of each round as

$$q_m = \frac{[r-\delta]^{(\sum_{i=1}^m y_i)}[1-[r-\delta]]^{(m-\sum_{i=1}^m y_i)}}{[r+\delta]^{(\sum_{i=1}^m y_i)}[1-[r+\delta]]^{(m-\sum_{i=1}^m y_i)}}.$$

The variables $y_1, y_2 \ldots$ signify a sequence of Bernoulli random variables which correspond to the set of trajectories with an assigned value of 1 if a trajectory k satisfies the property ψ or 0, if it does not satisfy. When sufficient samples are drawn, the test terminates. Otherwise, the test proceeds to draw more samples until the statistical guarantee defined by the error bounds and the indifference region are met. We define our stopping criterion as follows: We accept the null hypothesis H_0 if $q_m \geq \widehat{A}$ and accept the alternate hypothesis H_1 if $q_m \leq \widehat{B}$. Otherwise, we update q_m and sample a new random trajectory. For the thresholds of the sequential hypothesis test, we set $\widehat{A} = \frac{1-\beta}{\alpha}$ and $\widehat{B} = \frac{\beta}{1-\alpha}$.

4.2   The GPU Implementation

In this section, we first describe the design of our online method that overcomes the stringent memory restrictions imposed by the GPU platform to evaluate large number of trajectories as they are numerically generated. We then discuss how the SMC procedure is implemented in our setting using latency hiding.

Our online approach uses the automaton constructed in Sect. 3, which eliminates the need for handling different formulas explicitly. Recall that running an automaton $\mathcal{A}$ is equivalent to evaluating the corresponding BLTL formula under a series of valuations at different time points until a final state is reached. To efficiently implement this on GPU, branch divergence should be avoided as much as possible. Our solution is to index states, variables and the atomic propositions as defined in Sect. 3, and encode the transitions and the operations on the valuations into an array A_T. This array represents transitions and operations on the valuations, in which each row corresponds to an input state, and each column to an atomic proposition. Each element of the array consists of an output state and the operations on the valuations associated to the transition. Each GPU thread has access to A_T which is pre-computed and stored in the shared memory. A step in the run of the automaton is performed by all threads of a warp executing in lock-step updating the state and the variables according to A_T. Note that dummy self-loops for the terminal states are added so that once one of them is reached, the automaton stays there forever. This avoids explicit checking for termination, which induces branch divergence.

Example 3. For the fragment of the automaton A_ψ defined in Fig. 1, the array

$$
A_T = \begin{array}{c} \\ s_{in} \\ s_1 \\ s_2 \\ \top \\ \bot \end{array}
\begin{array}{c} \sigma_1 \qquad\quad \sigma_2 \\
\left[\begin{array}{cc}
s_1, a_{01} & s_2, a_{02} \\
s_1, a_{11} & s_2, a_{12} \\
s_1, a_{21} & s_2, a_{22} \\
\top, \{\} & \top, \{\} \\
\bot, \{\} & \bot, \{\}
\end{array}\right]
\end{array}
$$

encodes the automaton, where $\sigma_1 = \{p \mapsto false\}$ and $\sigma_2 = \{p \mapsto true\}$, and a_{ij} updates the set of time variables for the jth transition out of the ith state.

Our code generation scheme for the multi-thread based numerical simulation of an ODEs system is similar to the method developed by Hagiescu *et al.* [13]. During simulation, we generate a number of blocks of trajectories in parallel where the blocks are distributed across a number of GPU cores. At each time step, for each trajectory, we update the current state of the constructed deterministic automaton. We also periodically check if all the trajectories in a given block have hit a final state in the automaton. When this is the case we update this state information for all the trajectories in the block to the global memory.

If threads from other warps are also scheduled for such long latency global memory accesses, the memory access delay due to control flow divergence will impact performance. To get around this, we use a latency hiding technique where by the global memory accesses are pre-fetched by threads in a separate warp at the same time as when the other threads carry out the numerical integration.

At the global memory level we first pick the terminal state of a trajectory belonging to a block uniformly at random and use it to update the current SPRT score. When the SMC procedure reaches a decision we stop the concurrent numerical integration.

4.3   Parameter Estimation

As the first step, we describe how experimental data can be encoded as BLTL formulas. To do so we first mildly extend the syntax of BLTL with the formulas of type $\psi_1 U^t \psi_2$ with the semantics: ψ_1 will hold exactly up to t time units from now at which point ψ_2 will hold. The construction of the automaton presented in Sect. 3 can be easily extended to handle this case. Assume, without loss of generality, that $O \subseteq \{x_1, x_2, \ldots, x_k\}$ is the set of variables for which experimental data is available, and which has been allotted as training data to be used for parameter estimation. Assume $\mathcal{T}_i = \{\tau_1^i, \tau_2^i, \ldots, \tau_{T_i}^i\}$ are the time points at which the concentration level of x_i has been measured and reported as $[\ell_t^i, u_t^i]$ for each $t \in \mathcal{T}_i$. The interval $[\ell_t^i, u_t^i]$ is chosen to reflect the noisiness, the limited precision and the cell-population based nature of the experimental data. For each $t \in \mathcal{T}_i$, we define the formula $\psi_i^t = \mathcal{F}^t(i, \ell_t^i, u_t^i)$. Then $\psi_{exp}^i = \bigwedge_{t \in \mathcal{T}_i} \psi_i^t$. We then set $\psi_{exp} = \bigwedge_{i \in O} \psi_{exp}^i$. In case the species x_i has been measured under multiple experimental conditions, the encoding scheme is extended in the obvious way.

Often qualitative dynamic trends will be available for some of the molecular species in the pathway. For instance, we may know that a species shows transient activation, in which its level rises in the early time points, and later falls back to initial levels. Similarly, a species may be known to show oscillatory behavior with certain characteristics. Such information can be described as BLTL formulas that we term to be *trend* formulas. We let ψ_{qlty} to be the conjunction of all the trend formulas. We assume $\Theta_u = \{\theta_1, \theta_2, \ldots, \theta_K\}$ as the set of unknown parameters. For convenience we will assume that the other parameter values are known and that their nominal values do not fluctuate across the cell population. We will also assume nominal values for the initial concentrations and the range of their fluctuations of the form $[L_i^{init}, U_i^{init}]$ for each variable x_i. Again, for convenience, we fix a constant δ'' so that if the current estimate of the values of the unknown parameters is $\mathbf{w} \in \prod_{1 \leq j \leq K} [L^j, U^j]$ then this value will fluctuate in the range $[\mathbf{w}(j) - \delta'', \mathbf{w}(j) + \delta'']$. Setting $L_{init,\mathbf{w}}^j = \mathbf{w}(j) - \delta''$ and $U_{init,\mathbf{w}}^j = \mathbf{w}(j) + \delta''$ we define $INIT_{\mathbf{w}} = (\prod_i [L_i^{init}, U_i^{init}]) \times (\prod_j [L_{init,\mathbf{w}}^j, U_{init,\mathbf{w}}^j])$. The set of trajectories $TRJ_{\mathbf{w}}$ is defined accordingly.

To estimate the quality of $\mathbf{w}$, we run our parallel SMC procedure – using $INIT_{\mathbf{w}}$ – to verify $P_{\geq r}(\psi_{exp} \wedge \psi_{qlty})$. Depending on the outcome of the test for the various conjuncts in the specification, we assign a score to $\mathbf{w}$ using an objective function detailed below. This evaluation is done at the global memory level. We then iterate this scheme for various values of $\mathbf{w}$ generated using a suitable search strategy. For each such $\mathbf{w}$ we launch a fresh instance of the parallel SMC procedure on the GPU network. Using a cloud service, one can launch as many parallel sets of SMC procedures as there are GPU instances available.

The objective function is formed as follows. Let J_{exp}^i $(= T_i)$ be the number of conjuncts in ψ_{exp}^i, and J_{qlty} the number of conjuncts in ψ_{qlty}. Let $J_{exp}^{i,+}(\mathbf{w})$ be the number of formulas of the form ψ_i^t (a conjunct in ψ_{exp}^i) such that the statistical test for $P_{\geq r}(\psi_i^t)$ accepts the null hypothesis (that is, $P_{\geq r}(\psi_i^t)$ holds) with the strength $(\frac{\alpha}{J}, \beta)$, where $J = \sum_{i \in O} J_{exp}^i + J_{qlty}$. Similarly, let $J_{qlty}^+(\mathbf{w})$ be the number of conjuncts in ψ_{qlty} of the form $\psi_{\ell,qlty}$ that pass the statistical test $P_{\geq r}(\psi_{\ell,qlty})$ with the strength $(\frac{\alpha}{J}, \beta)$. Then $\mathcal{G}(\mathbf{w})$ is computed via

$$\mathcal{G}(\mathbf{w}) = J_{qlty}^+(\mathbf{w}) + \sum_{i \in O} \frac{J_{exp}^{i,+}}{J_{exp}^i}. \tag{1}$$

Thus the goodness of fit of $\mathbf{w}$ is measured by how well it agrees with the qualitative properties as well as the number of experimental data points with which there is acceptable agreement. To avoid over-training the model, we do not insist that every qualitative property and every data point must fit well with the dynamics predicted by $\mathbf{w}$.

The search strategy over the parameter space will use $\mathcal{G}(\mathbf{w})$ as an objective function. Global search methods including the Stochastic Ranking Evolutionary Strategy (SRES) [30] are much better at avoiding local minima than local methods but are computationally more intensive. We use the SRES algorithm in our work since it is known to perform well in the context of pathway models [26]. During the iterative optimization process, SRES maintains a *population* of parameter value vectors, and each search iteration is called a *generation* after which selection is performed over the population. We note, however, that the choice of search algorithm is orthogonal to our proposed method.

5   Experimental Evaluation

We applied our method[1] to three ODE based pathway models taken from the BioModels database [20]. We first verified properties of interest on each of the three pathway models. Using our parallelized SMC framework, we then performed parameter estimation on these models. The GPU implementation was based on CUDA 5.0 runtime and tested on four NVidia Tesla K20 m GPUs with 4.8 GB global memory, clocked at 706 MHz each. We compared the performance of our algorithm with that of a CPU based implementation on a PC with 3.4 Ghz Intel Core i7 processor with 8 GB of memory. The model checker and the numerical solver for the CPU implementation were written in C++. The CPU implementation uses the highly optimized stiff solvers in the CVODE package [14] for integrating ODEs. On the GPU, we implemented the fourth order Runge-Kutta

---
[1] Source code available at https://www.comp.nus.edu.sg/~rpsysbio/smcgpu/.

method (used for the EGF-NGF and segmentation clock model) and the adaptive step-size Runge-Kutta-Fehlberg method [19] (used for the thrombin pathway model). For the cloud implementation, we ported our single node implementation to 25 Amazon Web Service (AWS) cloud `g2.8xlarge` GPU nodes. Each such node has two Intel Xeon E5-2670 CPUs of 8 cores each and four NVIDIA GK104 GPUs with 60 GB host memory and 4 GB global memory on each GPU device. The nodes are connected by AWS Enhanced Networking and communicate using CUDA-aware OpenMPI. The NVIDIA GK104 GPUs have 1536 cores clocked at 797 MHz each with 4 GB global memory and a memory bandwidth of 160 GB/s.

5.1   Case Studies: Property Verification

Thrombin Dependent MLC-Phosphorylation Pathway. Thrombin plays an important role in the contraction of endothelial cells through multiple pathways leading to the phosphorylation of MLC [24]. The pathway model has 105 differential equations and 197 kinetic parameters. Simulation time was fixed at 1000 s divided into 20 equally spaced time points. We used the nominal model (all rate parameter values known) to verify if it conformed to a property with a high probability expressed in BLTL. It is known experimentally that the concentration of MLC* (phosphorylated MLC) starts at a low level, and then reaches a high steady state value. The corresponding formula is

$$P_{\geq 0.9}(([\text{MLC}^* \leq 1]) \land F^{\leq 5}(G^{\leq 20}([\text{MLC}^* \geq 3]))).$$

Our SMC analysis concluded that the nominal model does not satisfy this property, and we found that phosphorylated MLC shows a transient profile. This discrepancy has been studied in [25], where it was attributed to missing components in the proposed model.

Our online procedure for this case achieves significant speedup (4.6× in a single GPU setting) compared to an offline GPU based model checker which first generates trajectories in parallel, stores them in the global memory and then carries out the model checking procedure on the CPU.

EGF-NGF Pathway. The EGF-NGF signaling pathway captures the differential response to two growth factors, EGF and NGF in the PC12 neuro-endocrine cell line [6]. EGF induces cell proliferation while NGF promotes cell differentiation. The difference in cell fate is attributed to the duration of Erk activation. For studying this model, simulation time was set to 61 min divided into equally spaced intervals of 1 min each. We checked whether starting from a low value, the concentration of Erk* (active Erk) reaches a high value and then begins to fall. This property can be formalized as

$$P_{\geq 0.9}([0 \leq \text{Erk}^* \leq 2.2 \cdot 10^5] \land F^{\leq 10}([4.8 \cdot 10^5 \leq \text{Erk}^* \leq 5.6 \cdot 10^5])$$
$$\land F^{\leq 20}(G^{\leq 30}([2.2 \cdot 10^5 \leq \text{Erk}^* \leq 4.8 \cdot 10^5]))).$$

The property was confirmed to be true by our SMC method suggesting that Erk shows sustained activation upon EGF stimulation.

Segmentation Clock Pathway. The segmentation pattern of the spine in developing embryos is controlled by oscillations in Notch, Wnt and FGF signaling due to coupled feedback loops [11]. The ODE model representing this pathway was simulated up to 200 min with observations assumed to be available every 5 min. We formulated the oscillations observed in the concentration profile of Dusp6-mRNA as a BLTL property as follows

$$P_{\geq 0.9}([\text{Dusp6 mRNA} \leq 1] \wedge (F^{\leq 10}([\text{Dusp6 mRNA} \geq 5.5] \wedge$$
$$F^{\leq 10}([\text{Dusp6 mRNA} \leq 1] \wedge F^{\leq 10}([\text{Dusp6 mRNA} \geq 5.5]))))).$$

This property was verified to be true suggesting oscillations in Dusp6 mRNA with a period of approximately 100 min.

5.2   Case Studies: Parameter Estimation

We next evaluated our method for estimating unknown model parameters based on a combination of quantitative time series data and qualitative specifications of dynamical trends. Using the nominal model we generated training data to be used for parameter estimation and an independent set of test data not used for fitting. To generate time series data points, we simulated random trajectories on the GPU by sampling initial concentration from a ±5 % range around the nominal values. We also encoded the dynamic trends of a few species as properties in BLTL. Later, for each BLTL property, its respective symbolic automaton was constructed. We allowed 0.5 % parameter variability around the current estimate of parameters in each iteration of the search procedure. Table 1 summarizes the key features of the models including the number of variables (N_x), the number of parameters (N_Θ), the number of parameters assumed to be unknown (N_{Θ_u}), the number of equally spaced time points (T) and for SRES, the total number of individuals (λ) and the number of generations (G).

Table 1. Parameter estimation setup and model specifications

Bio-pathway model	N_x	N_Θ	N_{Θ_u}	T	λ	G
EGF-NGF	32	48	20	61	200	100
Segmentation clock	16	75	39	40	200	300
Thrombin	105	197	100	20	100	500

For the thrombin pathway all training data and test data were quantitative time course data with one exception. Namely, for Thrombin R*, a dynamical trend formulated in BLTL was used as training data expressing that it reaches a high level within 200 s and then falls to a low level (Fig. 2). We used only quantitative data for the EGF-NGF pathway and found a good fit to both training and test data by the fitted model (Fig. 3). For the segmentation clock model only

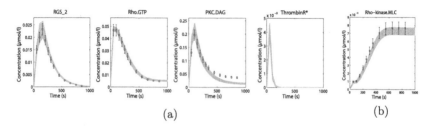

(a)   (b)

Fig. 2. Parameter estimation of the thrombin pathway, showing model fit to (a) training data and (b) test data.

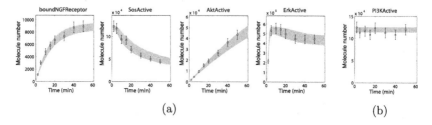

(a)   (b)

Fig. 3. Parameter estimation of the EGF-NGF pathway, showing fit to (a) training data and (b) test data.

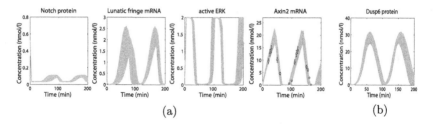

(a)   (b)

Fig. 4. Parameter estimation of the segmentation clock pathway, showing fit to (a) training data and (b) test data.

Axin2 mRNA was assumed to have quantitative time course data available, and dynamical trends were given as training and test data for the remaining species. For instance, the test data for Dusp6 protein expresses that at least two peaks and troughs are reached within 200 min – this test property was satisfied by the fitted model as seen in Fig. 4. In each case, the simulated dynamics of the fitted model is plotted by sampling randomly from the initial conditions while using the fitted parameter values.

5.3 Performance

We measured the runtime of the parameter estimation procedure with different combinations of SPRT error bounds (α and β), indifference regions (δ), and threshold probability (r) used within the SMC procedure. We found that for all three models, while GPU runtimes stayed roughly constant across all SPRT parameter combinations, runtimes for the CPU based implementation increased significantly for more stringent statistical tests (see Fig. 5). For instance with the most stringent statistical test, the GPU implementation took just 42 min for finding the best parameter set for the EGF-NGF model on a 4-GPU node, a 24.6× speed-up compared to the 17.2 h taken by the CPU implementation.

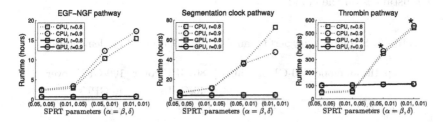

Fig. 5. Comparison of CPU and GPU runtimes on parameter estimation with different combinations of SPRT parameters (error bounds $\alpha = \beta$, indifference region δ and probability threshold r). *Estimated values based on shorter runs.

Next, Table 2 shows the performance of our parameter estimation method on a range of parallel architectures with the SPRT parameters set to $\alpha = \beta = \delta = 0.01$ and $r = 0.9$. In the 4-GPU server setup, for every generation in our single node parallel implementation, we divided the total number of individuals across 4 GPUs equally. For the cloud based implementation, the set of individuals were divided across 100 GPU instances in 25 machines with 4 GPUs per node.

Table 2. Performance of our scheme across different architectures (*Estimated values based on shorter runs.)

Model	CPU [hr]	4-GPU node [hr]	100-GPU cloud [hr]	4-GPU node over CPU
EGF-NGF	17.22	0.69	0.05	24.6×
Segmentation clock	47.5	4.01	0.45	11.9×
Thrombin	556.8*	111.1	5	5×*

While the 4-GPU server implementation took 42 min to complete the EGF-NGF parameter estimation task, the same took only 3 min on the 100-GPU cloud. For the segmentation clock pathway, the 4-GPU implementation took 4 h, a speed-up of approximately 11.9× over the CPU implementation. Finally, parameter estimation for the thrombin model would take an estimated 23.2 days using a CPU based implementation. (Note that this estimate was obtained by running an initial number of generations in the parameter search, calculating the average time taken for a generation, and then extrapolating the run time for the maximal generation number.) The cloud based implementation on the other hand is able to estimate the parameters in about 5 h.

Finally, Table 3 presents the scaled performance of our parameter estimation method applied on the EGF-NGF and the segmentation clock pathway models on the cloud. As might be expected our method achieves near perfect linear scaling when all the individuals in each round of the SRES procedure are launched on unique instances on the cloud.

Table 3. Strong scaling performance of the cloud based implementation

Bio-pathway model	40-GPUs Time[s]	80-GPUs over 40-GPUs	100-GPUs over 40-GPUs
EGF-NGF	445.28	1.62x	2.36x
Segmentation clock	3864.74	1.74x	2.35x

6   Conclusion

In this paper we proposed a technique for studying the dynamics of large models of biological pathways that utilizes the power of commodity graphics processors. In particular, starting with a model consisting of a system of ordinary differential equations, we developed a parallel, online procedure for checking if the trajectories of this model satisfy a bounded linear temporal logic formula. Our procedure works around various architectural constraints of the graphics processor execution model to achieve significant performance both on local systems as well as in the cloud. We believe that this opens the door for studying large pathway models in a scalable and cost-effective manner.

We used the parameter estimation problem to illustrate the applicability of our method, which consists of a parallel SMC procedure whose core is a deterministic online model checking procedure that determines if the trajectory under construction satisfies a given BLTL formula. Many analysis questions can be tackled by assuming a distribution over the set of initial concentrations and parameter values, which will then induce a probability measure on the set of trajectories satisfying a given BLTL formula. For instance, sensitivity analysis of a model can be carried out in this fashion as shown in [29] and we are currently constructing a GPU based implementation using the framework presented here.

When constructing dynamical models to explain experimental observations, one often ends up with a population of models with different structures corresponding to different hypotheses about the underlying system. With sufficient GPU units available, one can evaluate the quality of a large number of these models in parallel using our method. One can also explore the parameter landscape to identify regions most likely to induce the desired pathway responses to chosen stimuli. Our future work will involve exploring such issues in the context of model comparison.

References

1. Barnat, J., Brim, L., Cerná, I., Drazan, S., Fabriková, J., Láník, J., Safránek, D., Ma, H.: Biodivine: A framework for parallel analysis of biological models. In: Proceedings Second International Workshop on Computational Models for Cell Processes, COMPMOD 2009, Eindhoven, The Netherlands, 3 November 2009, pp. 31–45 (2009)
2. Barnat, J., Brim, L., Ceska, M., Lamr, T.: Cuda accelerated LTL model checking. In: 2009 15th International Conference on Parallel and Distributed Systems (ICPADS), pp. 34–41. IEEE (2009)
3. Barre, B., Klein, M., Soucy-Boivin, M., Ollivier, P.-A., Hallé, S.: MapReduce for parallel trace validation of LTL properties. In: Qadeer, S., Tasiran, S. (eds.) RV 2012. LNCS, vol. 7687, pp. 184–198. Springer, Heidelberg (2013)
4. Biere, A., Cimatti, A., Clarke, E., Zhu, Y.: Symbolic model checking without BDDs. In: Cleaveland, W.R. (ed.) TACAS 1999. LNCS, vol. 1579, p. 193. Springer, Heidelberg (1999)
5. Bortolussi, L., Sanguinetti, G.: Learning and designing stochastic processes from logical constraints. In: Joshi, K., Siegle, M., Stoelinga, M., D'Argenio, P.R. (eds.) QEST 2013. LNCS, vol. 8054, pp. 89–105. Springer, Heidelberg (2013)
6. Brown, K.S., Hill, C.C., Calero, G.A., Myers, C.R., Lee, K.H., Sethna, J.P., Cerione, R.A.: The statistical mechanics of complex signaling networks: nerve growth factor signaling. Phys. Biol. $1(3)$, 184 (2004)
7. Bulychev, P., David, A., Larsen, K.G., Mikučionis, M., Poulsen, D.B., Legay, A., Wang, Z.: Uppaal-smc: Statistical model checking for priced timed automata. arXiv preprint arXiv:1207.1272 (2012)
8. Clarke, E.M., Faeder, J.R., Langmead, C.J., Harris, L.A., Jha, S.K., Legay, A.: Statistical model checking in *BioLab*: applications to the automated analysis of T-cell receptor signaling pathway. In: Heiner, M., Uhrmacher, A.M. (eds.) CMSB 2008. LNCS (LNBI), vol. 5307, pp. 231–250. Springer, Heidelberg (2008)
9. David, A., Du, D., Guldstrand Larsen, K., Legay, A., Mikučionis, M.: Optimizing control strategy using statistical model checking. In: Brat, G., Rungta, N., Venet, A. (eds.) NFM 2013. LNCS, vol. 7871, pp. 352–367. Springer, Heidelberg (2013)
10. Dean, J., Ghemawat, S.: MapReduce: simplified data processing on large clusters. Commun. ACM $\mathbf{51}(1)$, 107–113 (2008)
11. Goldbeter, A., Pourquié, O.: Modeling the segmentation clock as a network of coupled oscillations in the Notch, Wnt and FGF signaling pathways. J. Theoret. Biol. $\mathbf{252}(3)$, 574–585 (2008)
12. Gyori, B.M., Liu, B., Paul, S., Ramanathan, R., Thiagarajan, P.: Approximate probabilistic verification of hybrid systems. In: Abate, A., Sǎfránek, D. (eds.) HSB 2015. LNCS (LNBI), vol. 9271, pp. 96–116. Springer, Heidelberg (2015)
13. Hagiescu, A., Liu, B., Ramanathan, R., Palaniappan, S.K., Cui, Z., Chattopadhyay, B., Thiagarajan, P., Wong, W.F.: GPU code generation for ode-based applications with phased shared-data access patterns. ACM Trans. Archit. Code Optim. (TACO) $\mathbf{10}(4)$, 55 (2013)

14. Hindmarsh, A., Brown, P., Grant, K., Lee, S., Serban, R., Shumaker, D., Woodward, C.: SUNDIALS: Suite of nonlinear and differential/algebraic equation solvers. ACM T. Math. Softw. **31**(3), 363–396 (2005)

15. Hirsch, M., Smale, S., Devaney, R.: Differential equations, dynamical systems, and an introduction to chaos. Academic Press, New York (2012)

16. Jha, S.K., Langmead, C.J.: Synthesis and infeasibility analysis for stochastic models of biochemical systems using statistical model checking and abstraction refinement. Theoret. Comput. Sci. **412**(21), 2162–2187 (2011)

17. Klipp, E., Herwig, R., Kowald, A., Wierling, C., Lehrach, H.: Systems Biology in Practice: Concepts, Implementation and Application. Wiley-VCH, Weinheim (2005)

18. Kuhtz, L., Finkbeiner, B.: Efficient parallel path checking for linear-time temporal logic with past and bounds. arXiv preprint arXiv:1210.0574 (2012)

19. Lambert, J.D.: Numerical Methods for Ordinary Differential Systems: The Initial Value Problem. Wiley, Chichester (1991)

20. Le Novere, N., Bornstein, B., Broicher, A., Courtot, M., Donizelli, M., Dharuri, H., Li, L., Sauro, H., Schilstra, M., Shapiro, B., Snoep, J., Hucka, M.: BioModels database: a free, centralized database of curated, published, quantitative kinetic models of biochemical and cellular systems. Nucleic Acids Res. **34**, D689–D691 (2006)

21. Legay, A., Delahaye, B., Bensalem, S.: Statistical model checking: an overview. In: Barringer, H., et al. (eds.) RV 2010. LNCS, vol. 6418, pp. 122–135. Springer, Heidelberg (2010)

22. Lindholm, E., Nickolls, J., Oberman, S., Montrym, J.: Nvidia tesla: a unified graphics and computing architecture. IEEE Micro **28**(2), 39–55 (2008)

23. Liu, B., Hagiescu, A., Palaniappan, S.K., Chattopadhyay, B., Cui, Z., Wong, W.F., Thiagarajan, P.: Approximate probabilistic analysis of biopathway dynamics. Bioinformatics **28**(11), 1508–1516 (2012)

24. Maeda, A., Ozaki, Y.I., Sivakumaran, S., Akiyama, T., Urakubo, H., Usami, A., Sato, M., Kaibuchi, K., Kuroda, S.: Ca2+-independent phospholipase A2-dependent sustained Rho-kinase activation exhibits all-or-none response. Genes to Cells **11**(9), 1071–1083 (2006)

25. Maedo, A., Ozaki, Y., Sivakumaran, S., Akiyama, T., Urakubo, H., Usami, A., Sato, M., Kaibuchi, K., Kuroda, S.: Ca^{2+}-independent phospholipase A2-dependent sustained Rho-kinase activation exhibits all-or-none response. Genes Cells **11**, 1071–1083 (2006)

26. Moles, C.G., Mendes, P., Banga, J.R.: Parameter estimation in biochemical pathways: a comparison of global optimization methods. Genome Res. **13**(11), 2467–2474 (2003)

27. Murray, L.: GPU acceleration of Runge-Kutta integrators. IEEE Trans. Parallel Distrib. Syst. **23**(1), 94–101 (2012)

28. Oshima, K., Matsumoto, T., Fujita, M.: Hardware implementation of BLTL property checkers for acceleration of statistical model checking. In: Proceedings of the International Conference on Computer-Aided Design, pp. 670–676. IEEE Press (2013)

29. Palaniappan, S.K., Gyori, B.M., Liu, B., Hsu, D., Thiagarajan, P.S.: Statistical model checking based calibration and analysis of bio-pathway models. In: Gupta, A., Henzinger, T.A. (eds.) CMSB 2013. LNCS, vol. 8130, pp. 120–134. Springer, Heidelberg (2013)

30. Runarsson, T.P., Yao, X.: Stochastic ranking for constrained evolutionary optimization. IEEE Trans. Evol. Comput. **4**(3), 284–294 (2000)

31. Sen, K., Viswanathan, M., Agha, G.: On statistical model checking of stochastic systems. In: Etessami, K., Rajamani, S.K. (eds.) CAV 2005. LNCS, vol. 3576, pp. 266–280. Springer, Heidelberg (2005)
32. Wald, A.: Sequential tests of statistical hypotheses. Ann. Math. Stat. **16**, 117–186 (1945)
33. Younes, H.L., Kwiatkowska, M., Norman, G., Parker, D.: Numerical vs. statistical probabilistic model checking. Int. J. Softw. Tools Technol. Transfer **8**(3), 216–228 (2006)
34. Younes, H.L., Simmons, R.G.: Statistical probabilistic model checking with a focus on time-bounded properties. Inf. Comput. **204**(9), 1368–1409 (2006)
35. Zhou, Y., Liepe, J., Sheng, X., Stumpf, M.P., Barnes, C.: GPU accelerated biochemical network simulation. Bioinformatics **27**(6), 874–876 (2011)

High-Performance Discrete Bifurcation Analysis for Piecewise-Affine Dynamical Systems

Luboš Brim, Martin Demko, Samuel Pastva, and David Šafránek[(✉)]

Systems Biology Laboratory, Faculty of Informatics, Masaryk University,
Botanická 68a, 602 00 Brno, Czech Republic
{brim,xdemko,xpastva,xsafran1}@fi.muni.cz

Abstract. Analysis of equilibria, their stability and instability, is an unavoidable ingredient of model analysis in systems biology. In particular, bifurcation analysis which focuses on behaviour of phase portraits under variations of parameters is of great importance. We propose a novel method for bifurcation analysis that employs coloured model checking to analyse phase portraits bifurcation in rectangular abstractions of piecewise-affine systems. The algorithm works on clusters of workstations and multi-core computers to allow scalability. We demonstrate the method on a repressilator genetic regulatory network.

1 Introduction

Many dynamical systems appearing in computational systems biology fall into the class of time-autonomous non-linear systems. Mathematical models focus on specific parts of studied biological mechanisms. This can be thought of as isolating a part from the whole system in a similar way as can be done in wet-lab experimentation (e.g., isolation of photosystem protein complexes in [16]) and is used as a methodology in synthetic biology (e.g., synthetic metabolic pathway modules [12]). Many models are constructed at the level of positive/negative feed-backs among system variables (e.g., enzyme kinetics or Hill kinetics).

A significant class of isolated non-linear systems displays kinetic functions that combine kinetic laws where each depends on a single variable, e.g., kinetics of first-order reactions, single-variable Hill functions, Michaelis-Menten kinetics in signalling pathways or in sequences of metabolic reactions. This restriction limits the complexity of the local behaviour in the systems phase space. Respective dynamical systems can be very well approximated by means of *piecewise-affine* systems.

Despite the limitations, piecewise-affine systems constitute a well-accepted modelling formalism. They have been found to be a valuable tool for practical analysis of complex genetic regulatory networks at the qualitative level [5,21] which would be difficult to handle with complex nonlinear systems.

This work has been supported by the Czech Science Foundation grant No. GA15-11089S.

© Springer International Publishing Switzerland 2015
A. Abate and D. Šafránek (Eds.): HSB 2015, LNBI 9271, pp. 58–74, 2015.
DOI: 10.1007/978-3-319-26916-0_4

The system dynamics is significantly affected by kinetic parameters. In order to obtain a precise (quantitative) analysis of piecewise-affine systems, various parameters, such as reaction rates or concentration values, need to be taken into account. For a typical model, some of the parameter values can be determined from the literature or experimental data, many parameters values are unknown.

In this paper we utilise the approach of abstracting the continuous phase space of the piecewise-affine systems dynamics into a discrete finite set of rectangles on which the system becomes *locally linear*. The *rectangular abstraction* procedure has two steps: (i) non-linear kinetic functions are approximated by piecewise linear functions resulting in a piecewise-affine system (the algorithm has been introduced in [15]), (ii) a finite-state automaton is defined on rectangles of the piecewise-affine system [9]. This approach enables qualitative analysis of the respective class of biochemical dynamical systems by means of methods based on model checking and allows to identify the system equilibria and approximate the character of stability in near proximity of individual equilibria.

Key Contributions. We propose a parallel (distributed-memory) algorithm for dynamic flow and stability analysis of continuous piecewise-affine models with respect to kinetic parameters. The main idea is to use special kind of atomic propositions, called *direction propositions*, which define elementary directions of flow in the discrete vectorfield of the rectangular abstraction and build various flow patterns from these propositions, possibly using temporal operators. Examples of such patterns might be sink, saddle, source, equilibrium, flow etc. Pattern formulas are analysed using parallel coloured parameter synthesis technique which is built on model checking. The novelty of our approach is thus in a high-performance automated method that analyses how patterns change (appear, disappear, move, etc.) depending on kinetic parameters. For piecewise-affine systems we can give guarantees on some of the computed results. The method keeps the simplicity and practicality of the pure qualitative analysis.

Related Work. The dynamical properties of the class of piecewise-affine systems have been the subject of active research for more than three decades (e.g., [14,17,21], see [19] for a review). Some attention has been also paid to the clarification of the correspondence between phase portraits in piecewise-affine systems and in the corresponding discrete models [3,8,17,22]. The theoretical background of our approach is based on [6,9] which we further generalise to parameterised dynamical systems by adopting the parameter uncertainty model for piecewise multi-affine systems [4]. Our setting differs from [4,15] in that we strictly focus on a class of continuous piecewise-affine systems where the rectangular abstraction takes the advantage of exact characterisation of systems equilibria. Since the rectangularisation is done by means of closed sets [9], we get an abstraction that completely covers the systems dynamics for a fixed parameterisation. However, in the similar way as in [4], parameter space is partitioned by open sets to avoid equilibria to occur on threshold planes and to avoid trajectories to slide over threshold planes. This assumption significantly simplifies the theoretical framework and allows us to characterise stability by means of single-state phase portrait patterns. In contrast to this paper, the work [10] provides characterisation of stability by means of a qualitative abstraction for a fixed symbolic parameterisation and in [23] the authors employ model checking for discrete-time piecewise-affine systems for a fixed parameterisation.

2 Parameterised Rectangular Abstraction

We give the theoretical background that combines the closed set rectangular abstraction for piecewise-affine systems summarised in [9] with the framework of parameter uncertainty that has been defined in [4] for piecewise multi-affine systems rectangularised on open sets. This combination is unique in the sense it allows us to take the advantage of complete and sound characterisation of equilibria by means of rectangularisation and brings the results of [9] to the domain of piecewise-affine systems with parameter uncertainty.

Another extension is in the form of the rectangular transition system resulting from the abstraction. In contrast with previous work, we encode the information of *flow direction* into transitions.

We define a class of parameterised non-linear systems covering dynamical systems with positive and negative regulations that do not multiply on any systems variable. The only allowed combination is summation. Such limitation still covers many interesting models of genetic regulatory networks [1] or metabolic and signalling pathways relevant in systems and synthetic biology [12]. Optimal approximation of regulation functions by means of piecewise-affine functions defined in [15] is employed.

Let $\mathbb{P} \subset \mathbb{R}^m$ denote the *parameter space* of m uncertain *parameters*. For fixed $m_1, ..., m_n \in \mathbb{N}_0$ such that $\sum_{i=1}^n m_i = m$, a *parameterisation* $p \in \mathbb{P}$ is defined as a tuple $p = (p_1^1, ..., p_1^{m_1}, p_2^1, ..., p_2^{m_2}, ..., ..., p_n^1, ..., p_n^{m_n})$ where $p_i^j \in \mathbb{R}^+$. The meaning of m_i is the number of uncertain parameters in the ith systems dimension.

We consider as admissible a class of dynamical systems in the form $\dot{x} = f(x, p)$ where $x = (x_1, ..., x_n)$ is a vector of system *variables*, $n \in \mathbb{N}$ the systems *dimension*, and for every fixed $p \in \mathbb{P}, f(p) = (f_1(p), ..., f_n(p)) : \mathbb{R}^n \rightarrow \mathbb{R}^n$ is a vector of *kinetic functions* satisfying the following constraints:

1. Variables $x_1, ..., x_n$ are linearly independent.
2. For every $p \in \mathbb{P}, f(p)$ is completely defined and continuous on an n-dimensional rectangle $\mathcal{D} = \prod_{i=1}^n [0, max_i] \subset \mathbb{R}^n$ where max_i is the upper bound assumed for $x_i, \nu_\mathcal{D} \overset{\mathrm{df}}{=} \prod_{i=1}^n \{0, max_i\}$ is the set of *boundary vertices*, and $\mathcal{D}_i$ denotes the *domain* of x_i.
3. $\forall p \in \mathbb{P}, \forall x \in \nu_\mathcal{D}, \forall i \in \{1, ..., n\}.(x_i = 0 \Rightarrow f_i(x, p) > 0) \wedge (x_i = max_i \Rightarrow f_i(x, p) < 0)$.
4. For each $i \in \{1, ..., n\}, p \in \mathbb{P}, f_i$ has the form

$$f_i(x, p) = \sum_{j \in I^+} \kappa_i^j \varrho_i^j(x_{l_{ij}}) - \sum_{j \in I^-} \gamma_i^j \varrho_i^j(x_{l_{ij}})$$

where I^+ and I^- are finite index sets such that $I^+ \cap I^- = \emptyset$, $0 \neq \kappa_i^j, \gamma_i^j \in \mathbb{R}^+$ are *kinetic coefficients*, and $\varrho_i^j(x_{l_{ij}})$ is an arbitrary monotonic and continuous *regulatory function* defined for the variable $x_{l_{ij}}$ where $l_{ij} \in \{1, ..., n\}$ that is total on $\mathcal{D}_{l_{ij}}$. Any kinetic coefficient κ_i^j (resp. γ_i^j) can be *uncertain* provided that there exists $P \subseteq \mathbb{P}$ and a unique k such that $\kappa_i^j = p_i^k$ (resp. $\gamma_i^j = p_i^k$) for any $p \in \mathbb{P}$.

5. We assume $\mathbb{P}$ such that every component p_i^k of $p \in \mathbb{P}$ corresponds to some unique uncertain kinetic coefficient appearing in ith kinetic function.

The set $\Omega = \{X_i \mid i = 1, ..., m\}$ is called *rectangular partitioning* of $\mathcal{D}$ if (i) for all $i = 1, ..., m$: X_i is a closed full-dimensional rectangle in $\mathbb{R}^n$, (ii) $\cup_{i=1}^m X_i = \mathcal{D}$, and (iii) for all $i, j = 1, ..., m$, $i \neq j$, the intersection $X_i \cap X_j$ is either empty, or a common face of X_i and X_j. We use the notation ν_{X_i} to denote the set of *boundary vertices* of X_i.

A mapping $g : \mathcal{D} \to \mathbb{R}^n$ is called *piecewise-affine* on Ω if (i) g is continuous on $\mathcal{D}$, and (ii) for all $i = 1, ..., m$ there exist $A_i \in \mathbb{R}^{n \times n}$ and $a_i \in \mathbb{R}^n$ such that for all $x \in X_i$: $g(x) = A_i x + a_i$, i.e., $g \mid_{X_i}$ is an affine mapping.

For a given $p \in \mathbb{P}$, a *piecewise-affine system (on a rectangle $\mathcal{D}$)* is a tuple $\Sigma = (\mathcal{D}, \Omega, x_0, t_0, g)$ where Ω is a rectangular partitioning of $\mathcal{D}$, $x_0 \in \mathcal{D}$ is the initial continuous state, $t_0 \in \mathbb{R}_0^+$ is the initial time, and $g(p) : X \to \mathbb{R}^n$ is a piecewise-affine function on Ω. A *trajectory* $x : \mathbb{R}_0^+ \to \mathcal{D}$ of system Σ is a solution of the differential equation $\dot{x}(t) = g(x(t), p)$ for the initial condition $x(t_0) = x_0$.

For every parameterisation $p \in \mathbb{P}$, an admissible (non-linear) dynamical system of dimension n given by the equation $\dot{x} = f(x, p)$, initial condition $x(t_0) = x_0$, and bounded in the domain $\mathcal{D}$, can be approximated by a piecewise-affine system $\Sigma = (\mathcal{D}, \Omega, x_0, t_0, g(p))$ using the following procedure:

1. For each $i \in \{1, ..., n\}$, g_i is constructed from f_i by replacing every $\varrho_i^j(x_{l_{ij}})$ with a piecewise-affine function $R_i^j(x_{l_{ij}})$ obtained in the algorithm [15] which for a given number of requested affine segments approximates $\varrho_i^j(x_{l_{ij}})$ by an optimal polygonal chain (to simplify the notation let $\lambda := l_{ij}$ in the following definition):

$$R_i^j(x_\lambda) \overset{\text{df}}{=} \sum_{k=1}^{w_\lambda - 1} r(x_\lambda, \mu_i^k, \mu_i^{k+1}, \varrho_i^j(\mu_i^k), \varrho_i^j(\mu_i^{k+1}))$$

where
  - w_λ is number of *thresholds* defined on x_λ including 0 and $max_\lambda(w_\lambda \geq 2)$, $\mu_i^0 = 0$ and $\mu_i^k < \mu_i^{k+1}$ for each $k \in \{0, ..., w_\lambda - 2\}$, $\mu_i^{w_\lambda - 1} = max_\lambda$,
  - $r(x, \sigma, \sigma', y, y') = \begin{cases} y + \frac{x - \sigma}{\sigma' - \sigma} \cdot (y' - y) & \text{if } \sigma \leq x \leq \sigma' \wedge \varrho_i^j(x_\lambda) \text{ increasing,} \\ y + \frac{x - \sigma}{\sigma' - \sigma} \cdot (y - y') & \text{if } \sigma \leq x \leq \sigma' \wedge \varrho_i^j(x_\lambda) \text{ decreasing,} \\ 0 & \text{otherwise.} \end{cases}$

2. For each $\iota = (j_1, ..., j_n) \in \prod_{i=1}^n \{0, ..., w_i - 2\}$ we define a rectangle $X_\iota = \prod_{i=1}^n [\mu_i^{j_i}, \mu_i^{j_i+1}]$. Note that $\forall 1 \leq i \leq n$. $max_i = \mu_i^{w_i - 1}$.

3. Finally, the partitioning Ω is constructed: $\Omega = \{X_\iota | \iota \in \prod_{i=1}^n \{0, ..., w_i - 2\}\}$.

Every pair $X_\iota, X_{\iota'} \in \Omega$ such that $\exists! j. \iota'_j = \iota_j + \delta, \delta \in \{-1, 1\}$ and $\forall i, i \neq j. \iota'_i = \iota_i$ satisfies $||\iota - \iota'|| = 1$ where $||.||$ denotes the Euclid norm. For such ι, ι' we say that $X_{\iota'}$ is a *neighbouring rectangle of* X_ι *in direction* $\delta \cdot j$ and denote $ex(X_\iota, j, \delta) \overset{\mathrm{df}}{=} X_\iota \cap X_{\iota'}$ the $(n-1)$-dimensional *exiting face of* X_ι *in direction* $\delta \cdot j$. Note that $||\iota - \iota'|| \leq 1$ means that either $X_\iota = X_{\iota'}$ or $X_\iota, X_{\iota'}$ are neighbouring rectangles. The notation $ex_V(X_\iota, j, \delta)$ then denotes the set of vertices of the respective exiting face, $ex_V(X_\iota, j, \delta) \overset{\mathrm{df}}{=} \nu_{X_\iota} \cap \nu_{X_{\iota'}}$.

Note that in every rectangle $X_\iota \in \Omega$ the system is affine, i.e., it has the form $\dot{x}(t) = Ax(t) + a$. According to [6, Theorem 3.1] it holds that there exists a trajectory starting in X_ι which never leaves X_ι if and only if there exists an equilibrium in X_ι. Moreover, there exists [9, Theorem 9] an equilibrium in X_ι if and only if $\mathbf{0} \in hull(\{A_\iota x + a_\iota \mid x \in \nu_{X_\iota}\})$ where *hull* denotes convex hull.

The *parameterised rectangular abstraction transition system* for $\mathbb{P}$ and Σ, written $RATS(\mathbb{P}, \Sigma)$, is a quadruple $(\mathbb{P}, S, T, I)$ where $S = \Omega$ is the *set of states*, $I \subseteq S$ the *set of initial states*, and $T subseteq S \times \{-n, ..., n\} \times 2^\mathbb{P} \times S$ the (parameterised) *transition relation*. The relation T contains only those tuples $\langle X_\iota, i, P, X_{\iota'} \rangle$, $P \subseteq \mathbb{P}$, denoted as $X_\iota \overset{P}{\to}_i X_{\iota'}$ for which $||\iota - \iota'|| \leq 1$ and either of the following conditions holds:

1. $||\iota - \iota'|| = 1$, $i = j \cdot \delta$ with $1 \leq j \leq n$, $\delta \in \{-1, 1\}$ such that $\iota'_j = \iota_j + \delta$ and there exists $x \in ex_V(X_\iota, j, \delta)$ satisfying $\forall p \in P. f_j(x, p) \cdot \delta > 0$.
2. $||\iota - \iota'|| = 0, i = 0 \wedge \forall p \in P. \mathbf{0} \in hull\{f(x, p) \mid x \in \nu_{X_\iota}\}$.

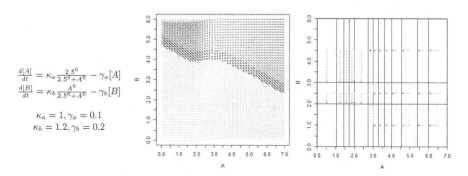

$$\frac{d[A]}{dt} = \kappa_a \frac{2.5^6}{2.5^6 + A^6} - \gamma_a[A]$$
$$\frac{d[B]}{dt} = \kappa_b \frac{A^6}{2.5^6 + A^6} - \gamma_b[B]$$

$\kappa_a = 1, \gamma_a = 0.1$
$\kappa_b = 1.2, \gamma_b = 0.2$

Fig. 1. Example of an admissible system (left). The vector field (middle) and the corresponding RATS (right) were obtained under the given parameterisation.

We denote the rectangle associated with a state $s \in S$ as $X(s)$. An example of a RATS is given in Fig. 1 for a 2-dimensional system.

Similarly to [4], we assume $\mathbb{P}$ includes *almost all* parameterisations excluding singular cases for which some component p_i^j leads to $f_i(\mu, p) = 0$ for some threshold intersection point $\mu \in \mathcal{D}$. Additionally, we exclude all parameterisations that would allow a trajectory to slide along a threshold plane.

The parameter space $\mathbb{P}$ is finitely discretised by *parameter partitioning* $\mathcal{P} = \{P_i | i \in \{1, ..., l\}\}$ such that for each P_i, $RATS(\{p\}, \Sigma)$ is the same for all $p \in P_i$. Moreover, $\mathcal{P}$ is a rectangular partitioning if $\forall 1 \leq i \leq n.m_i \leq 1$ (there is at most a single uncertain kinetic coefficient per a kinetic function).

In [9] it is shown (global sufficiency) that for any $p \in \mathbb{P}$ and every continuous trajectory x of Σ there is a path in $RATS(P_i, \Sigma)$ that corresponds to x and $P_i \in \mathcal{P}$ is a class of parameterisations such that $p \in P_i$. The converse (global necessity) is not guaranteed. This conservatively affects reachability analysis.

For a given $p \in \mathbb{P}$, the results referred above enable a complete identification of fixed points (equilibria) by means of RATS. In particular, every state $s \in S$ for which there is a self-transition (see condition (2) in def. of T) must contain an *equilibrium* $\hat{x} \in X$ such that $f(\hat{x}, p) = 0$ where $X \in \Omega$ is the rectangle associated with s. Note that by the restriction to almost all parameterisations stated above $\hat{x}$ must be situated in the interior of X. Moreover, linear independence of systems variables implies there can be at most a single equilibrium per a rectangle. In particular, we deal with systems with finitely many equilibrium points all of which are correctly represented in the abstraction.

3   Phase Portraits Patterns

In this section we suppose a rectangular abstraction representing the finite discrete (over)approximation of the continuous state space of the given dynamical system. The trajectories in the rectangular abstraction thus (over)approximate the flow in the original vectorfield. A representative set of trajectories in the continuous system is called a *phase portrait*.

Phase portraits can have many shapes. Typical shapes that can appear in a rectangle of a piecewise-affine system in plane are the following (see also Fig. 2):

- *source*: a point away from which all nearby trajectories flow;
- *sink*: a point into which all nearby trajectories flow;
- *saddle*: a point near which two trajectories flow in, two flow out and the rest come close but then move away again;
- *stable spiral*: a point to which trajectories converge in a spiral;
- *unstable spiral*: a point near which trajectories diverge out in a spiral; and
- *centre*: infinite number of orbits.

The phase portraits in Fig. 2 also illustrate the notion of *stability*. A state of a dynamical system is stable if the system returns to that state after a small disturbance, or perturbation. Otherwise, it is unstable. In a phase portrait, a state is stable if all nearby trajectories point towards it, and unstable otherwise. In Fig. 2, the sink and limit cycle are stable, but the source and saddle (which has both types of arrows) are unstable. All listed shapes characterise systems stability (resp. instability) around an *equilibrium*.

If we limit ourselves to linear systems in the form $\dot{x}(t) = Ax(t) + a$ (as is the case of systems on individual rectangles), stability of an equilibrium is characterised by means of eigenvalues of the systems matrix A [2]. Figure 2 shows

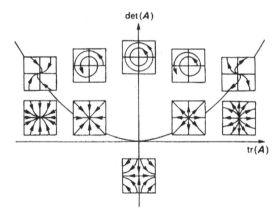

Fig. 2. Phase portraits with respect to the trace and determinant of the two-dimensional region linear system matrix.

the equilibria realised in plane for different values of the determinant and trace of matrix A, i.e., $det(A) = \lambda_1 \cdot \lambda_2$ and $tr(A) = \lambda_1 + \lambda_2$ where λ_1, λ_2 are the real parts of eigenvalues. An equilibrium is considered to be a *sink* if it is asymptotically stable in all directions and its eigenvalues are only real and negative. If the equilibrium is unstable in all directions, it is a *source*. If the eigenvalues include both positive and negative numbers, the equilibrium is of *saddle* type. Other kinds of equilibria are possible in a phase space with dimension $n \geq 3$, e.g., an equilibrium state called a saddle-focus which is unstable according to Lyapunov. They arise from a variety of combinations of systems flow around the equilibrium in all dimensions.

If the topological structure of phase portraits changes with parameters, we refer to this as a *bifurcation*. In this paper, we would like to analyse how phase portraits changes depend on parameters. To do this we formulate temporal queries representing topological structure of phase portraits and compute parameters values that guarantee the queries are satisfied. In such a way we can analyse, how parameter values influence the portraits topology in the phase space.

When projecting phase portraits into rectangular abstraction we get abstract characterisation of portraits called *portrait patterns*. Some patterns can be described as single-state patterns (Fig. 3) while other require several states to be represented. Here we consider single-state patterns only. For such patterns we can guarantee that the abstracted pattern represents the original phase portrait. In the rectangular abstraction, single-state patterns are built from transitions between neighbouring states that are axis parallel. For each state we thus can have incoming and outgoing transitions facing a particular direction.

The individual patterns are defined by transitions (vectors) associated with a state. To characterise patterns in the rectangular abstraction of an n-dimensional model Σ we consider a set

$$DP = \{\theta_j^i \mid i \in \{-n, \ldots, n\} \setminus \{0\}, j \in \{in, out\}\} \cup \{\theta^0\}$$

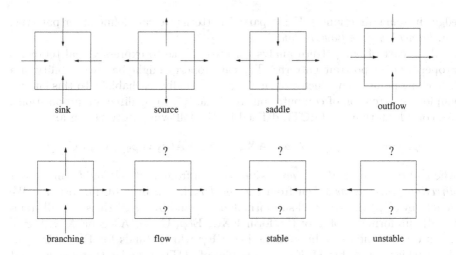

Fig. 3. Single-state portrait patterns. The question mark denotes either situation: incoming only, outgoing only, both or none.

of *direction propositions* evaluated in states. The direction proposition θ_{in}^i holds in a state s if there is a transition from an adjacent state in the *direction* i to s (incoming edge) and θ_{out}^i holds in a state s if there is a transition to an adjacent state in the *direction* i from s (outgoing edge). Directions i and $-i$ are called *opposite directions*. The *zero* direction θ^0 corresponds to a "cyclic transition" from a state to itself, which can be traversed in "any" direction. Note that there are $4n + 1$ direction propositions for an n-dimensional model.

Using direction propositions we can define portrait patterns. Examples of patterns are:

$$
\begin{aligned}
sink &\equiv (\theta_{in}^{-n} \wedge \ldots \wedge \theta_{in}^n) \wedge (\neg\theta_{out}^{-n} \wedge \ldots \wedge \neg\theta_{out}^n) \\
source &\equiv (\theta_{out}^{-n} \wedge \ldots \wedge \theta_{out}^n) \wedge (\neg\theta_{in}^{-n} \wedge \ldots \wedge \neg\theta_{in}^n) \\
stable_i &\equiv (\theta_{in}^i \wedge \neg\theta_{out}^i \wedge \theta_{in}^{-i} \wedge \neg\theta_{out}^{-i}) \text{ (for } i : 1 \leq |i| \leq n) \\
unstable_i &\equiv (\theta_{out}^i \wedge \neg\theta_{in}^i \wedge \theta_{out}^{-i} \wedge \neg\theta_{in}^{-i}) \text{ (for } i : 1 \leq |i| \leq n) \\
outflow_i &\equiv (\theta_{in}^{-n} \wedge \ldots \wedge \theta_{in}^{i-1} \wedge \theta_{in}^{i+1} \wedge \ldots \wedge \theta_{in}^n) \wedge \\
&\quad (\neg\theta_{out}^{-n} \wedge \ldots \wedge \neg\theta_{out}^{i-1} \wedge \neg\theta_{out}^{i+1} \wedge \ldots \wedge \neg\theta_{out}^n) \wedge \\
&\quad (\theta_{out}^i \wedge \neg\theta_{in}^i) \text{ for } i : 1 \leq |i| \leq n \\
flow_i &\equiv (\theta_{out}^i \wedge \neg\theta_{in}^i \wedge \theta_{in}^{-i} \wedge \neg\theta_{out}^{-i}) \text{ (for } i : 1 \leq |i| \leq n) \\
branching_{i,j} &\equiv (\theta_{in}^{-n} \wedge \ldots \wedge \theta_{in}^{i-1} \wedge \theta_{in}^{i+1} \wedge \ldots \wedge \theta_{in}^{j-1} \wedge \theta_{in}^{j+1} \wedge \ldots \wedge \theta_{in}^n) \wedge \\
&\quad (\neg\theta_{out}^{-n} \wedge \ldots \wedge \neg\theta_{out}^{i-1} \wedge \neg\theta_{out}^{i+1} \wedge \ldots \wedge \neg\theta_{out}^{j-1} \wedge \neg\theta_{out}^{j+1} \wedge \ldots \wedge \neg\theta_{out}^n) \wedge \\
&\quad (\theta_{out}^i \wedge \neg\theta_{in}^i \wedge \theta_{out}^j \wedge \neg\theta_{in}^j) \text{ (for } i, j : 1 \leq |i| \leq |j| \leq n) \\
2d-saddle &\equiv \theta_{out}^1 \wedge \theta_{in}^2 \wedge \theta_{out}^{-1} \wedge \theta_{in}^{-2} \text{ (for } n = 2) \\
equilibrium &\equiv \theta^0
\end{aligned}
$$

Note that all the patterns can be rotated along a particular axis which gives additional examples. We can also consider "partial" patterns like a sink without any edges in some direction(s). Using partial patterns we can define other patterns, e.g. $branching'_{i,j} \equiv flow_i \wedge flow_j$.

As a part of our pattern analysis we would like to express global temporal properties over portrait patterns. Typical property might be "reachability of a branching point from which two sinks are eventually reachable". To this end we employ an extension of computation tree logic (CTL) by direction propositions. We consider formulae of dCTL defined by the following abstract syntax:

$$\varphi ::= \theta \mid Q \mid \neg\varphi \mid \varphi_1 \wedge \varphi_2 \mid \mathbf{AX}\varphi \mid \mathbf{EX}\varphi \mid \mathbf{A}(\varphi_1 \, \mathbf{U} \, \varphi_2) \mid \mathbf{E}(\varphi_1 \, \mathbf{U} \, \varphi_2)$$

where Q ranges over *atomic propositions* taken from a set AP and θ ranges over *direction propositions* taken from the set DP. Let φ be a dCTL formula. We denote by $cl(\varphi)$ the set of all sub-formulae of φ and by $tcl(\varphi)$ the set of all (temporal) sub-formulae of φ of the form $\mathbf{EX}\varphi$, $\mathbf{E}(\varphi_1 \, \mathbf{U} \, \varphi_2)$, $\mathbf{AX}\varphi$ or $\mathbf{A}(\varphi_1 \, \mathbf{U} \, \varphi_2)$. We use the standard abbreviations like $\mathbf{EF}\varphi$ which stands for $\mathbf{E}(true \, \mathbf{U} \, \varphi)$ or $\mathbf{AG}\varphi$ which stands for $\neg\mathbf{EF}\neg\varphi$. Examples of dCTL formulae that speak about patterns are:

- $\mathbf{EF}sink$ expresses reachability of a sink pattern,
- $\mathbf{AG}((E2F1 \geq 4 \wedge E2F1 \leq 7.5) \Rightarrow flow_i)$ expresses that in a specified region of concentration levels of a protein *E2F1* the system moves in the direction i,
- $\mathbf{AG}(saddle \Rightarrow \mathbf{E}(flow_i \mathbf{U} sink))$ expresses that from a saddle there is reachable a sink in the direction i.

Note that we have not added directions to temporal operators. It would be meaningful to consider for example an $\mathbf{EX}_\delta$ operator for expressing the property "exists next in direction δ". Such an extension leads to a more richer framework based on doubly labelled transition systems [11] that allow to combine information on states and transitions. However, such multi-state patterns suffer from the extent of over-approximation in rectangular abstraction and therefore we do not consider them.

The dCTL semantics is defined on a special kind of Kripke structures, that in addition to standard Kripke structures have directions associated with transitions. We will call them *rectangular Kripke structures* (RKS). A rectangular Kripke structure of *dimension n* (over AP and DP) is a tuple $\mathcal{K} = (S, I, \rightarrow, L)$, where S is the finite set of states, $I \subseteq S$ the set of initial states, $L : S \rightarrow 2^{AP}$ is a labelling of states by atomic propositions, $\rightarrow \subseteq S \times \{-n, \ldots, n\} \times S$ is a total transition relation, an element $(s, \delta, s') \in \rightarrow$ is a *transition facing direction* δ, also written as $s \rightarrow_\delta s'$. Satisfaction of a dCTL formula φ in a state s, written as $s \models \varphi$, is defined inductively in the same way as in CTL, for the new formulas we define:

$$s \models \theta^i_{in} \text{ iff } \exists s' \in S.s' \rightarrow_i s$$
$$s \models \theta^i_{out} \text{ iff } \exists s' \in S.s \rightarrow_i s'$$
$$s \models \theta^0 \text{ iff } s \rightarrow_0 s$$

To effectively compute the bifurcation analysis we suppose a *finite set* of parameter values (parameterisations) obtained as the parameter partitioning $\mathcal{P}$

(see Sect. 2). We define *parameterised rectangular Kripke structure* $\mathcal{K}_{\mathcal{P}}$ *of dimension* n (PRKS) as a general structure that encapsulates the family of rectangular Kripke structures for all parameterisations. Formally, $\mathcal{K}_{\mathcal{P}} = (\mathcal{P}, S, I, \rightarrow, L)$, where for each $P \in \mathcal{P}$ the tuple $\mathcal{K}(P) = (S, I, \overset{P}{\rightarrow}, L)$ is a rectangular Kripke structure of dimension n.

It is evident that a parameterised rectangular abstraction $RATS(\mathbb{P}, \Sigma) = (\mathbb{P}, S, T, I)$ can be directly turned into a PRKS $\mathcal{K}_{\mathcal{P}} = (\mathcal{P}, S, I, \rightarrow, L)$ of the same dimension with directions DP, labelled by AP, and the set of parameters $\mathbb{P}$ finitely discretised by parameter partitioning $\mathcal{P}$ as described in Sect. 2.

It is important to note some implications that follow from the discretisation we use (defined in Sect. 2). For each parameterisation $p \in \mathbb{P}$ the rectangular Kripke structure over-approximates the original continuous piecewise-affine system provided that every trajectory of the system has a corresponding run (sequence of transitions) in the RKS, but not *vice versa*. This implies that any phase portrait in piecewise-affine system thus has a corresponding portrait pattern in RKS. The following characterisation can be inferred from selected patterns:

1. $s \models \theta^0$ if and only if there exists an equilibrium in $X(s)$ interior;
2. $s \models sink$ implies there exists a stable equilibrium $\hat{x} \in X(s)$ in the interior of $X(s)$ such that every continuous trajectory starting in any point of $X(s)$ does not leave $X(s)$ and reaches $\hat{x}$;
3. $s \models source$ implies there exists an unstable equilibrium $\hat{x} \in X(s)$ in the interior of $X(s)$ such that for each $x_0 \in X(s), x_0 \neq \hat{x}$ a continuous trajectory starting at x_0 leaves $X(s)$ in finite time and no trajectory enters $X(s)$ from outside;
4. $s \models outflow_i$ for some i implies there is no equilibrium in $X(s)$ and every trajectory starting in $X(s)$ leaves $X(s)$ in finite time at the face $ex(X(s), j, \delta)$ where $i = j \cdot \delta$, $j \geq 1$ and $\delta = -1$, if $i < 0$, or $\delta = 1$, if $i > 0$;
5. $s \models flow_i$ for some i implies there is no equilibrium in $X(s)$ and there exists a trajectory that leaves $X(s)$ in finite time at the face $ex(X(s), j, \delta)$ such that $i = j \cdot \delta$, $j \geq 1$ and $\delta = -1$, if $i < 0$, or $\delta = 1$, if $i > 0$, and a trajectory that enters $X(s)$ at $ex(X(s), j, -\delta)$, additionally, no trajectory leaves $X(s)$ at the face $ex(X(s), j, -\delta)$ and no trajectory enters $X(s)$ at the face $ex(X(s), j, \delta)$.

The characterisation is inferred from the assumptions and facts discussed in Sect. 2: (1) follows from the complete characterisation of a fixed point from convex hull of vectors in rectangle vertices, (2,3) besides the assumptions in Sect. 2 follow from linear systems stability analysis [2] and the fact that vectorfield in $X(s)$ is a linear combination of vectors in vertices (note that $s \models sink \lor source$ implies $s \models \theta^0$), pattern (4) gives a necessary condition for a flow to exit in finite time at the face $en(X(s), j, \delta)$ and again follows from the linear vectorfield in $X(s)$ characterised by $X(s)$ vertices (Sect. 2). Similarly, (5) provides characterisation of flow in direction i (here we know nothing about other directions). Note that (5) also holds in a state where there is a trajectory entering at $ex(X(s), j, -\delta)$ and leaving at $ex(X(s), j, \delta)$, but this cannot be guaranteed by the pattern due to the extent of over-approximation. We consider the characterisation above just for the given selection of most significant patterns. In this paper we focus on computational aspects of the pattern analysis. Full characterisation of all patterns will be given in an extended version of the paper.

We say that a formula is *universal* (or in dACTL) if it only contains universal temporal operators and no negations. The rectangular abstraction provides an over-approximation of the system dynamics and preserves thus the truth of universal formulae. This guarantees that all parameterisations synthesised from such formulae are correct representatives of the original ones. Bifurcations are represented by universal formulae, any bifurcation identified in a PRKS has a counter-part in the piecewise-affine system. We call a *bifurcation point* a parameter $P \in \mathcal{P}$ such that in P the satisfaction of a given portrait pattern is changing.

Our goal is to analyse how phase portraits change depending on a change in parameters. The problem is formally defined in the following way. Suppose we are given a PRKS $\mathcal{K}$ and a dCTL formula φ. The goal is to compute the function $\mathcal{B} : \mathcal{P} \rightarrow 2^S$ such that $\mathcal{B}(P) = \{s \mid s \models_{\mathcal{K}(P)} \varphi\}$. Since $\mathcal{P}$ is finite, the function will be represented as a table serving as a source for additional tasks like visualisation, identifying bifurcation points, etc.

4   Parallel Portrait Patterns Analysis

We suppose a parameterised n-dimensional piecewise-affine system with at most one uncertain parameter per systems equation. Moreover, we assume a given dCTL formula. The analysis algorithm is an adaptation of the parallel coloured model-checking algorithm for parameter synthesis from CTL formulae [7].

The analysis is supposed to be performed on a cluster of workstations which allows not only to speed-up the analysis, but also to accommodate larger models using the cumulative memory of workstations. The distributed algorithm uses one master node responsible for processing input and output of the analysis and for initialisation of the distributed computing. The state space is distributed among N nodes using a *partitioning function* $f : S \rightarrow \{1, \ldots, N\}$. After partitioning, each node owns only a part of the original state space.

In our approach to partitioning, we utilise the regular structure of the state space for biochemical models [18]. We define the partition function f so that the rectangular state space is divided into N almost equally sized rectangular sub-spaces. The construction of the discrete state space as described in Sect. 2 ensures there are only transitions between the adjacent states with respect to the hyper-rectangular structure. Therefore, our partitioning naturally provides almost the minimal number of cross transitions, since the only cross transitions are the ones occurring between the rectangular sub-spaces.

The portrait patterns analysis for a given dCTL formula φ works in three phases. In the *initial phase* the given piecewise-affine system is transformed into a RATS and the RATS into a PRKS. Each computational node performs this operation only for states it owns. This can be distributed almost trivially since computation of the transitions between adjacent states depends only on the information provided in the piecewise-affine model (each process has a full copy of the piecewise-affine system). We also need to compute the labelling function for all atomic propositions and evaluate all direction propositions present in the formula. The labelling function does not depend on model parameters. However, since the transition relation depends on parameter values, direction propositions are also parameter-dependent. Therefore, in evaluating direction propositions we have to consider all possible parametrisations (which can be done efficiently due to the coloured approach). If required by the user, the analysis can also return additional global information, e.g. about numbers of various patterns together with their corresponding state "addresses". This additional information can be afterwards used in the second phase of the analysis. For example, if there is more than one *sink state* in the state space for some parametrisation, the user can rewrite the formula with atomic propositions which are true in different sinks (using their different addresses). This way, the user can synthesise parameters for this kind of multi-stability. Using command-line options we can instruct the algorithm to compute such analysis automatically.

The *second phase* uses the parallel coloured model checking algorithm for computing the function $\mathcal{F}_\varphi^{\mathcal{K}} : S \to 2^{\mathcal{P}}$ such that $\mathcal{F}_\varphi^{\mathcal{K}}(s) = \{P \in \mathcal{P} \mid s \models_{\mathcal{K}(P)} \varphi\}$, where $s \models_{\mathcal{K}(P)} \varphi$ denotes that φ is satisfied in the state s of $\mathcal{K}(P)$. We only provide the basic outline of the main algorithm (Algorithm 1), a full description can be found in [7].

Once the system is partitioned, the Kripke structure on each network node, called a *fragment*, can contain states that represent "border" states, which are those states that in fact belong to some other network node. Whenever the model checking algorithm reaches a border state it uses information provided by other network nodes about the truth of formulas in that state – *assumptions*. As the assumptions can change, a re-computation is necessary in general. More precisely, we consider an assumption function $\mathcal{A} : \mathcal{P} \times S \times cl(\varphi) \to Bool$. For each parameter P, state s and formula φ, the function $\mathcal{A}$ returns true if we can assume that $s \models_{\mathcal{K}(P)} \varphi$. Symmetrically, if $\mathcal{A}$ returns false, we can assume that $s \not\models_{\mathcal{K}(P)} \varphi$. If we cannot assume anything (the information has not been computed yet), $\mathcal{A}$ returns undefined value $\perp$. If the assumption function is defined for all arguments then $\mathcal{F}_\varphi^{\mathcal{K}}(s) = \{P \in \mathcal{P} \mid \mathcal{A}(P, s, \varphi) = true\}$.

On each computation node we first initialise the assumption function with $\mathcal{A}_\perp$, which is undefined for all inputs. Afterwards, we iterate over all sub-formulas of ψ, starting from the smallest ones. For each formula we modify the assumption function so that it correctly represents states and parameters where the formula holds (based on previous assumptions). This is achieved using a function $\mathcal{C}_{\mathcal{K}_i}^\varphi : AS_{\mathcal{K}_i}^\varphi \to AS_{\mathcal{K}_i}^\varphi$ ($AS_{\mathcal{K}_i}^\varphi$ is the set of all possible assumption functions over fragments $\mathcal{K}_i$ and sub-formulas of φ) which takes an initial assumption function and returns, intuitively, the model checking results relative to the initial assumptions. The full formal definition of $\mathcal{C}_{\mathcal{K}_i}^\varphi$ can be found in [7].

As the system is partitioned, neither of the computation nodes has all information required to complete the whole computation. Therefore, each computation node first computes everything it can from available data and then sends

---

Algorithm 1. Main Idea of the Distributed Algorithm

 1: **procedure** CHECK-dCTL($\varphi, \mathcal{K} = (\mathcal{P}, S, I, \rightarrow, L)$)
 2:    Partition $\mathcal{K} = (\mathcal{P}, S, I, \rightarrow, L)$ into fragments $\mathcal{K}_1, \cdots, \mathcal{K}_N$ (initial phase)
 3:    **for all** $\mathcal{K}_i$ where $i \in \{1, \cdots, N\}$ **do in parallel**
 4:        $\mathcal{A} \leftarrow \mathcal{A}_\perp$ ▷ Initialise assumption function as undefined
 5:        **for all** $i \leq |\varphi|$ **do**
 6:            **for all** ψ in $cl(\varphi)$ **where** $|\psi| = i$ **do**
 7:                **repeat**
 8:                    $\mathcal{A} \leftarrow \mathcal{C}^\varphi_{\mathcal{K}_i}(\mathcal{A})$ ▷ Update $\mathcal{A}$ based on available information
 9:                    exchange assumptions with other processes
10:                **until** all processes reach fixpoint
11:        **end parallel**
12:    output $\mathcal{F}^{\mathcal{K}}_\varphi$

---

relevant assumptions to other nodes that might need it. This is repeated until all computation nodes reach fixpoint (no new information can be computed). Such situation is detected using suitable distributed termination detection algorithm.

Finally, in the *third phase*, data from all workstations is collected, $\mathcal{B}(P) = \{s \in S \mid P \in \mathcal{F}^{\mathcal{K}}_\varphi(s)\}$ is computed and eventually post-processed and presented to the user.

5   Case Study

Applicability of our approach is demonstrated on a model of genetic repressilator given in [13]. In particular, we deal with a scalable genetic regulatory network with various genes repressing each other in a circle (Fig. 4 (left)). Let us denote the individual models as Rep^i where i is the number of genes.

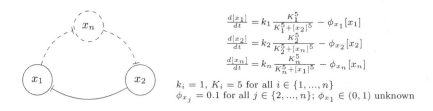

$$\frac{d[x_1]}{dt} = k_1 \frac{K_1^5}{K_1^5 + [x_2]^5} - \phi_{x_1}[x_1]$$

$$\frac{d[x_2]}{dt} = k_2 \frac{K_2^5}{K_2^5 + [x_n]^5} - \phi_{x_2}[x_2]$$

$$\frac{d[x_n]}{dt} = k_n \frac{K_n^5}{K_n^5 + [x_1]^5} - \phi_{x_n}[x_n]$$

$k_i = 1$, $K_i = 5$ for all $i \in \{1, ..., n\}$
$\phi_{x_j} = 0.1$ for all $j \in \{2, ..., n\}$; $\phi_{x_1} \in (0,1)$ unknown

Fig. 4. Schematic model (left) and mathematical model (right) of n-dimensional repressilator (Rep^n).

According to [20], there is a bistability in Rep^2 (general model is given in Fig. 4 (right)). Our algorithm has been able to discover the bistability (two distinct sink nodes) guaranteed for $\phi_{x_1} \in (0.119, 0.120) \cup (0.138, 1.0)$. Bifurcation points are predicted at the borders of both parameter intervals as the bistability appears/disappears there. Figure 6 visualises three vector fields for three different parameter values of the parameter ϕ_{x_1} and the corresponding RATS that

Table 1. The second column shows intervals of ϕ_{x_1} where just a single sink has been discovered. The third column shows parameterisations where both sinks are guaranteed to occur simultaneously.

Rep^i	single stability	bistability
$i = 2$	$(0.119, 0.120) \cup (0.138, 1.0)$	$(0.022, 0.119) \cup (0.120, 0.138)$
$i = 4$	$(0.119, 0.120) \cup (0.128, 0.129) \cup$ $(0.138, 1.0)$	$(0.022, 0.119) \cup (0.120, 0.128) \cup$ $(0.129, 0.138)$
$i = 6$	$(0.087, 0.088) \cup (0.106, 0.107) \cup$ $(0.123, 0.124) \cup (0.140, 1.0)$	$(0.021, 0.087) \cup (0.088, 0.106) \cup$ $(0.107, 0.123) \cup (0.124, 0.14)$
$i = 8$	$(0.101, 0.103) \cup (0.131, 1.0)$	$(0.015, 0.101) \cup (0.103, 0.131)$

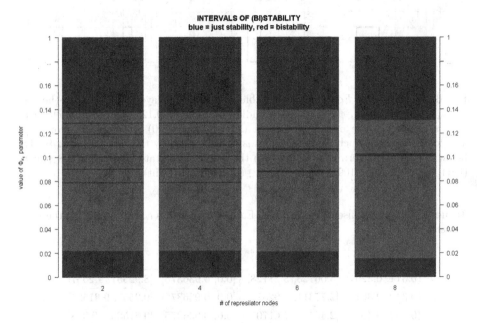

Fig. 5. Visualisation of the results shown in Table 1.

reveal the pattern bifurcation. For models Rep^i where $i \in \{4, 6, 8\}$, multistable behaviour has been also discovered (see Table 1 for summary of the results) (Fig. 5).

In the model Rep^3, we have discovered equilibrium points for specific intervals of ϕ_{x_1} that have character of a 3d-sink (Table 2). All obtained results are in agreement with the analysis provided in [20]. We have also considered and successfully checked several temporal properties over patterns, like $\mathbf{AG}(source \Rightarrow \mathbf{AF}sink)$, that express some global characteristics of the vector field.

Additionally, we have evaluated the scalability of the algorithm. The results confirmed the same level of scalability as reported in [7] for the main procedure

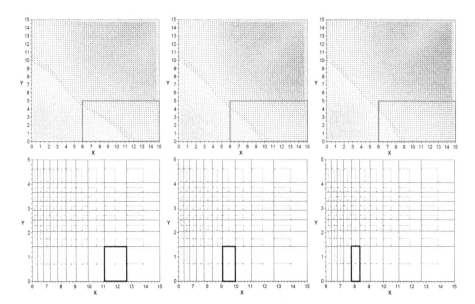

Fig. 6. Vectorfield of Rep^2 with two stable equilibria displayed in figures on the top for $\phi_{x_1} = 0.085$ (left), $\phi_{x_1} = 0.105$ (middle), and $\phi_{x_1} = 0.125$. The axis labelling is $x = x_1$, $y = x_2$. The upper fixed point ($x = 0.5$, $y = 10$) is independent on ϕ_{x_1}. The focused zone in each vectorfield plot is characterised by the corresponding RATS below. The *sink* pattern is guaranteed in the emphasised state for $\phi_{x_1} \in (0.08, 0.089)$ (left), $\phi_{x_1} \in (0.101, 0.109)$ (middle), $\phi_{x_1} \in (0.12, 0.128)$ (right).

Table 2. Sink nodes discovered in Rep^3. Each row identifies a state satisfying the *sink* pattern.

ϕ_{x_1}	x_1	x_2	x_3
(0.311, 0.326)	[3.06723, 3.21128]	[0.0, 0.936375]	[8.92557, 9.2557]
(0.342, 0.361)	[2.77311, 2.92317]	[0.0, 0.936375]	[9.2557, 9.61585]
(0.382, 0.407)	[2.45498, 2.61705]	[0.0, 0.936375]	[9.61585, 10.018]
(0.407, 0.438)	[2.28091, 2.45498]	[0.0, 0.936375]	[9.61585, 10.018]
(0.438, 0.477)	[2.08884, 2.28091]	[0.0, 0.936375]	[9.61585, 10.018]
(0.479, 0.532)	[1.87875, 2.08884]	[0.0, 0.936375]	[9.61585, 10.018]
(0.532, 0.612)	[1.63265, 1.87875]	[0.0, 0.936375]	[9.61585, 10.018]
(0.613, 0.747)	[1.33854, 1.63265]	[0.0, 0.936375]	[9.61585, 10.018]
(0.747, 1.000)	[0.936375, 1.33854]	[0.0, 0.936375]	[9.61585, 10.018]

(second phase). All experiments have been conducted on a homogeneous cluster with 12 nodes each equipped with 16 GB of RAM and a quad-core Intel Xeon 2 GHz processor.

6 Conclusions

We have proposed a parallel algorithm that can assist the user in analysing bifurcation of piecewise-affine systems. The framework uses a rich temporal logic language extended by directions propositions and very effective model checking technology for representing and analysing the vector field topology changes. Most of our approach can be extended to more general (piecewise)-multiaffine systems and we leave this for our future work.

References

1. Alon, U.: An Introduction to Systems Biology: Design Principles of Biological Circuits. Chapman & Hall/CRC Mathematical and Computational Biology, Boca Raton (2006)
2. Anishchenko, V.S., Vadivasova, T.E., Strelkova, G.I.: Deterministic Nonlinear Systems: A Short Course. Springer, Heidelberg (2014)
3. Bagley, R.J., Glass, L.: Counting and classifying attractors in high dimensional dynamical systems. J. Theor. Biol. **183**(3), 269–284 (1996)
4. Batt, G., Belta, C., Weiss, R.: Model checking genetic regulatory networks with parameter uncertainty. In: Bemporad, A., Bicchi, A., Buttazzo, G. (eds.) HSCC 2007. LNCS, vol. 4416, pp. 61–75. Springer, Heidelberg (2007)
5. Batt, G., Besson, B., Ciron, P.-E., de Jong, H., Dumas, E., Geiselmann, J., Monte, R., Monteiro, P.T., Page, M., Rechenmann, F., Ropers, D.: Genetic network analyzer: A tool for the qualitative modeling and simulation of bacterial regulatory networks. In: van Helden, J., Toussaint, A., Thieffry, D. (eds.) Bacterial Molecular Networks: Methods and Protocols. Methods in Molecular Biology, vol. 804, pp. 439–462. Springer, Heidelberg (2012)
6. Belta, C., Habets, L.C.G.J.M.: Controlling a class of nonlinear systems on rectangles. IEEE Trans. Automat. Contr. **51**(11), 1749–1759 (2006)
7. Brim, L., Češka, M., Demko, M., Pastva, S., Šafránek, D.: Parameter synthesis by parallel coloured CTL model checking. In: Roux, O., Bourdon, J. (eds.) CMSB 2015. LNCS, vol. 9308, pp. 251–263. Springer, Heidelberg (2015)
8. Chaves, M., Tournier, L., Gouzé, J.-L.: Comparing boolean and piecewise affine differential models for genetic networks. Acta Biotherica **58**(2–3), 217–232 (2010)
9. Collins, P., Habets, L.C.G.J.M., van Schuppen, J.H., Černá, I., Fabriková, J., Šafránek, D.: Abstraction of biochemical reaction systems on polytopes. In: IFAC World Congress, pp. 14869–14875. IFAC (2011)
10. de Jong, H., Page, M.: Search for steady states of piecewise-linear differential equation models of genetic regulatory networks. IEEE/ACM Trans. Comput. Biol. Bioinform. **5**(2), 208–222 (2008)
11. De Nicola, R., Vaandrager, F.: Three logics for branching bisimulation. In: Proceedings of the Fifth Annual IEEE Symposium on Logic in Computer Science, pp. 118–129 (1990)
12. Dvorak, P., Bidmanova, S., Damborsky, J., Prokop, Z.: Immobilized synthetic pathway for biodegradation of toxic recalcitrant pollutant 1,2,3-trichloropropane. Environ. Sci. Technol. **48**(12), 6859–6866 (2014)
13. Elowitz, M.B., Leibler, S.: A synthetic oscillatory network of transcriptional regulators. Nature **403**(6767), 335–338 (2000)

14. Glass, L., Kauffman, S.A.: The logical analysis of continuous, non-linear biochemical control networks. J. Theor. Biol. **39**(1), 103–129 (1973)
15. Grosu, R., Batt, G., Fenton, F.H., Glimm, J., Le Guernic, C., Smolka, S.A., Bartocci, E.: From cardiac cells to genetic regulatory networks. In: Gopalakrishnan, G., Qadeer, S. (eds.) CAV 2011. LNCS, vol. 6806, pp. 396–411. Springer, Heidelberg (2011)
16. Holzwarth, A.R., Müller, M.G., Reus, M., Nowaczyk, M., Sander, J., Rögner, M.: Kinetics and mechanism of electron transfer in intact photosystem II and in the isolated reaction center: Pheophytin is the primary electron acceptor. Proc. Nat. Acad. Sci. **103**(18), 6895–6900 (2006)
17. Jamshidi, S., Siebert, H., Bockmayr, A.: Comparing discrete and piecewise affine differential equation models of gene regulatory networks. In: Lones, M.A., Smith, S.L., Teichmann, S., Naef, F., Walker, J.A., Trefzer, M.A. (eds.) IPCAT 2012. LNCS, vol. 7223, pp. 17–24. Springer, Heidelberg (2012)
18. Jha, S., Shyamasundar, R.K.: Adapting biochemical Kripke structures for distributed model checking. In: Priami, C., Ingólfsdóttir, A., Mishra, B., Riis Nielson, H. (eds.) Transactions on computational systems biology vii. LNCS (LNBI), vol. 4230, pp. 107–122. Springer, Heidelberg (2006)
19. De Jong, H.: Modeling and simulation of genetic regulatory systems: a literature review. J. Comput. Biol. **9**, 67–103 (2002)
20. Müller, S., Hofbauer, J., Endler, L., Flamm, C., Widder, S., Schuster, P.: A generalized model of the repressilator. J. Math. Biol. **53**(6), 905–937 (2006)
21. Snoussi, E.H.: Qualitative dynamics of piecewise-linear differential equations: a discrete mapping approach. Dyn. Stab. Syst. **4**(3–4), 565–583 (1989)
22. Veliz-Cuba, A., Arthur, J., Hochstetler, L., Klomps, V., Korpi, E.: On the relationship of steady states of continuous and discrete models arising from biology. Bull. Math. Biol. **74**(12), 2779–2792 (2012)
23. Yordanov, B., Belta, C., Batt, G.: Model checking discrete time piecewise affine systems: application to gene networks. In: European Control Conference (ECC) (2007)

Integrating Time-Series Data in Large-Scale Discrete Cell-Based Models

Louis Fippo Fitime[1], Christian Schuster[2], Peter Angel[2], Olivier Roux[1],
and Carito Guziolowski[1]([✉])

[1] LUNAM Université, École Centrale de Nantes, IRCCyN UMR CNRS 6597
(Institut de Recherche En Communications Et Cybernétique de Nantes),
1 rue de la Noë, B.P. 92101, 44321 Nantes Cedex 3, France
{louis.fippo-fitime,carito.guziolowski}@irccyn.ec-nantes.fr
http://www.irccyn.ec-nantes.fr/en/,
http://www.irccyn.ec-nantes.fr/en/research-teams/meforbio
[2] Division of Signal Transduction and Growth Control (A100), DKFZ-ZMBH
Alliance, Deutsches Krebsforschungszentrum, Heidelberg, Germany

Abstract. In this work we propose an automatic way of generating and
verifying formal hybrid models of signaling and transcriptional events,
gathered in large-scale regulatory networks. This is done by integrating
temporal and stochastic aspects of the expression of some biological
components. The hybrid approach lies in the fact that measurements
take into account both times of lengthening phases and discrete switches
between them. The model proposed is based on a real case study of keratinocytes differentiation, in which gene time-series data was generated
upon Calcium stimulation.

To achieve this we rely on the Process Hitting (PH) formalism that
was designed to consider large-scale system analysis. We first propose an
automatic way of detecting and translating biological motifs from the
Pathway Interaction Database to the PH formalism. Then, we propose a
way of estimating temporal and stochastic parameters from time-series
expression data of action on the PH. Simulations emphasize the interest
of synchronizing concurrent events.

Keywords: Time-series data · Large-scale network · Hybrid models ·
Compositional approach · Stochastic simulation

1 Introduction

Unraveling and describing the mechanisms involved in the regulation of a cell-
based biological system is a fundamental issue. These mechanisms can be modeled as biological regulatory networks, whose analysis requires to preliminary
build a mathematical or computational model. By just considering qualitative
regulatory effects between components, biological regulatory networks depict
fairly well biological systems, and can be built upon public repositories such as
the Pathways Interaction Database [23] and hiPathDB [30] for human regulatory

© Springer International Publishing Switzerland 2015
A. Abate and D. Šafránek (Eds.): HSB 2015, LNBI 9271, pp. 75–95, 2015.
DOI: 10.1007/978-3-319-26916-0_5

knowledge. In this work we built a hybrid model of signaling and transcriptional events, gathered in large-scale regulatory networks, for which stochastic simulation parameters were inferred from gene expression time-series data.

High-throughput experimental data has been used since more than one decade ago to infer biological regulatory models. A variety of methods were proposed to infer dynamic or static models of protein signaling or gene regulations depending on the nature of the experimental data. We can cite methods that infer static gene regulatory models from steady-state gene expression datasets of over-expression or knock-down perturbations using statistical models generating small-scale (ten species) models [8] or middle-scale (maximum 100 species) models [19]. Additionally, we can cite methods that recovered gene regulatory dynamic models from time-series data using kinetic modeling [5,20] generating small-scale models. Recently, static boolean models for middle and large-scale (over 100 species) signaling protein networks have been derived from a *prior* network and fitted to steady-state multiple perturbation phosphoproteomics data [10,16] using combinatorial optimization through logic and integer linear programing to explore the vast search space of candidate boolean models. When using time-series multi-perturbation phosphoproteomics data, results can be extended to reconstruct middle-scale dynamic signaling models via the use of stochastic search approaches [14] that do not guarantee an exhaustive exploration of the search space of candidate models. The approach presented in this work confronts a prior signaling and gene regulatory large-scale network, obtained from publicly curated databases, to time-series gene expression data, by using discrete automaton models and stochastic simulations. Our built model verifies the agreement of expression traces over time given a signed (activations/inhibitions), directed and cyclic prior graph.

The advantages and complementariness of our method with respect to the afore cited approaches are that it allows us to define a logic that integrates signaling and transcription events (imposing different regulatory rules on these events), it also integrates multi-valued states of the system components, and importantly it deals with the complexity of large-scale dynamic models.

Several conceptually different approaches are available for modeling Biological Regulatory Network (BRN) dynamics. The most common approach is ordinary differential equations (ODE) that describe deterministic (population average) behavior in a continuous manner. Even for simple models including a simple interaction between two components, the analytical solution is impossible. Thus we must refer to simulation as the only practical method. Furthermore, continuous models require quantitative knowledge in terms of kinetic coefficients, which are unknown and very difficult to measure. Thereby, various abstraction approaches have been developed to make BRN models more convenient for analysis. Synchronous Boolean model was first proposed by Kauffman [12] and an alternative asynchronous model was proposed by Thomas [27]. Following these two papers, many other models have been proposed [6,7,25,26] for modeling dynamic of BRN. All of these models are purely qualitative and discrete, thus do not incorporate quantitative time or other quantities. As well, discrete models have been extended to integrate quantitative aspects. Time aspect have been introduced by [1,4,24,29]. It relies on timed automaton implementations. These models, however, do not take into account the stochastic aspects of the influences of a BRN.

In the context of modeling and analyzing stochastic and concurrent biological systems various formalisms have been introduced such as Stochastic Petri Nets which is suitable for the representation of parallel systems [17]. They have been successfully applied in many areas; in particular, the specification of Petri Nets allows an accurate modeling of a wide range of systems including biological systems [11]. The major problem of Stochastic Petri Nets is that, generally, they do not lead to compact models. In addition, they do not provide results to deal with the state space explosion and are thus computationally expensive when modeling large-scale biological networks. The Stochastic pi-calculus formalism was introduced by [21] and used in [15] for the modeling of biological systems. Stochastic pi-calculus has a rich expressiveness and is well adapted for the use of compositional approach. In this work we rely on this formalism through the Process Hitting (PH) framework [18], since it is especially useful for studying systems composed of biochemical interactions, and provides stochastic simulation as well as efficient algorithms, based on the verification of state reachability, to study dynamical properties of the system. The PH framework uses qualitative and discrete information of the system without requiring enormous parameter estimation tasks for its stochastic simulation. This framework has been previously used to verify dynamical properties on biological systems without integrating high-throughput experimental data.

In this work we provide a method to build a time-series data integrated PH model and we evaluate the prediction power of this model concerning the simultaneously predicted traces of 12 mRNA expression components of the system upon system stimulation. The main results of this work are: (1) automatic generation of PH models integrating gene transcription and signaling events, with and without synchronization of concurrent events, from the Pathways Interaction Database, (2) parameter estimation from time-series data and parameter integration in the PH model, and (3) comparison of the PH model predictions and experimental results. To illustrate our approach, we used a time-series dataset of human keratinocytes cells, which shows the fluctuations of mRNA expression across time upon Calcium stimulation. This dataset was built to study keratinocytes differentiation, a time-dependent process in which the sequence of activation of signaling proteins is not yet completely understood. The method proposed in this paper remains general and can be applied to other case-studies.

2   Data and Methods

The general workflow for integrating time-series data in a PH model is depicted in Fig. 1, in the following sections we detail some of the workflow steps.

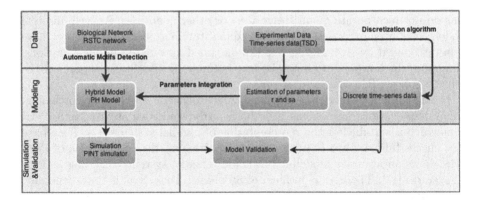

Fig. 1. Integrating stochastic and temporal information in a large-scale discrete biological model. The parameters rate (r) and stochasticity absorption factor (sa) will be presented later in Sect. 2.3.

2.1 Data

Interaction Graph

Definition 1 (Terminal Transient Interaction Graph (TTIG)). *A* TTIG *N is a couple (V, E), where:*

- *$V = V_T \bigcup V_I$ is the finite set of* nodes; *with $V_T = \{v_{1t}, v_{2t}, \ldots, v_{n1t}\}$ the set of terminal nodes; $V_I = \{v_{1i}, v_{2i}, \ldots, v_{n2i}\}$ the set of transient nodes.*
- *$E = \{e_1, e_2, \ldots, e_m\}$ is the set of edges. $E \subseteq (V_T \times V_T) \bigcup (V_T \times V_I) \bigcup (V_I \times V_T)$*

In this definition terminal nodes can be either mRNA expression, proteins, complexes, cellular states, biological processes or positive conditions. On the other side, transient nodes can be either transcriptions or translocations or modifications or compounds. Edges are of different types: activation (agent), inhibition, output, input and protein-family-member.

Definition 2 (Multi-layer Receptor-Signaling-Transcription-Cell State (RSTC)). *A RSTC network is a TTIG where nodes are linked to a layer (Receptor, Signaling, Transcription, Cell state) according to their position in the cell. The position in the cell usually induces specific behavior that has to be modeled differently.*

The interactions of the biological system under study were represented in a RSTC network, which stands for multi-layer Receptor-Signaling-Transcription-Cell state network and that was generated from the Pathway Interaction Database (PID). In order to build this network one needs to select a set of seed nodes related to the biological process studied. For our case study, the seed nodes were: (1) *E-cadherin*, which is a protein having Calcium binding domains and which plays an important role in cell adhesion; (2) the 12 significantly differentially expressed genes across the 10 time-points; and (3) the cell states of keratinocytes-differentiation and cell-cycle-arrest. The network was extracted automatically from the whole content of the PID database by using a subgraph algorithm to link the seed nodes [9]. In Fig. 2 we show the RSTC network obtained.

Definition 3 (Pattern). *A pattern can be defined as an atomic set of biological components and their interacting roles.*

The first column of Table 1 shows some examples of patterns that can be found in a RSTC Network.

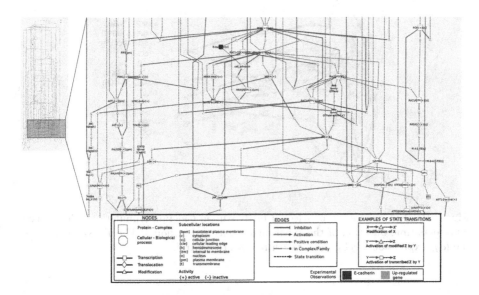

Fig. 2. Interaction graph linking E-cadherin with 12 genes of the time-series dataset. Blue nodes correspond to E-cadherin entities, red or green, to time-series genes, and cyan nodes to cellular processes. The graph is composed of 293 nodes and 375 edges (interactions). The set of nodes are composed of terminal nodes (proteins, complexes, mRNA expression, cellular state, biological processes and positive conditions) and of transient nodes (transcriptions, translocations, modifications and compounds). The set of edges are composed of interactions of type activation, inhibition, output, input and protein-family-member (Color figure online).

Time-Series Microarray Dataset. We use the time-series microarray data from Calcium stimulated human keratinocyte cells measured at 10 time-points (1h, 2h, 3h, 4h, 5h, 6h, 8h, 12h, 18h, 24h). The expression levels were measured in log_2; the expression of a gene at an specific time point is compared with respect to a control condition (gene expression in a kerationocyte cell without Calcium stimulation). We selected genes, which mRNA expression e was significantly

($log_2(e) \geq 1$) up-regulated or significantly ($log_2(e) \leq -1.0$) down-regulated in at least one time point compared to control. From this procedure 200 mRNA expression transcripts were selected. We included in our model a subset of 12 of the 200 selected (see Fig. 3) because these 12 genes had upstream regulatory mechanisms when querying the PID database and therefore were connected in the interaction graph to the E-cadherin node.

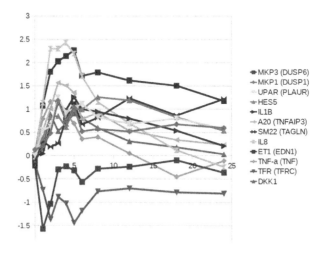

Fig. 3. Relative expression of selected mRNA upon Calcium stimulation. The X axis represents time duration of the experiment measured in hours. The Y axis represents the log_2 expression level of genes with respect to control.

2.2 The Process Hitting Framework

In order to model the dynamics of the system, we use the Process Hitting framework [18]. The Process Hitting (PH) gathers a finite number of concurrent processes grouped into a finite set of sorts. A sort stands for a component of a biological system while a process, which belongs to a unique sort, stands for one of its expression levels. At any time exactly one process of each sort is present. A state of the PH corresponds to such a set of processes. We denote here a process by a_i where a is the sort and i is the process identifier within the sort a. The concurrent interactions between processes are defined by a set of *actions*. Actions describe the replacement of a process by another of the same sort conditioned by the presence of at most one other process in the current state. An action is denoted $a_i \rightarrow b_j \,\sqcap\, b_k$, which is read as "$a_i$ *hits* b_j to make it bounce to b_k", where a_i, b_j, b_k are processes of sorts a and b, called respectively *hitter*, *target* and *bounce* of the action.

Definition 4 (Process Hitting). *A* Process Hitting *is a triple* $(\Sigma, L, \mathcal{H})$, *where:*

- $\Sigma = \{a, b, \dots\}$ *is the finite set of* sorts;
- $L = \prod_{a \in \Sigma} L_a$ *is the set of states with* $L_a = \{a_0, \dots, a_{l_a}\}$ *the finite set of processes of sort* $a \in \Sigma$ *and* l_a *a positive integer, with* $a \neq b \Rightarrow L_a \cap L_b = \emptyset$;
- $\mathcal{H} = \{a_i \rightarrow b_j \,\Gamma\, b_k \in L_a \times L_b \times L_b \mid (a, b) \in \Sigma^2 \wedge b_j \neq b_k \wedge a = b \Rightarrow a_i = b_j\}$ *is the finite set of* actions.

Given a state $s \in L$, the process of sort $a \in \Sigma$ present in s is denoted by $s[a]$. An action $h = a_i \rightarrow b_j \,\Gamma\, b_k \in \mathcal{H}$ is *playable* in $s \in L$ if and only if $s[a] = a_i$ and $s[b] = b_j$. In such a case, $(s \cdot h)$ stands for the state resulting from playing the action h in s, with $(s \cdot h)[b] = b_k$ and $\forall c \in \Sigma, c \neq b, (s \cdot h)[c] = s[c]$. In order to model the fact that a molecule in the interaction graph is influenced by various molecules, two types of modeling-scenarios can be proposed: cooperation and synchronization.

Modeling Cooperation. The cooperation between processes to make another process bounce can be expressed in PH by building a *cooperative sort* [18]. Figure 4 shows an example of a cooperative sort ab between sorts a and b, which is composed of 4 processes (one for each sub-state of the presence of processes in a and b). For the sake of clarity, processes of ab are indexed using the sub-state they represent. Hence, ab_{01} represents the sub-state $\langle a_0, b_1 \rangle$, and so on. Each process of sort a and b hits ab, which makes it bounce to the process reflecting the status of the sorts a and b (e.g., $a_1 \rightarrow ab_{00} \,\Gamma\, ab_{10}$ and $a_1 \rightarrow ab_{01} \,\Gamma\, ab_{11}$). Then, to represent the cooperation between processes a_1 and b_1, the process ab_{11} hits c_1 to make it bounce to c_2 instead of independent hits from a_1 and b_1. The same cooperative sort is used to make a_0 and b_0 cooperate to hit c_1 and make it bounce to c_0. Cooperation sort allows to model the fact that two components cooperate to hit another component.

Modeling Synchronization. The synchronization sort implements another type of cooperation. If we refer to the example of Fig. 4 left, we can similarly construct a *synchronization sort* ab between sorts a and b, defined with also 4 processes. Then, component c is activated (c_1 bounces to c_2 or c_0 bounces to c_1) if either a or b are activated. Therefore, each one of these processes ab_{01}, ab_{10}, ab_{11} can activate c. In order to inhibit c, both sorts, a and b, need to be in the sub-state 0, i.e. ab_{00}. Notice that this rule is a combination of OR logical gates for activation and AND logical gates for inhibition. Imposing the synchronization sort to model a target component regulated independently by multiple predecessors avoids oscillations in the behavior of the target component over time. These oscillations appear because each predecessor can independently activate the target component when it is active, but when one predecessor is inhibited, it inhibits the target component. This competition between the predecessors generates oscillations on the target component.

Example 1. Figure 4 represents a PH $(\Sigma, L, \mathcal{H})$ with $\Sigma = \{a, b, c, ab\}$, and:

$$L_a = \{a_0, a_1\},$$
$$L_b = \{b_0, b_1\},$$
$$L_{ab} = \{ab_{00}, ab_{01}, ab_{10}, ab_{11}\},$$
$$L_c = \{c_0, c_1, c_2\}.$$

This example models a Biological Regulatory Network (BRN) where the component c has three qualitative levels, components a and b are Boolean and ab is a cooperative sort. In this BRN, ab inhibits c at level 2 through the cooperative sort ab (e.g. $ab_{00} \rightarrow c_2 \upharpoonright c_1$, $ab_{00} \rightarrow c_1 \upharpoonright c_0$) while a and b activate c through the cooperative sort ab (e.g. $ab_{11} \rightarrow c_0 \upharpoonright c_1$ $ab_{11} \rightarrow c_1 \upharpoonright c_2$). Indeed, the reachability of c_2 and c_0 is conditioned by a cooperation of a and b as explained above.

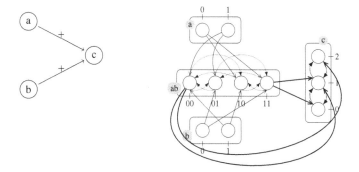

Fig. 4. (Left) biological pattern example. Nodes represent molecules (components) and edges, interactions. In this pattern components a and b cooperate to activate c. (Right) equivalent PH model with four sorts: three components (a, b and c) and a cooperative sort (ab). Actions targeting processes of c are drawn as thick lines.

2.3   Model Construction (from RSTC to PH)

Modeling the RSTC Network as a PH Model. In order to model the RSTC network as a PH model we select known biological regulatory patterns (atomic set of biological components and their interacting roles), represented as biochemical reactions in the RSTC network and we propose their PH representation. Table 1 shows some examples of this transformation. The automatic pattern selection and PH model generation algorithms use two procedures. The first one takes the graph as parameter argument and automatically browses it node by node and detects all the patterns in the graph. For each node (output node of the pattern) we call a recursive procedure, that allows to detect a minimal set of nodes (input node of the pattern) that has a direct influence over that node. This set of nodes plus the output node and the way input and output are linked form a pattern. The type of a pattern is determined by the type of the output node, the type of regulations that come to that node and the type of input nodes of the pattern. Consequently, the algorithm of patterns detection returns the pattern and its type to another procedure which translates the pattern into the PH formalism. This transformation takes care of different cases (cooperation, synchronization, simple activation, simple inhibition, etc.) For example a molecule a cooperating with a molecule b to activate a molecule c (Fig. 4, left), is a regulatory pattern because it is a protein-complex biochemical reaction that appears at recurrent times. We model this pattern by four sorts (Fig. 4, right) a, b, c and ab. Sorts a, b and c stand for components a, b and c. The cooperative sort ab is introduced in order to characterize constraints on the components a and b. In the RSTC network, we find 25 regulatory patterns. We show some examples in Table 1.

Table 1. Examples of patterns

Biological Patterns	PH Transformations	Descriptions
Simple activation		This pattern model the activation of the component b by the component a.
Simple inhibition		This pattern model the inhibition of the component b by the component a
Activation or inhibition		This pattern model either the activation of the component c by the component a or the inhibition of the component c by the component b

Estimating the Parameters for the PH-simulation from Time-Series Gene Expression Data. Since the simulation of the execution of the PH actions is done stochastically, we need to relate each action with temporal and stochastic parameters introduced into the PH framework to achieve dynamic refinement [18]. To fire an action in the PH framework we need to provide two parameters: (1) the rate $r = t^{-1}$, where t is the mean time for firing an action, and (2) the stochasticity absorption factor sa, which is introduced to control the variance of firing time of an action.

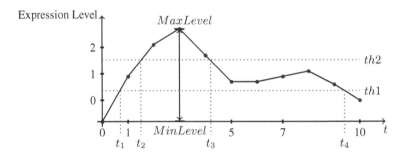

Fig. 5. Estimating temporal parameters from time series data: The mean firing time of an action that makes a component (mRNA expression) change of sub-state is estimated as $r_i = \frac{1}{t_i - t_{i-1}}$. $MaxLevel$ represents the maximum expression of a mRNA expression, while $MinLevel$, its minimum expression. The thresholds $th1$ and $th2$ define the PH discrete sub-states (e.g. 0,1,2) of a component according to its gene expression data.

For the model components which have a measurement in the time-series data we estimate the r and sa parameters and they are introduced in the PH model. The other components are assigned default parameters. In order to estimate r_i and sa_i for each action $h_i \in \mathcal{H}$, we need to know the different times t_i when the action could be fired as illustrated in Fig. 5. Each t_i represents the time at which we assume that a component moves from one process to another. Therefore the action that leads this change must be played at the rate $r_i = \frac{1}{t_i - t_{i-1}}$. The integer sa represents the window of firing the action at rate r: the larger the sa is, the smaller the variance around r is. Studies [2,3,22] have proposed more elaborated methods for parameters estimation from gene expression data. These methods are well adapted in the case of biochemical reactions where the concept of threshold is implicit. In the proposed case we assume an explicit threshold. Thus a basic estimation algorithm can be used for temporal and stochastic parameter estimation.

Discretization of Time-Series Data. Because the outputs of a PH simulation are discrete traces of PH components, we discretized continuous experimental data to facilitate the comparison with simulation outputs. When looking at the time-series data (see Fig. 3) one can distinguish a high level of activity in early hours [0h-5h] and a low level in late hours [5h-10h]. This trend was confirmed by the SMA (Simple moving average) function of the R package TTR which allows us to smooth time-series data. We used the SMA function with parameter $n = 2$ and we observed that more than 50 % of the time-series data presented these two levels of activities. We implemented a discretization method to capture these two activation times. For each time-series, we introduced two thresholds $th1$ and $th2$ (see Fig. 5) were introduced: $th1 = \frac{1}{3}(MaxLevel - MinLevel)$ and $th2 = \frac{2}{3}(MaxLevel - MinLevel)$. In this way, the expression level in the range $[0 - th1]$ is at level 0, the one in the range $[th1 - th2]$ is at level 1, and the one in the last range is at level 2.

2.4 Simulation

We set the same initial conditions to PH components belonging to the same network layer, chosen from the RSTC structure. These initial conditions are detailed below and summarized in Fig. 6.

- **Receptor Layer: E-cadherin.** We choose the pulse signal for the input node E-cadherin to be active for a duration of 5 time units in average. This choice was made in orders to take into account the average time of the Calcium stimuli effect.
- **Signaling Layer: Signaling Proteins.** The components in this layer are activated and inhibited with the same rate and the same stochasticity absorption factor. The actions between a controller component A and a controlled component B are constrained so that B is first activated by A and then inhibited. That is, the time interval in which an inhibition action from A to B fires is greater than the time interval in which an activation action from A to B fires. Additionally, these two time intervals must not overlap. These constraints can be seen as reachability constraints from the entry node (E-cadherin) to the output nodes (mRNA expression). The values of these parameters are selected by considering the delay of signal transduction from the entry node to the output nodes.
- **Transcription Layer: Transcription Factors.** In this layer, the activation/inhibition over a transcription factor (TF) comes from signaling proteins; however, for all TFs we introduced an auto-inhibition action that represents their degradation over time.
- **mRNA Expression.** The mRNA expression are activated or inhibited according to the estimated values from time-series data.

2.5 Automatic Analysis of Simulation Traces

Due to the stochastic and concurrent aspects of the system, each execution of the model can generate a different dynamic trace. Therefore, to validate the proposed model we analyzed the traces generated by each component for a set of simulations of the model. The idea was to calculate the percentage of traces that reproduced the expected dynamic of the system. To achieve this goal, we take each trace generated at each simulation for a given component and passed it to an automaton ($\mathcal{A}_i$) that recognize the experimental trace of that component. Thus we can count the number of accepted traces ($Trace_{accp}$); the percentage of accepted traces is $\frac{Trace_{accp}}{Trace_N}$ if $Trace_N$ is the total number of simulations. Following this we introduce the concept of tolerance in accepting traces. It means that an automaton can accept a trace with a difference of one or more levels at each state. In our case study we used a tolerance T_1 that allows accepting a difference of one between the simulated trace and the expected trace at each state.

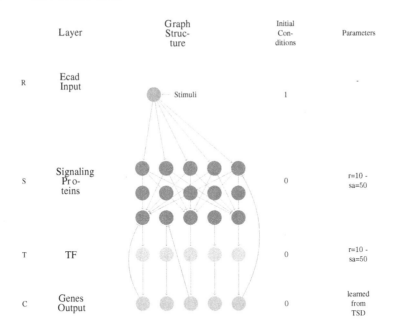

Fig. 6. RSTC network structure and initial conditions assigned to each node in the layer

3   Results

3.1   Automatic Generation of PH Model from the PID Network

PH models are written using the PINT[1] format. PINT implements stochastic simulations and static analyses for computing dynamical properties on very large-scale PH models. For the PINT code generation two procedures were used. The first procedure detects motifs (controller and controlled components) in graphs from the Pathway Interaction Database; the second, generates the PINT code by choosing an adequate concurrency rule, based on synchronization sorts, to represent the motif dynamic in PH. With this method it is possible to convert the whole content of the PID database into a PH model, as well as individual pre-selected pathways, as is the case for the system under study. It is implemented in Java and available upon request.

3.2   Simulation of Calcium Stimulated Biological System

We simulated the model with and without the inclusion of the synchronization sort. In the following, we present the results of the simulation.

---

[1] http://process.hitting.free.fr.

Without the Introduction of the Synchronization Sort. One can notice in Fig. 7 the occurrence of oscillations. Whereas it is not the expected behavior from the biological system, it is coherent with the choice of the modeling and the way the simulator works as explained in Sect. 2.2. In this simulation, cooperation sorts were used to model multiple controllers of a common controlled (target) component. It is important to notice that the intensity of the oscillation is linked with the size of the concurrence, i.e. the number of controllers a controlled component has. Despite the presence of the oscillations, the model reproduces expected dynamical behaviors namely the dynamics of components, the signal transduction and takes into account the stochastic and time aspect of the model.

With the Introduction of the Synchronization Sort. In Fig. 8 we can see that the introduction of the synchronization sort significantly reduces the impact of concurrency. The result shows a clear elimination of the previously observed oscillations (Fig. 7). Comparing the simulated results with the ones observed experimentally, we found four different cases. We found 5 simulation traces (IL8, uPAR, IL1_beta, ET1, A20) that matched the sequence of all the component expression levels perfectly. In this case, delays exist among simulation and experiment but these delays are not comparable since experimental time-points are measured in hours and simulation-units for the simulated PH model. We found 6 simulation traces (MKP1, MKP3, Hes5, SM22, TfR, DKK1) that matched the sequence of experimental discrete expression levels missing one expression-level. We found 1 components (TNF-alpha) in which at least 2 expression levels are missed.

Simulating Biological Processes. To validate our model, we studied the prediction of non-observed components of such a system and we focused on biological processes linked to Calcium stimulation, such as keratinocyte-differentiation, cell-adhesion and cell-cycle arrest. Our results are shown in Fig. 9 and confirm literature experimental evidences on these processes. In the case of keratinocyte-differentiation, this was a functional behavior measured on the cultured cells upon Calcium stimulation, so there was experimental evidences of this effect before measuring the gene expression. In the case of cell-cycle arrest, the switch-on of this component represents the fact that the E-cadherin stimulated model predicts the stop of growth, as confirmed by literature in human and mouse keratinocytes [13]. Finally, the cell-adhesion component is predicted to switch-on, also in according to published evidence [28] in human and mouse keratinocytes.

3.3 Model Validation: Traces Analysis

To validate the results of the simulations, we automatically analyzed the traces generated by a set of 100 simulations. Table 2 shows the results of the percentage of acceptance for the traces of each of the 12 mRNA expressions. One can observe that there are 4 components with a good acceptance rate ($> 76\,\%$), which are:

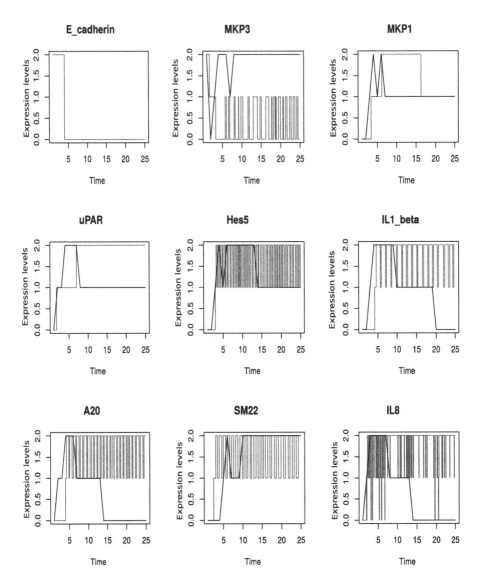

Fig. 7. Results of system simulations without introducing the synchronization sort for 9 genes. The traces representing the discretized time-series data are shown as black lines. The traces representing the simulated traces are shown as blue lines (Color figure online).

A20, IL1_beta, IL8, uPar; 4 traces with a good acceptance rate (> 94 %) when considering 1 level of tolerance, which are: MKP1, MKP3, SM22, and TfR; and finally 4 traces, for which the model failed to predict their expressions: ET1, Hes5, DKK1, and TNFa. All in all, for this case study our model predicts

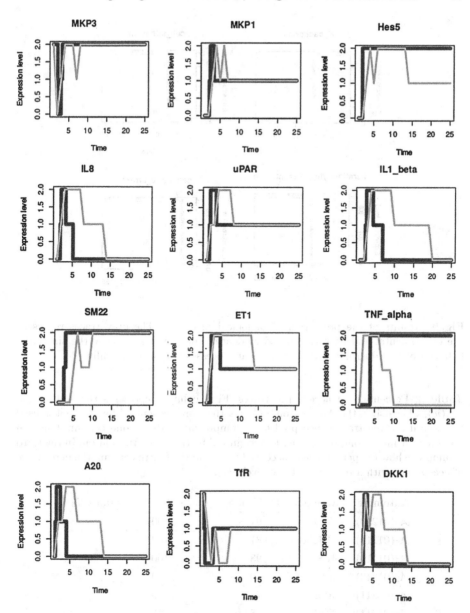

Fig. 8. Results of simulations by introducing the synchronization sort. The gray traces represent the experimental expected behaviors from the discretization of the time-series data. The blue traces show the simulated behavior (Color figure online).

relatively well, 8 out of 12, the experimental traces. Errors on the prediction of the missing 4 components may be because of missing regulatory interactions.

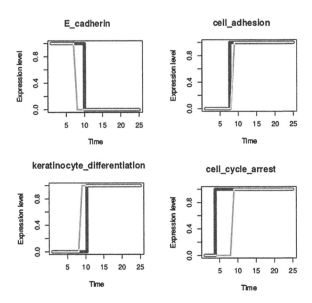

Fig. 9. Results of the prediction of biological processes. The gray traces represent the experimental and literature-based evidence. The blue traces show the simulated behavior of E-cadherin and three biological processes (Color figure online).

Table 2. Percentage of acceptance traces. First column represents the Automaton $(\mathcal{A}_i(w)$, where $\mathcal{A}_i$ is the Automaton and w is the word recognized by $\mathcal{A}_i$) that is used to check if a given trace is accepted for a component in the second column. One can observe that many components can be recognized by the same Automaton. In the third column we show the percentage of accepted traces; in the fourth column, the percentage of acceptance with a tolerance of one level (T_1).

Automate	Components	% of acceptance	% of acceptance T_1
$\mathcal{A}_2(01210)$	A20	91	100
$\mathcal{A}_2(01210)$	IL1_beta	81	100
$\mathcal{A}_2(01210)$	IL8	93	100
$\mathcal{A}_2(01210)$	TNF_alpha	0	0
$\mathcal{A}_3(01211)$	uPar	76	99
$\mathcal{A}_3(01211)$	ET1	8	19
$\mathcal{A}_4(0121210)$	DKK1	13	43
$\mathcal{A}_5(0121211)$	Hes5	0	17
$\mathcal{A}_5(0121211)$	MKP1	9	97
$\mathcal{A}_6(0212)$	SM22	11	100
$\mathcal{A}_7(02010)$	MKP3	11	98
$\mathcal{A}_8(02121)$	Tfr	0	94

4   Conclusion

This work describes the steps towards the integration of time-series data in large-scale cell-based models. We proposed an automatic method to build a timed and stochastic PH model from pathways of biochemical reactions present in the Pathway Interaction Database (PID). As a case-study we built a model combining signaling and transcription events relevant to keratinocyte differentiation induced by Calcium, which linked E-cadherin nodes and 12 genes, which expression profiles was measured upon Calcium stimulation over time. The interaction graph represented by the model had 293 nodes and 375 edges. We proposed a method to discretize time-series gene expression data, so they can be integrated to the PH simulations and logically explained by the PH stochastic analyses. Additionally, we implemented a method to automatically estimate the temporal and stochastic parameters for the PH simulation, so this estimation process will not be biased by over fitting. Our results show that we can observe dynamic effects on 11 out of 12 genes, for which 5 of them represent accurate predictions, and 6 of them missed few dynamic levels. This error may be also a result from the incompleteness of the regulatory information in PID. Moreover, when observing the predicted behavior of biological processes linked to Calcium stimulation, our predictions agreed with experimental and literature-based evidences. Overall, with this work we show the feasibility of modeling and simulating large-scale networks with very few parameter estimation and having good quality predictions. As perspectives of this work we intend to study the effects of computing automatically the concurrent rules on this system. Also, we intend to improve the model prediction quality by empirically obtaining the dynamics of the system components by performing large stochastic simulations, as well as by implementing static analysis of quantitative properties by adding probabilistic features to the PH static solver.

Acknowledgements. This work was supported by a PhD grant from the CNRS and the French region *Pays de la Loire* and grants from the German Ministry for Research and Education (BMBF) funding program MedSys (grant number FKZ0315401A) and AGENET (FKZ0315898).

A   Algorithm of Patterns Detection

Here are the algorithms that allow to detect and construct a process hitting model from an RSTC network. These algorithms have a polynomial time running that correspond to the running time of the procedure 2.

Proposition 1. *Algorithm 2 has a time complexity of $\mathcal{O}(|V| \log (h))$. Where h is the average height of the patterns in the RSTC network. In the worst case $h = \log_V (|V|)$.*

Algorithm 1. Algorithm for Pattern detection in an RSTC Network in order to generate the equivalent model in the PH formalism

---

Require: *Net* {The RSTC network}
Ensure: generate the PH Model associated to *Net*
1: **for all** Node n in *Net*.getSetOfNodes() **do**
2:     Pat = detectPattern (*Net*, n)
3:     patternInPHModel (*out*, Pat)
4: **end for**

Algorithm 2. Algorithm for pattern detection, function detectPattern (*Net*, n)

Require: *Net*, n {*Net* is the network and n is the current node}
Ensure: Build a set of nodes associated to node n that we call pattern.
 1: **switch** (n)
 2: **case** TerminalNode:
 3:     add node n to the pattern Pat
 4:     numberPredecessor= n.getNumberOfPredecessor()
 5:     **switch** (numberPredecessor)
 6:     **case** 1:
 7:       **for all** p in setOfPredecessor (n) **do**
 8:         **switch** (p)
 9:         **case** TerminalNode:
10:             add node n to the pattern Pat
11:         **case** TransientNode:
12:             detectPattern (*Net*, p);
13:         **end switch**
14:       **end for**
15:       Set the code of pattern Pat;
16:       return Pat;
17:     **case** 2:
18:
19:     **end switch**
20: **case** TransientNode:
21:     numberPredecessor= n.getNumberOfPredecessor()
22:     **switch** (numberPredecessor)
23:     **case** 1:
24:       **for all** p in setOfPredecessor (n) **do**
25:         **switch** (p)
26:         **case** TerminalNode:
27:             added node to the pattern Pat;
28:         **case** TransientNode:
29:             detectPattern (*Net*, p);
30:         **end switch**
31:       **end for**
32:       Set the code of pattern Pat;
33:       return Pat;
34:     **case** 2:
35:
36:     **end switch**
37: **end switch**

Algorithm 3. Algorithm for writing a given pattern into a file, function patternInPHModel (*out*, *Pat*)

Require: *out*, *Pat* {Pat is The pattern to be translated into the PH Model, out is the output file}

Ensure: The correspondent PH Model of the given pattern Pat will write into the file *out*

$nocp = Pat$.getNumberOfComponents() {Number of the components of the pattern *Pat*}

$tabPat = Pat$.getTableOfPattern() {return the components of the pattern in *tabPat*}

switch (*nocp*)

  case 2:

    switch (*code*)

    case A:

      out.write ($tabPat[1]$ 1 → $tabPat[0]$ 0 1 r_a sa_a); {Component $tabPat[1]$ activates component $tabPat[0]$ with $r = r_a$ and $sa = sa_a$ }

    case I:

      out.write ($tabPat[1]$ 0 → $tabPat[0]$ 1 0 r_i sa_i); {Component $tabPat[1]$ inhibits component $tabPat[0]$ with $r = r_i$ and $sa = sa_i$ }

    end switch

  case 3:

    switch (*code*)

    case C:

      out.write (coop ($[tabPat[2];tabPat[1]]$) → $tabPat[0]$ 0 1); {Cooperation between $tabPat[1]$ and $tabPat[2]$ to activate $tabPat[0]$}

    case S:

    out.write (coop ($[tabPat[2];tabPat[1]]$) → $tabPat[0]$ 0 1); {Synchronization between $tabPat[1]$ and $tabPat[2]$ to activate $tabPat[0]$}

    default:

      out.write ((*unknow pattern*));

    end switch

  end switch

References

1. Ahmad, J., Roux, O., Bernot, G., Comet, J.-P., Richard, A.: Analysing formal models of genetic regulatory networks with delays. Int. J. Bioinform. Res. Appl. (IJBRA) **4**(3), 240–262 (2008)
2. Andreychenko, A., Mikeev, L., Spieler, D., Wolf, V.: Approximate maximum likelihood estimation for stochastic chemical kinetics. EURASIP J. Bioinform. Syst. Biol. **2012**(1), 9 (2012)
3. Batt, G., Page, M., Cantone, I., Goessler, G., Monteiro, P., de Jong, H.: Efficient parameter search for qualitative models of regulatory networks using symbolic model checking. Bioinformatics **26**(18), i603–i610 (2010)
4. Batt, G., Ben Salah, R., Maler, O.: On timed models of gene networks. In: Raskin, J.-F., Thiagarajan, P.S. (eds.) FORMATS 2007. LNCS, vol. 4763, pp. 38–52. Springer, Heidelberg (2007)

5. Busch, H., Camacho-Trullio, D., Rogon, Z., Breuhahn, K., Angel, P., Eils, R., Szabowski, A.: Gene network dynamics controlling keratinocyte migration. Mol. Syst. Biol. **4**(1), 199 (2008)
6. Chaouiya, C., Remy, E., Mossé, B., Thieffry, D.: Qualitative analysis of regulatory graphs: a computational tool based on a discrete formal framework. In: Benvenuti, L., De Santis, A., Farina, L. (eds.) Positive Systems. LNCIS, vol. 294, pp. 119–126. Springer, Heidelberg (2003)
7. De Jong, H., Geiselmann, J., Hernandez, C., Page, M.: Genetic network analyzer: qualitative simulation of genetic regulatory networks. Bioinformatics **19**(3), 336–344 (2003)
8. Gardner, T.S., Di Bernardo, D., Lorenz, D., Collins, J.J.: Inferring genetic networks and identifying compound mode of action via expression profiling. Science **301**(5629), 102–105 (2003)
9. Guziolowski, C., Kittas, A., Dittmann, F., Grabe, N.: Automatic generation of causal networks linking growth factor stimuli to functional cell state changes. FEBS J. **279**(18), 3462–3474 (2012)
10. Guziolowski, C., Videla, S., Eduati, F., Thiele, S., Cokelaer, T., Siegel, A., Saez-Rodriguez, J.: Exhaustively characterizing feasible logic models of a signaling network using answer set programming. Bioinformatics **29**(18), 2320–2326 (2013)
11. Heiner, Monika, Gilbert, David, Donaldson, Robin: Petri nets for systems and synthetic biology. In: Bernardo, Marco, Degano, Pierpaolo, Zavattaro, Gianluigi (eds.) SFM 2008. LNCS, vol. 5016, pp. 215–264. Springer, Heidelberg (2008)
12. Kauffman, S.A.: Metabolic stability and epigenesis in randomly constructed genetic nets. J. Theor. Biol. **22**(3), 437–467 (1969)
13. Kolly, C., Suter, M.M., Muller, E.J.: Proliferation, cell cycle exit, and onset of terminal differentiation in cultured keratinocytes: pre-programmed pathways in control of C-Myc and Notch1 prevail over extracellular calcium signals. J. Invest. Dermatol. **124**(5), 1014–1025 (2005)
14. MacNamara, A., Terfve, C., Henriques, D., Bernabé, B.P., Saez-Rodriguez, J.: State-time spectrum of signal transduction logic models. Phys. Biol. **9**(4), 045003 (2012)
15. Maurin, M., Magnin, M., Roux, O.: Modeling of genetic regulatory network in stochastic π-calculus. In: Rajasekaran, S. (ed.) BICoB 2009. LNCS, vol. 5462, pp. 282–294. Springer, Heidelberg (2009)
16. Mitsos, A., Melas, I.N., Siminelakis, P., Chairakaki, A.D., Saez-Rodriguez, J., Alexopoulos, L.G.: Identifying drug effects via pathway alterations using an integer linear programming optimization formulation on phosphoproteomic data. PLoS Comput. Biol. **5**(12), e1000591 (2009)
17. Molloy, M.K.: Performance analysis using stochastic petri nets. IEEE Trans. Comput. **100**(9), 913–917 (1982)
18. Paulevé, L., Magnin, M., Roux, O.: Refining dynamics of gene regulatory networks in a stochastic π-calculus Framework. In: Priami, C., Back, R.-J., Petre, I., de Vink, E. (eds.) Transactions on Computational Systems Biology XIII. LNCS, vol. 6575, pp. 171–191. Springer, Heidelberg (2011)
19. Pinna, A., Soranzo, N., de la Fuente, A.: From knockouts to networks: Establishing direct cause-effect relationships through graph analysis. PLoS ONE **5**(10), e12912 (2010)
20. Porreca, R., Cinquemani, E., Lygeros, J., Ferrari-Trecate, G.: Identification of genetic network dynamics with unate structure. Bioinformatics **26**(9), 1239–1245 (2010)

21. Priami, C.: Stochastic π-calculus. Comput. J. **38**(7), 578–589 (1995)
22. Altman, R., Reinker, S., Timmer, J.: Parameter estimation in stochastic biochemical reactions. IEE Pro. Syst. Biol. **153**, 168–178 (2006)
23. Schaefer, C.F., Anthony, K., Krupa, S., Buchoff, J., Day, M., Hannay, T., Buetow, K.H.: Pid: the pathway interaction database. Nucleic Acids Res. **37**(suppl 1), D674–D679 (2009)
24. Siebert, H., Bockmayr, A.: Incorporating time delays into the logical analysis of gene regulatory networks. In: Priami, C. (ed.) CMSB 2006. LNCS (LNBI), vol. 4210, pp. 169–183. Springer, Heidelberg (2006)
25. Snoussi, E.H.: Qualitative dynamics of piecewise-linear differential equations: a discrete mapping approach. Dyn. Stab. Syst. **4**(3–4), 565–583 (1989)
26. Thieffry, D., Thomas, R.: Dynamical behaviour of biological regulatory networks immunity control in bacteriophage lambda. Bull. Math. Biol. **57**(2), 277–297 (1995)
27. Thomas, R.: Boolean formalization of genetic control circuits. J. Theor. Biol. **42**(3), 563–585 (1973)
28. Tu, C.L., Chang, W., Bikle, D.D.: The calcium-sensing receptor-dependent regulation of cell-cell adhesion and keratinocyte differentiation requires Rho and filamin A. J. Invest. Dermatol. **131**(5), 1119–1128 (2011)
29. Van Goethem, S., Jacquet, J.-M., Brim, L., Šafránek, D.: Timed modelling of gene networks with arbitrarily precise expression discretization. Electron. Notes Theoret. Comput. Sci. **293**, 67–81 (2013)
30. Namhee, Y., Seo, J., Rho, K., Jang, Y., Park, J., Kim, W.K., Lee, S.: hipathdb: a human-integrated pathway database with facile visualization. Nucleic Acids Res. **40**(D1), D797–D802 (2012)

Approximate Probabilistic Verification
of Hybrid Systems

Benjamin M. Gyori[1], Bing Liu[2], Soumya Paul[3], R. Ramanathan[4],
and P.S. Thiagarajan[1($\boxtimes$)]

[1] Laboratory of Systems Pharmacology, Harvard Medical School, Boston, USA
benjamin_gyori@hms.harvard.edu, psthiagu@gmail.com
[2] Department of Computational and Systems Biology, University of Pittsburgh,
Pittsburgh, USA
[3] Institute de Recherche En Informatique de Toulouse, 31062 Toulouse, France
[4] Department of Computer Science, National University of Singapore,
Singapore, Singapore

Abstract. Hybrid systems whose mode dynamics are governed by non-linear ordinary differential equations (ODEs) are often a natural model for biological processes. However such models are difficult to analyze. To address this, we develop a probabilistic analysis method by approximating the mode transitions as stochastic events. We assume that the probability of making a mode transition is proportional to the measure of the set of pairs of time points and value states at which the mode transition is enabled. To ensure a sound mathematical basis, we impose a natural continuity property on the non-linear ODEs. We also assume that the states of the system are observed at discrete time points but that the mode transitions may take place at any time between two successive discrete time points. This leads to a discrete time Markov chain as a probabilistic approximation of the hybrid system. We then show that for BLTL (bounded linear time temporal logic) specifications the hybrid system meets a specification iff its Markov chain approximation meets the same specification with probability 1. Based on this, we formulate a sequential hypothesis testing procedure for verifying–approximately–that the Markov chain meets a BLTL specification with high probability. Our case studies on cardiac cell dynamics and the circadian rhythm indicate that our scheme can be applied in a number of realistic settings.

Keywords: Hybrid systems · Markov chains · Dynamical systems · Statistical model checking

1 Introduction

Hybrid systems are often used to model biological processes [6,10,11]. The analysis of these models is difficult due to the high expressive power of the mixed

S. Paul—The author would like to thank ERC grant 269427 that supported a part of this work.

© Springer International Publishing Switzerland 2015
A. Abate and D. Šafránek (Eds.): HSB 2015, LNBI 9271, pp. 96–116, 2015.
DOI: 10.1007/978-3-319-26916-0_6

dynamics [22]. Various lines of work have explored ways to mitigate this problem with a common technique being to restrict the mode dynamics [2,3,14,18,20,23]. However, for many of the models arising in systems biology the mode dynamics will be governed by a system of non-linear ordinary differential equations (ODEs). To analyze such systems, we develop a scheme under which such systems can be approximated as a discrete time Markov chain.

A key difficulty in analyzing a hybrid system's behavior is that the time points and value states at which a trajectory meets a guard will depend on the solutions to the ODE systems associated with the modes. For high-dimensional systems these solutions will not be available in closed form. To get around this, we approximate the mode transitions as stochastic events by fixing the probability of a mode transition to be proportional to the measure of the set of value state and time point pairs at which this transition is enabled. More sophisticated hypotheses could be considered. For instance one could tie the mode transition probability to how long the guard has been continuously enabled or how deeply within a guard region the current state is. To bring out the main ideas we will postpone exploring such approximations to our future work.

To secure a sound mathematical basis for our approximation, we further assume: (i) The vector fields associated with the ODEs are Lipschitz continuous functions. (ii) The states of the hybrid system are observable only at discrete time points. (iii) The set of initial states and the guard sets are bounded open sets. (iv) The hybrid dynamics is strictly non-Zeno in the sense that there is a uniform upper bound on the number of transitions that can take place in a unit time interval. For technical convenience we in fact assume that time discretization is so chosen that at most one mode transition takes place between two successive discrete time points.

Under these assumptions, we show that the dynamics of the hybrid system H can be approximated as an infinite state Markov chain M. Given the application domain we have in mind, namely, biological pathway dynamics, we focus on the behavior of the hybrid system over a finite time horizon and BLTL (bounded linear time temporal logic) [15] to specify dynamic properties of interest. The maximum discrete time point we fix will be determined by the BLTL specification. Our probabilistic approximation is such that the set of trajectories satisfying a BLTL formula will correspond to a measurable set of paths of the Markov chain and hence can be assigned a probability value. We then show that H meets the specification ψ–i.e. every trajectory of H is a model of ψ–iff M meets the specification ψ with probability 1. This allows us to approximately verify interesting properties of the hybrid system using its Markov chain approximation. However, even a bounded portion of M can not be constructed effectively. This is because the transition probabilities of the Markov chain will depend on the solutions to the ODEs associated with the modes, which will not be available in a closed form. In addition, the structure of M itself will be unknown since the states of the chain will be those that can reached with non-zero probability from the initial mode and we can not determine which transitions have non-zero probabilities. To cope with this, we design a statistical model checking procedure to approximately verify that the chain (and hence the hybrid system) almost certainly meets the specification. One just needs to ensure that the dynamics of the Markov chain is being sampled according to underlying probabilities. We achieve this by randomly generating trajectories of H through numerical simulations in a way that corresponds to randomly sampling the paths of the Markov chain according to its underlying structure and transition probabilities. We note that a simple minded Monte Carlo simulations based strategy consisting of sampling an initial state (according to the given initial distribution) followed by a random generation of trajectory will flounder on the issue of how one "randomly" picks a mode transition during the generation of a trajectory in the presence of the non-linear dynamics captured by the ODEs systems. Our approximation technique instead establishes a principled way of achieving this.

In establishing these results, we assume that the atomic propositions in the specification are interpreted over the modes of the hybrid system. Consequently one can specify patterns of mode visitations while quantitative properties can be inferred only indirectly and in a limited fashion. Our results however can be extended to handle quantitative atomic propositions ("the current concentration of protein X is greater than 2 μM"). Due to space limitations, we present the details of this extension in a technical report available online [4].

To demonstrate the applicability of our method, we first study the electrical activity of cardiac cells represented by a hybrid model. By varying parameters we analyze key dynamical properties on multiple cell types, in healthy and disease conditions, and under different input stimuli. We also analyze a hybrid model of the circadian rhythm, and find distinct roles of multiple feedback loops in maintaining oscillatory properties of the dynamics.

1.1   Related Work

Mode transitions have been approximated as random events in the literature. In [1] the dynamics of a hybrid system is approximated by substituting the guards with probabilistic barrier functions. Our transition probabilities are constructed using similar but simpler considerations. We have done so in order to be able to carry out temporal logic based verification based on simulations. An alternative approach to approximately verifying non-linear hybrid systems is one based on δ-reals [19]. Here one verifies bounded reachability properties that are robust under small perturbations of the numerical values mentioned in the specification. Since the approximation involved is of a very different kind, it is difficult to compare this line of work with ours. However, it may be fruitful to combine the two approaches to verify a richer set of reachability properties.

The present work may be viewed as an extension of [31] where a *single* system of ODEs is considered. This method however, breaks down in the multi-mode hybrid setting and one needs to construct–as we do here–an entirely new machinery. Finally, a wealth of literature is available on the analysis of stochastic automata [5,8,13,25]. It will be interesting to explore if these methods can be transported to our setting.

2   Hybrid Automata

We fix n real-valued variables $\{x_i\}_{i=1}^n$ viewed as functions of time $x_i(t)$ with $t \in \mathbb{R}_+$, the set of non-negative reals. A valuation of $\{x_i\}_{i=1}^n$ is $\mathbf{v} \in \mathbb{R}^n$ with $\mathbf{v}(i) \in \mathbb{R}$ representing the value of x_i. The language of *guards* is given by: (i) $a < x_i$ and $x_i < b$ are guards where a, b are rationals and $i \in \{1, 2, \ldots, n\}$. (ii) If g and g' are guards then so are $g \wedge g'$ and $g \vee g'$.

$\mathcal{G}$ denotes the set of guards. We define $\mathbf{v} \models g$ (i.e. $\mathbf{v}$ satisfies the guard g) via: $\mathbf{v} \models a < x_i$ iff $a < \mathbf{v}(i)$ and similarly for $x_i < b$. The clauses for conjunction and disjunction are standard. We let $\|g\| = \{\mathbf{v} \mid \mathbf{v} \models g\}$. We note that $\|g\|$ is an open subset of $\mathbb{R}^n$ for every guard g. We will abbreviate $\|g\|$ as g.

Definition 1. *A hybrid automaton is a tuple* $H = (Q, q_{in}, \{F_q(\boldsymbol{x})\}_{q \in Q}, \mathcal{G}, \to,$ INIT), *where*

- Q *is a finite set of* modes *and* $q_{in} \in Q$ *is the* initial *mode.*
- *For each* $q \in Q$, $d\boldsymbol{x}/dt = F_q(\boldsymbol{x})$ *is a system of ODEs, where* $\boldsymbol{x} = (x_1, x_2, \ldots, x_n)$ *and* $F_q = (f_q^1(\boldsymbol{x}), f_q^2(\boldsymbol{x}), \ldots, f_q^n(\boldsymbol{x}))$. *Further,* f_q^i *is Lipschitz continuous for each* i.
- $\to \subseteq (Q, \mathcal{G}, Q)$ *is the mode transition relation. If* $(q, g, q') \in \to$ *we shall often write it as* $q \xrightarrow{g} q'$.
- INIT $= (L_1, U_1) \times (L_2, U_2) \ldots \times (L_n, U_n)$ *is the set of initial states where* $L_i < U_i$ *and* L_i, U_i *are rationals.*

We have not associated invariant conditions with the modes or reset conditions with the mode transitions. They can be introduced with some additional work.

Fixing a suitable unit time interval Δ, we discretize the time domain as $t = 0, \Delta, 2\Delta, \ldots$. We assume the states of the system are observed only at these discrete time points. Furthermore, we shall assume that only a bounded number of mode changes can take place between successive discrete time points. Both in engineered and biological processes this is a reasonable assumption. Given this, we shall in fact assume that Δ is such that at most one mode change takes place within a Δ time interval. We note that there can be multiple choices for Δ that meet this requirement and in practice one must choose this parameter carefully. (Our method can be extended to handle a bounded number of mode transitions in a unit time interval but this will entail notational complications that will obscure the main ideas.) In what follows, for technical convenience we also assume the time scale has been normalized so that $\Delta = 1$. As a result, the discretized set of time points will be $\{0, 1, 2, \ldots\}$.

2.1   Trajectories

We have assumed that for every mode q, the right hand side of the ODEs, $F_q(\mathbf{x})$, is Lipschitz continuous for each component. As a result, for each initial value $\mathbf{v} \in \mathbb{R}^n$ and in each mode q, the system of ODEs $d\mathbf{x}/dt = F_q(\mathbf{x})$ will have a unique solution $Z_{q,\mathbf{v}}(t)$ [24]. We are also guaranteed that $Z_{q,\mathbf{v}}(t)$ is Lipschitz and hence measurable [24]. It will be convenient to work with two sets of functions derived from solutions to the ODE systems.

The (unit interval) *flow* $\Phi_q : (0,1) \times \mathbb{R}^n \to \mathbb{R}^n$ is given by $\Phi_q(t,\mathbf{v}) = Z_{q,\mathbf{v}}(t)$. Φ_q will also be Lipschitz. Next we define the parametrized family of functions $\Phi_{q,t} : \mathbb{R}^n \to \mathbb{R}^n$ given by $\Phi_{q,t}(\mathbf{v}) = \Phi_q(t,\mathbf{v})$. Applying once again the fact that the RHS of the ODEs are Lipschitz continuous functions, we can conclude that these parametrized functions $\Phi_{q,t}$ (which will be $1-1$) as well as their inverses will be Lipschitz.

A (finite) *trajectory* is a sequence $\tau = (q_0, \mathbf{v}_0)(q_1, \mathbf{v}_1) \ldots (q_k, \mathbf{v}_k)$ such that for $0 \le j < k$ the following conditions are satisfied: (i) For $0 \le j < k$, $q_j \xrightarrow{g_j} q_{j+1}$ for some guard g_j. (ii) there exists $t \in (0,1)$ such that $\Phi_{q_j,t}(\mathbf{v}_j) \in g$. Furthermore $\mathbf{v}_{j+1} = \Phi_{q_{j+1},1-t}(\Phi_{q_j,t}(\mathbf{v}_j))$.

We say that the trajectory τ as defined above *starts* from q_0 and *ends* in q_k. Further, its initial value state is $\mathbf{v}_0$ and its final value state is $\mathbf{v}_k$. We let TRJ denote the set of all finite trajectories that start from the initial mode q_{in} with an initial value state in INIT.

3   The Markov Chain Approximation

A (finite) path in H is a sequence $\rho = q_0 q_1 \ldots q_k$ such that for $0 \le j < k$, there exists a guard g_j such that $q_j \xrightarrow{g_j} q_{j+1}$. We say that this path starts from q_0, ends at q_k and is of length $k+1$. We let paths_H denote the set of all finite paths that start from q_{in}.

In what follows μ will denote the standard Lebesgue measure over finite dimensional Euclidean spaces. We will construct $M_H = (\Upsilon, \Rightarrow)$, the Markov chain approximation of H inductively. Each state in Υ will be of the form $(\rho, X, \mathbf{P}_X)$ with $\rho \in \mathsf{paths}_H$, X an open subset of $\mathbb{R}^n$ of non-zero, finite measure and $\mathbf{P}_X$ a probability distribution over $SA(X)$, the Borel σ-algebra generated by X.

We start with $(q_{in}, \text{INIT}, \mathbf{P}_{\text{INIT}}) \in \Upsilon$. Clearly, INIT is an open set of non-zero, finite measure since $\mu(\text{INIT}) = \prod_i (U_i - L_i)$. For technical convenience we shall assume $\mathbf{P}_{\text{INIT}}$ to be the uniform probability distribution, but other probability distributions with respect to which the Lebesgue-measure is absolutely continuous could also be chosen. Assume inductively that $(\rho, X, \mathbf{P}_X)$ is in Υ with X an open subset of $\mathbb{R}^n$ of non-zero, finite measure and $\mathbf{P}_X$ a probability distribution over $SA(X)$. Suppose ρ ends in q and there are m outgoing transitions $q \xrightarrow{g_1} q_1, \ldots, q \xrightarrow{g_m} q_m$ from q in H (Fig. 1 illustrates this inductive step).

Then for $1 \le j \le m$ we define the triples $(\rho q_j, X_j, \mathbf{P}_{X_j})$ as follows. In doing so we will assume the required properties of the objects involved in this construction. We will then establish these properties and thus the soundness of the construction. For convenience, through the remaining parts of this section j will range over $\{1, 2, \ldots, m\}$.

For each $\mathbf{v} \in X$ and each j we first define the set of time points $\mathbb{T}_j(\mathbf{v}) \subseteq (0,1)$ via

$$\mathbb{T}_j(\mathbf{v}) = \{t \mid \Phi_q(t, \mathbf{v}) \in g_j\}. \tag{1}$$

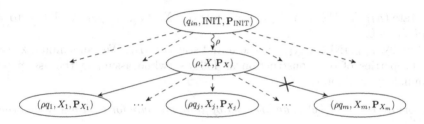

Fig. 1. The Markov chain construction. The edge from the state $(\rho, X, \mathbf{P}_X)$ to the state $(\rho q_m, X_m, \mathbf{P}_{X_m})$ marked with a '$\times$' represents the case where X_m has measure 0, and hence the probability of this transition is 0. Thus, $(\rho q_m, X_m, \mathbf{P}_{X_m})$ will not be a state of the Markov chain.

Thus $\mathbb{T}_j(\mathbf{v})$ is the set of time points in $(0, 1)$ at which the guard g_j is satisfied if the system starts from $\mathbf{v}$ in mode q and evolves according to dynamics of mode q up to time t. We next define X_j for each j as

$$X_j = \bigcup_{\mathbf{v} \in X} \{\Phi_{q_j}(1 - t, \Phi_q(t, \mathbf{v})) \mid t \in \mathbb{T}_j(\mathbf{v})\}. \tag{2}$$

Thus X_j is the set of all value states obtained by starting from some $\mathbf{v} \in X$ at time k, evolving up to $k+t$ according to the dynamics q, making an instantaneous mode switch to q_j at this time point, and evolving up to time $k + 1$ according to dynamics of mode q_j.

To complete the definition of the triples $(\rho q_j, X_j, \mathbf{P}_{X_j})$, we first denote by $\mathbf{P}_{\mathbb{T}_j(\mathbf{v})}$ a probability distribution over $\mathbb{T}_j(\mathbf{v})$. We shall choose this distribution to be uniform but it could be any other non-uniform probability distribution with respect to which the Lebesgue-measure is absolutely continuous. We now define the probability distributions $\mathbf{P}_{X_j}$ over $SA(X_j)$ as follows. Suppose Y is a measurable subset of X_j. Then

$$\mathbf{P}_{X_j}(Y) = \int_{\mathbf{v} \in X} \int_{t \in \mathbb{T}_j(\mathbf{v})} \mathbf{1}_{(\Phi_{q_j}(1-t, \Phi_q(t, \mathbf{v})) \cap Y)} d\mathbf{P}_{\mathbb{T}_j(\mathbf{v})} d\mathbf{P}_X. \tag{3}$$

As usual $\mathbf{1}_Z$ is the indicator function of the set Z while $d\mathbf{P}_{\mathbb{T}_j(\mathbf{v})}$ indicates that the inner integration over $\mathbb{T}_j(\mathbf{v})$ is w.r.t. the probability measure $\mathbf{P}_{\mathbb{T}_j(\mathbf{v})}$ and $d\mathbf{P}_X$ indicates that the outer integration over X is w.r.t. the probability measure $\mathbf{P}_X$. Thus $\mathbf{P}_{X_j}(Y)$ captures the probability that the value state $\Phi_{q_j}(1 - t, \Phi_q(t, \mathbf{v}))$ lands in $Y \subseteq X_j$ by taking the transition $q \xrightarrow{g_j} q_j$ at some time point in $\mathbb{T}_j(\mathbf{v})$ given that one started with some value state in X.

Next we define the triples $((\rho, X, \mathbf{P}_X), p_j, (\rho q_j, X_j, \mathbf{P}_{X_j}))$, where p_j is given by

$$p_j = \int_{\mathbf{v} \in X} \frac{\mu(\mathbb{T}_j(\mathbf{v}))}{\sum_{\ell=1}^{m} \mu(\mathbb{T}_\ell(\mathbf{v}))} d\mathbf{P}_X. \tag{4}$$

Thus p_j captures the probability of taking the mode transition $q \xrightarrow{g_j} q_j$ when starting from the value states in X and mode q. For every j we add the state $(\rho q_j, X_j, \mathbf{P}_{X_j})$ to Υ and the triple $((\rho, X, \mathbf{P}_X), p_j, (\rho q_j, X_j, \mathbf{P}_{X_j}))$ to $\Rightarrow$ iff $\mu(X_j) > 0$.

Finally, $(q_{in}, \text{INIT}, \mathbf{P}_{\text{INIT}})$ is the initial state of M_H. We can summarize the key properties of our construction as follows (while assuming the associated terminology and notations).

Theorem 1. 1. $\mathbb{T}_j(\boldsymbol{v})$ is an open set of finite measure for each $\boldsymbol{v} \in X$ and each j.
2. X_j is open and is of finite measure for each j.
3. If $(\rho q_j, X_j, \mathbf{P}_{X_j}) \in \Upsilon$ then $\mu(X_j) > 0$.
4. $\boldsymbol{P}_{X_j}$ is a probability distribution for each j.
5. $M_H = (\Upsilon, \Rightarrow)$ is an infinite state Markov chain whose underlying graph is a finitely branching tree.

Proof. To prove the first part, suppose $t \in \mathbb{T}_j(\mathbf{v})$. Then $\Phi_q(t, \mathbf{v}) = \mathbf{v}' \in g_j$ and g_j is open. Hence $\mathbf{v}'$ will be contained in an open neighborhood U contained in g_j. Since Φ_q is Lipschitz we can pick U such that $Y' = \Phi_q^{-1}(U)$ is an open set containing $(\mathbf{v}, t)$ with $Y' \subseteq (0, 1) \times X$. Thus every element of $\mathbb{T}_j(\mathbf{v})$ is contained in an open neighborhood in $(0, 1)$ and hence $\mathbb{T}_j(\mathbf{v})$ is open.

Using the definition of X_j, the fact that X and $\mathbb{T}_j(\mathbf{v})$ are open, and the continuity of the inverses of the flow functions it is easy to observe that X_j is open. To see that it is of finite measure, by the induction hypothesis, X is open and $\mu(X)$ is finite. Hence $((0, 1) \times X)$ is open as well and $\mu((0, 1) \times X)$ is finite. Since $\mathbb{R}^{n+1}$ is second-countable [33], there exists a countable family of disjoint open-intervals $\{I_i\}_{i \geq 1}$ in $\mathbb{R}^{n+1}$ such that $((0, 1) \times X) = \bigcup_i I_i$. Clearly each I_i has a finite measure. By the Lipschitz continuity of Φ_q we know that there exists a constant c such that $\mu(\Phi_q(I_i)) < c \cdot \mu(I_i)$ for all i. We thus have

$$\mu(\Phi_q((0, 1), X)) \leq \sum_i \mu(\Phi_q(I_i))$$
$$< c \sum_i \mu(I_i) = c\mu((0, 1) \times X) < \infty. \tag{5}$$

Therefore $\Phi_q((0, 1), X)$ has a finite measure. By a similar argument we can show that $\Phi_{q_j}((0, 1), \Phi_q((0, 1), X))$ has a finite measure as well. Since $X_j = \bigcup_t \Phi_{q_j, 1-t}(\Phi_{q, t}(X) \cap g) \subseteq \Phi_{q_j}((0, 1), \Phi_q((0, 1), X))$, it must have a finite measure.

The remaining parts of the theorem follow easily from the definitions and basic measure theory.

4   Relating the Behaviors of H and M_H

We shall use bounded linear-time temporal logic (BLTL) [15] to specify time bounded properties and use it to relate the behaviors of H and M_H. For convenience we shall write M instead of M_H from now on.

We assume a finite set of atomic propositions AP and a valuation function $Kr : Q \rightarrow 2^{AP}$. Formulas of BLTL are defined as: (i) Every atomic proposition as well as the constants *true, false* are formulas. (ii) If ψ, ψ' are formulas then $\neg\psi$ and $\psi \vee \psi'$ are formulas. (iii) If ψ, ψ' are formulas and ℓ is a positive integer then $\psi\mathbf{U}^{\leq\ell}\psi'$ is a formula. The derived operators $\mathbf{F}^{\leq\ell}$ and $\mathbf{G}^{\leq\ell}$ are defined as usual: $\mathbf{F}^{\leq\ell}\psi \equiv true\mathbf{U}^{\leq\ell}\psi$ and $\mathbf{G}^{\leq\ell}\psi \equiv \neg\mathbf{F}^{\leq\ell}\neg\psi$.

We shall assume through the rest of the paper that the behavior of the system is of interest only up to a maximum time point $K > 0$. This is guided by the fact that given a BLTL formula ψ there is a constant K_ψ that depends only on ψ so that it is enough to evaluate an execution trace of length at most K_ψ to determine whether ψ is satisfied [7]. Hence we assume that a sufficiently high K has been chosen to handle the specifications of interest. Having fixed K, we denote by TRJ^{K+1} the trajectories of length $K + 1$, and view this set as representing the time bounded non-deterministic behavior of H of interest.

To develop the corresponding notion for M, we first define a finite path in M to be a sequence $\eta_0\eta_1 \ldots \eta_k$ such that $\eta_j \in \Upsilon$ for $0 \leq j \leq k$. Furthermore for $0 \leq j < k$ there exists $p_j \in (0,1]$ such that $\eta_j \overset{p_j}{\Rightarrow} \eta_{j+1}$. Such a path is said to start from η_0 and its length is $k + 1$. We define paths_M to be the set of finite paths that start from the initial state of M while paths_M^{K+1} is the set of paths in paths_M of length $K + 1$.

The Trajectory Semantics: Let $\tau = (q_0, \mathbf{v}_0) (q_1, \mathbf{v}_1) \ldots (q_k, \mathbf{v}_k)$ be a finite trajectory, ψ a BLTL formula and $0 \leq j \leq K$. Then $\tau, j \models_H \psi$ is defined via:

- $\tau, j \models_H A$ iff $A \in Kr(q_j)$, where A is an atomic proposition.
- $\neg$ and $\vee$ are interpreted in the usual way.
- $\tau, j \models_H \psi\mathbf{U}^{\leq\ell}\psi'$ iff there exists j' such that $j' \leq \ell$ and $j + j' \leq k$ and $\tau, (j + j') \models_H \psi'$. Further, $\tau, (j + j'') \models_H \psi$ for every $0 \leq j'' < j'$.

We now define $models_H(\psi) \subseteq TRJ^{K+1}$ via: $\tau \in models_H(\psi)$ iff $\tau, 0 \models_H \psi$. We say that H *meets the specification* ψ, denoted $H \models \psi$, iff $models_H(\psi) = TRJ^{K+1}$.

The Markov chain semantics: Let $\pi = \eta_0\eta_1 \ldots \eta_k$ be a path in M with $\eta_j = (\rho q_j, X_j, \mathbf{P}_{X_j})$ for $0 \leq j \leq k$. Let ψ be a BLTL formula and $0 \leq j \leq k$. Then $\pi, j \models_M \psi$ is given by:

- $\pi, j \models_M A$ iff $A \in Kr(q_j)$, where A is an atomic proposition.
- The remaining clauses are defined just as in the case of $\models_H$.

Now we define $models_M(\psi) \subseteq \mathsf{paths}_M^{K+1}$ via: $\pi \in models_M(\psi)$ iff $\pi, 0 \models_M \psi$. We can now define the probability of satisfaction of a formula in M. Let $\pi = \eta_0\eta_1 \ldots \eta_K$ be in paths_M^{K+1}. Then $\Pr(\pi) = \prod_{0 \leq \ell < K} p_\ell$, where $\eta_\ell \overset{p_\ell}{\Rightarrow} \eta_{\ell+1}$ for $0 \leq \ell < K$. This leads to

$$\Pr(models_M(\psi)) = \sum_{\pi \in models_M(\psi)} \Pr(\pi).$$

We write $M \models \psi$ to denote $\Pr(models_M(\psi)) = 1$. For $p \in [0,1]$ we write as usual $\Pr_{\geq p}(\psi)$ instead of $\Pr(models_M(\psi)) \geq p$. We note that $Pr(\pi) > 0$ for every $\pi \in models_M(\psi)$. Furthermore $\sum_{\pi \in models_M(\psi)} \Pr(\pi) \leq 1$. Hence $\Pr_{\geq 1}(\psi)$ iff $models_M(\psi) = \mathsf{paths}_M^{K+1}$ iff $M \models \psi$.

4.1 The Correspondence Result

We wish to show that H meets the specification ψ iff $\Pr_{\geq 1}(\psi)$. To this end let $\pi = \eta_0 \eta_1 \ldots \eta_k$ be a path in M with $\eta_j = (q_0 q_1 \ldots q_j, X_j, \mathbf{P}_{X_j})$ for $0 \leq j \leq k$ and let $\tau = (q_0', \mathbf{v}_0)\,(q_1', \mathbf{v}_1) \ldots (q_{k'}', \mathbf{v}_{k'})$ be a trajectory. Then we say that π and τ are *compatible* iff $k = k'$ and $q_j = q_j'$ and $\mathbf{v}_j \in X_j$ for $0 \leq j \leq k$. The following three observations based on this notion will easily lead to the main result.

Lemma 1.   *1. Suppose the path $\pi = \eta_0 \eta_1 \ldots \eta_k$ in M and the trajectory $\tau = (q_0, \mathbf{v}_0)\,(q_1, \mathbf{v}_1) \ldots (q_k, \mathbf{v}_k)$ are compatible. Let $0 \leq j \leq k$ and ψ be a BLTL formula. Then $\pi, j \models_M \psi$ iff $\tau, j \models_H \psi$.*
  2. Suppose π is a path in M. Then there exists a trajectory τ such that π and τ are compatible. Furthermore if $\pi \in \mathsf{paths}_M$ then $\tau \in TRJ$.
  3. Suppose τ is a trajectory. Then there exists a path π in M such that τ and π are compatible. Furthermore if $\tau \in TRJ$ then $\pi \in \mathsf{paths}_M$.

Proof. The proof follows via a systematic application of the definitions. To prove the first part we note that if A is an atomic proposition then $\pi, j \models_M A$ iff $A \in Kr(q_j)$ iff $\tau, j \models_H A$. We next note that the suffix of length m of π will be compatible with the suffix of length m of τ whenever π and τ are compatible. The result now follows at once by structural induction on ψ.

To show the second part let $\pi = \eta_0 \eta_1 \ldots \eta_k$ be a path in M with $\eta_j = (q_0 q_1 \ldots q_j, X_j, \mathbf{P}_{X_j})$ for $0 \leq j \leq k$. Clearly X_j is non-empty for $0 \leq j \leq k$ since $\eta_j \in \Upsilon$ implies $\mu(X_j) > 0$. We proceed by induction on k. If $k = 0$ then we can pick $\mathbf{v}_0 \in X_0$ and the trajectory $(q_0, \mathbf{v}_0)$ will be compatible with τ. So assume $k > 0$. Then by the induction hypothesis there exists a trajectory $(q_1, \mathbf{v}_1)(q_2, \mathbf{v}_2) \ldots (q_k, \mathbf{v}_k)$ which is compatible with the path $\eta_1 \eta_2 \ldots \eta_k$. Let $q_0 \xrightarrow{g} q_1$. Since $\mathbf{v}_1 \in X_1$ there must exist $\mathbf{v}_0$ in X_0 and $t \in (0,1)$ such that $\Phi_{q_0, t}(\mathbf{v}_0) \in g$ and $\mathbf{v}_1 = \Phi_{q_1, 1-t}(\Phi_{q_0, t}(\mathbf{v}_0))$. Clearly $\mathbf{v}_0 \mathbf{v}_1 \ldots \mathbf{v}_k$ is a trajectory that is compatible with π. The fact that $\tau \in TRJ$ if $\pi \in \mathsf{paths}_M$ follows from the definition of compatibility.

To prove the third part let $\tau = (q_0, \mathbf{v}_0)\,(q_1, \mathbf{v}_1) \ldots (q_k, \mathbf{v}_k) \in TRJ$. Again we proceed by induction on k. Suppose $k = 0$. Then $(q_{in}, \mathrm{INIT}, \mathbf{P}_{\mathrm{INIT}})$ is in paths_M which is compatible with τ. So suppose $k > 0$. Then by the induction hypothesis there exits $\pi' = \eta_0 \eta_1 \ldots \eta_{k-1}$ such that π' is compatible with $\tau' = (q_0, \mathbf{v}_0)(q_1, \mathbf{v}_1) \ldots (q_{k-1}, \mathbf{v}_{k-1})$. Let $q_{k-1} \xrightarrow{g} q_k$. Since X_{k-1} is open there exists an open neighborhood $Y \subseteq X_{k-1}$ that contains $\mathbf{v}_{k-1}$. But then both $\Phi_{q_{k-1}}^{-1}$ and $\Phi_{q_k}^{-1}$ are continuous. Thus $\Phi_{q_{k-1}, t}(Y)$ is open and $\Phi_{q_{k-1}, t}(Y) \cap g$ should be open and non-empty (since g is open and $(q_k, \mathbf{v}_k)$ is part of the trajectory). Hence $Y' = \bigcup_{t \in (0,1)} \Phi_{q_k, 1-t}(\Phi_{q_{k-1}, t}(Y) \cap g)$ is a non-empty open set with a positive measure. This means there will be a state of the form $\eta_k = (\rho_k, X_k, \mathbf{P}_{X_k})$ in Υ

with $Y' \subseteq X_k$ and $\eta_{k-1} \xrightarrow{p} \eta_k$ for some $p \in (0,1]$. Clearly $\pi = \pi'\eta_k \in \mathsf{paths}_M$ and is compatible with τ. Again the fact that $\pi \in \mathsf{paths}_M$ if $\tau \in TRJ$ follows from the definition of compatibility.

Theorem 2. $H \models \psi$ iff $M \models \psi$.

Proof. Suppose H does not meet the specification ψ. Then there exists $\tau \in TRJ^{K+1}$ such that $\tau, 0 \not\models_H \psi$. By the third part of Lemma 1 there exists $\pi \in \mathsf{paths}_M^{K+1}$ which is compatible with τ. By the first part of Lemma 1 we then have $\pi \notin \mathsf{models}_M(\psi)$ which leads to $Pr_{<1}(\psi)$.

Next suppose that $Pr_{<1}(\psi)$. Then there exists $\pi \in \mathsf{paths}^{K+1}$ such that $\pi, 0 \not\models_M \psi$. By the second part of Lemma 1 there exists $\tau \in TRJ^{K+1}$ which is compatible with π. By the first part of Lemma 1 this implies $\tau, 0 \not\models_H \psi$ and this in turn implies that H does not meet the specification ψ.

4.2 Quantitative Atomic Propositions

The above results can be extended to handle atomic propositions of the form $\langle x_i < c \rangle$ and $\langle x_i > c \rangle$ where c is a rational constant. We partition $\mathbb{R}^n$ into hypercubes according to the constants appearing in the given set of quantitative atomic propositions in AP_{qt}. We then blow up the state space of the Markov chain to record which hypercube the current values of the variables fall in. We restrict our attention to robust trajectories and show that every robust trajectory of H meets a BLTL specification iff its Markov chain approximation meets the same specification with probability 1. Informally a robust trajectory is one which has an open neighborhood of trajectories under a natural topology over the space of $K+1$-length trajectories. Under an associated measure the set of non-robust trajectories will have measure 0. The details can be found in [4].

5   The SMC Procedure

To verify whether H meets the specification ψ, we solve the equivalent problem whether $Pr_{\geq 1}(\psi)$ on M. However, as discussed in Sect. 1, M cannot be constructed explicitly since both its structure and transition probabilities, defined in terms of the solutions to the ODEs, will not be available. Therefore we shall use randomly generated trajectories to sample the paths of M and formulate a sequential hypothesis test to decide with bounded error rate whether $Pr_{\geq 1}(\psi)$ holds. Algorithm 1 describes our trajectory sampling procedure.

Clearly Algorithm 1 generates a trajectory in TRJ^{K+1}. We now relate these trajectories to paths in M.

The initial value $\mathbf{v}_0$ is sampled uniformly on INIT, and we start in mode q_{in}, consistent with the initial state $(q_{in}, \mathrm{INIT}, \mathbf{P}_{\mathrm{INIT}})$ of M. Inductively, suppose $\eta = (\rho, X, \mathbf{P}_X)$ is a state of M with ρ ending in q. Suppose $\eta \xrightarrow{p_j} \eta_j$ is a transition in M such that $\eta_j = (\rho q_j, X_j, \mathbf{P}_{X_j})$.

Algorithm 1. Trajectory simulation

Input: Hybrid automaton $H = (Q, q_{in}, \{F_q(\mathbf{x})\}_{q \in Q}, \mathcal{G}, \rightarrow, \text{INIT})$, maximum time step K.
Output: Trajectory τ
1: Sample $\mathbf{v}_0$ from INIT uniformly, set $q_0 := q_{in}$ and $\tau := (q_0, \mathbf{v}_0)$.
2: **for** $k := 1 \dots K$ **do**
3:     Generate time points $T := \{t_1, \dots, t_J\}$ uniformly in $(0, 1)$.
4:     Simulate $\mathbf{v}^j := \Phi_{q_{k-1}}(t_j, \mathbf{v}_{k-1})$, for $j \in \{1, \dots, J\}$
5:     Let $\widehat{\mathbb{T}}_j := \{t \in T : \mathbf{v}^j \in g_j\}$ be the time points where g_j is enabled.
6:     Pick g_ℓ randomly according to probabilities $\{p_j := |\widehat{\mathbb{T}}_j| / \sum_{i=1}^{m} |\widehat{\mathbb{T}}_i|\}$.
7:     Pick t_ℓ uniformly at random from $\widehat{\mathbb{T}}_\ell$.
8:     Simulate $\mathbf{v}' := \Phi_{q'}(1 - t_\ell, \mathbf{v}^\ell)$, where q' is the target of g_ℓ.
9:     Set $q_k := q'$, $\mathbf{v}_k := \mathbf{v}'$, and extend $\tau := (q_0, \mathbf{v}_0) \dots (q_k, \mathbf{v}_k)$.
10: **end for**
11: **return** τ

Proposition 1. *Suppose, we obtain a sample $\mathbf{v} \sim \mathbf{P}_X$. The probability of choosing guard g_j whose target mode is q_j in Algorithm 1 tends to p_j as $J \rightarrow \infty$.*

Proof. According to Algorithm 1, the probability of picking guard g_j for a trajectory starting at $\mathbf{v} \in X$ is defined as $|\widehat{\mathbb{T}}_j| / \sum_{i=1}^{m} |\widehat{\mathbb{T}}_i|$, which, by the law of large numbers tends to

$$p_j(\mathbf{v}) := \frac{\mu(\mathbb{T}_j(\mathbf{v}))}{\sum_{i=1}^{m} \mu(\mathbb{T}_i(\mathbf{v}))} \tag{6}$$

as J tends to ∞.

Now if $\mathbf{v}$ is randomly sampled according to $\mathbf{P}_X$, then the probability of picking guard j can be expressed as the expected value of $p_j(\mathbf{v})$ under $\mathbf{v} \sim \mathbf{P}_X$ as

$$\mathbb{E}_{\mathbf{v} \sim \mathbf{P}_X}[p_j(\mathbf{v})] = \int_{\mathbf{v} \in X} p_j(\mathbf{v}) d\mathbf{P}_X = \int_{\mathbf{v} \in X} \frac{\mu(\mathbb{T}_j(\mathbf{v}))}{\sum_{i=1}^{m} \mu(\mathbb{T}_i(\mathbf{v}))} d\mathbf{P}_X, \tag{7}$$

which by (4) is equal to p_j, the corresponding transition probability in the Markov chain.

Similarly, picking the transition time t from $\widehat{\mathbb{T}}_j$ will approximate sampling $t \sim \mathbf{P}_{\mathbb{T}_j(\mathbf{v})}$, for sufficiently high J. Next, assume that we have picked $q \xrightarrow{g_j} q_j$ as the transition to take. We sample $t \sim \mathbf{P}_{\mathbb{T}_j(\mathbf{v})}$, and obtain $\mathbf{v}'$ by numerical simulation via:

$$\mathbf{v}' = \Phi_{q_j}(1 - t, \Phi_q(t, \mathbf{v})). \tag{8}$$

Proposition 2. *$\mathbf{v}'$ is distributed according to $\mathbf{P}_{X_j}$.*

Proof. Clearly it suffices to show that for a measurable subset $Y \subseteq X_j$, $\Pr(\mathbf{v}' \in Y) = \mathbf{P}_{X_j}(Y)$. We start with

$$\Pr(\mathbf{v}' \in Y \mid \mathbf{v}) = \int_{t \in \mathbb{T}_j(\mathbf{v})} \mathbf{1}_{(\Phi_{q_j}(1-t, \Phi_q(t, \mathbf{v})) \cap Y)} d\mathbf{P}_{\mathbb{T}_j(\mathbf{v})}.$$

Integrating now over all possible choices of $\mathbf{v}$ with respect to $\mathbf{P}_X$ we have

$$\Pr(\mathbf{v}' \in Y) = \int_{\mathbf{v} \in X} \Pr(\mathbf{v}' \in Y \mid \mathbf{v}) d\mathbf{P}_X.$$

From (3) it follows that $\Pr(\mathbf{v}' \in Y) = \mathbf{P}_{X_j}(Y)$ with $\mathbf{v} \sim \mathbf{P}_X$ and $t \sim \mathbf{P}_{\mathbb{T}_j(\mathbf{v})}$.

Consequently, the trajectory being generated will now be in mode q_j with $\mathbf{v}' \in X_j$ and $\mathbf{v}'$ distributed according to $\mathbf{P}_{X_j}$, compatible with the state $\eta_j = (\rho q_j, X_j, \mathbf{P}_{X_j})$ of M. Inductively it is hence guaranteed that each subsequent iteration of Algorithm 1 will produce values compatible with a path of M.

Whether the generated trajectory of length $K + 1$ (and hence the corresponding path of M) is a model of ψ can be determined using a standard BLTL model checker [15]. In fact this can be done on the fly which will often avoid generating the whole trajectory. Based on this, we can test whether $\Pr_{\geq 1}(\psi)$ on M by testing the following alternative pair of hypotheses: $H_0 : \Pr_{\geq 1}(\psi)$ and $H_1 : \Pr_{< 1 - \delta}(\psi)$, where $0 < \delta < 1$ is a parameter chosen by the user marking the interval $[1 - \delta, 1)$ as an indifference region in which accepting either hypothesis is fine. In our setting, whenever we encounter a sample (i.e. a randomly generated trajectory) that does not satisfy ψ, we can reject H_0 and accept H_1. Thus we only have to deal with false positives (when H_0 is accepted while H_1 happens to be true).

This leads to Algorithm 2 that repeatedly generates a random trajectory (using Algorithm 1), and decides after a finite number of tries between H_0 and H_1. For doing so we also fix a user-defined false positive rate α.

Algorithm 2. Sequential hypothesis test

Input: Markov chain M, BLTL property ψ, indifference parameter δ, false positive bound α.
Output: H_0 or H_1.
1: Set $N := \lceil \log \alpha / \log(1 - \delta) \rceil$
2: **for** $i := 1 \dots N$ **do**
3:     Generate a random trajectory τ using Algorithm 1
4:     **if** $\tau, 0, \models^H \psi$ **then** Continue
5:     **else return** H_1
6: **end for**
7: **return** H_0

The accuracy of Algorithm 2 is captured by the next result.

Theorem 3. *The probability of choosing H_1 when H_0 is true (false negative) is 0. Further, suppose $N \geq \log \alpha / \log(1 - \delta)$. Then the probability of choosing H_0 when H_1 is true (false positive) is no more than α.*

Proof. As observed earlier the first part is obvious. To prove the second part, if H_1 is true, then we know that $\Pr_{< 1 - \delta}(\psi)$. The probability of N sampled trajectories all satisfying ψ (and thus returning H_0, a false positive) is at most $(1 - \delta)^N$. Therefore we have $\alpha \leq (1 - \delta)^N$, leading to $N \geq \log \alpha / \log(1 - \delta)$.

Hence we use $N := \lceil \log \alpha / \log(1 - \delta) \rceil$ to set the sample size. For example for $\delta = 0.01$ and $\alpha = 0.01$ we get $N = 459$ while for $\delta = 0.001$ and $\alpha = 0.01$ we get $N = 4603$.

6   Case Studies

We first evaluated our method on a model of the electrical dynamics of the cardiac cell [12]. We also applied our method on a model of circadian rhythm network [28]. The Δ time step parameter for the cardiac cell model and the circadian rhythm model were both set to 0.1. The parameters used for the statistical model checking were $\delta = 0.01$ and $\alpha = 0.01$. We have implemented our method using MATLAB. The source code is available at http://github.com/bgyori/hybrid. The experiments were carried out on a PC with a 3.4 GHz Intel Core i7 processor with 8 GB RAM. Simulating one trajectory took, on average, 5.2 s for the circadian clock model and 18.3 s for the cardiac cell model. We note that when checking quantitative properties, the trajectories that hit corner points such as $u = 1.4$ will be non-robust and hence can be ignored. Our implementation exploits the parallelization enabled by statistical model checking, hence multiple trajectories can be simulated simultaneously. A summary of the results for the verification of all properties for both models along with the number of samples taken to complete the verification is given in Table 3 of the Appendix.

In our experiments, we used $J = 10$ as the number of intermediate time steps for choosing mode transitions. We investigated whether this choice is sufficient for accurate simulation. We simulated 1000 independent realizations of the cardiac cell system with $J = 10$ and $J = 100$, and compared the distributions of the modes that the system is in at a series of discrete time points. The Kolmogorov-Smirnov statistical test did not reject the hypothesis that the two distributions are the same (at confidence level 95 %). This indicates that using $J = 10$ is adequate.

6.1 Cardiac Cell Model

Heart rhythm depends on the organized opening and closing of gates–called ion channels–on the cell membrane, which govern the electrical activity of cardiac cells. Disordered electric wave propagation in heart muscle can cause cardiac abnormalities such as *tachycardia* and *fibrillation*. The dynamics of the electrical activity of a single human ventricular cell has been modeled as a hybrid automaton [12,21] shown in Fig. 2. The model contains 4 state variables and 26 parameters. Ventricular cells consist of three subtypes, namely epicardial, endocardial, and midmyocardial cells, which possess different dynamical characteristics. The cell-type-specific parameters of the model are summarized in Table 1.

An action potential (AP) is a change in the cell's transmembrane potential u, as a response to an external stimulus (current) ϵ. The flow of total currents is controlled by a fast channel gate v and two slow gates w and s.

In mode q_0, the "Resting mode", the cell is waiting for stimulation. We assume an external stimulus ϵ equal to 1 mV lasting for 1 millisecond. The stimulation causes u to increase which may trigger a mode transition to mode q_1. In mode q_1, gate v starts closing and the decay rate of u changes. The system will jump to mode q_2 if $u > \theta_w$. In mode q_2, gate w is also closing. When $u > \theta_v$, mode q_3 can be reached, which means a successful "AP initiation". In mode q_3, u reaches its peak due to the fast opening of a sodium channel. The cardiac muscle then contracts and u starts decreasing.

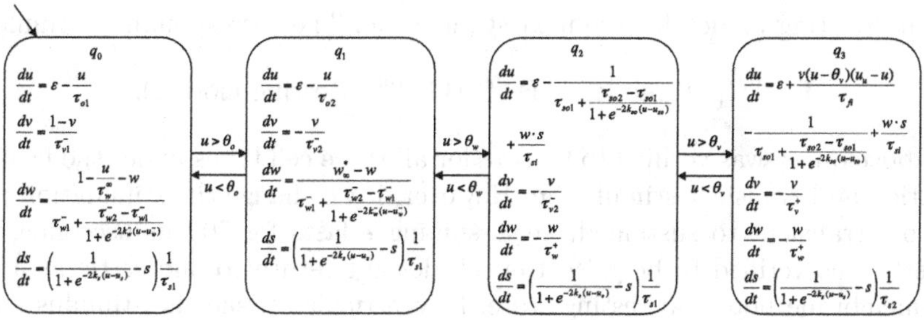

Fig. 2. The hybrid automaton model for the cardiac cell system [21].

Table 1. Parameter values of the cardiac model for epicardial (EPI), endocardial (ENDO), and midmyocardial (MID) cells under healthy condition.

Parameter	EPI	ENDO	MID	Parameter	EPI	ENDO	MID
θ_o	0.006	0.006	0.006	τ_{v1}^-	60	75	80
θ_w	0.13	0.13	0.13	τ_{v2}^-	1150	10	1.4506
θ_v	0.3	0.3	0.3	τ_{w1}^-	60	6	70
u_w^-	0.03	0.016	0.016	τ_{w2}^-	15	140	8
u_{so}	0.65	0.65	0.6	τ_{o1}	400	470	410
u_s	0.9087	0.9087	0.9087	τ_{o2}	6	6	7
u_u	1.55	1.56	1.61	τ_{so1}	30.0181	40	91
w_∞^*	0.94	0.78	0.5	τ_{so2}	0.9957	1.2	0.8
k_w^-	65	200	200	τ_{s1}	2.7342	2.7342	2.7342
k_{so}	2.0458	2	2.1	τ_{s2}	16	2	4
k_s	2.994	2.994	2.994	τ_{fi}	0.11	0.1	0.078
τ_v^+	1.4506	1.4506	1.4506	τ_{si}	1.8875	2.9013	3.3849
τ_w^+	200	280	280	$\tau_{w\infty}$	0.07	0.0273	0.01

Property C1. It is known that the cardiac cell can lose its excitability, which will lead to disorders such as ventricular tachycardia and fibrillation. We formulated the property for responding to stimulus by leaving the resting mode:

$$\mathbf{F}^{\leq 500}(\neg[\text{Resting mode}]).$$

The property was verified to be *true* for all three cell types under the healthy condition. However, under a disease condition (for example $\tau_{o1} = 0.004$ or $\tau_{o2} = 0.1$ [27]) the property was verified to be *false* no matter what stimulation value of ϵ was used. Consequently, a region of such unexcitable cells blocks the impulse conduction and can lead to cardiac disorders such as fibrillation. This is consistent with experimental results reported in [34].

Property C2. After successfully generating an AP (that is, reaching the "AP mode", q_3), the cardiac cell should return to a low transmembrane potential and wait in "Resting mode" for the next stimulation. The corresponding formula is

$$\mathbf{F}^{\leq 500}([\text{AP mode}]) \wedge \mathbf{F}^{\leq 500}(\mathbf{G}^{\leq 100}([\text{Resting mode}])).$$

The above query was verified to be *true* for all three cell types under the healthy condition and transient stimulation. However, if we change the stimulation profile from transient to sustained, i.e. assuming ϵ lasts for 500 milliseconds, the property was verified to be *false*–the cell doesn't return to and settle at a low transmembrane potential resting state. In ventricular tissue the stimulus ϵ can be delivered from neighboring cells [12]. Thus, our results suggest that the transient activation of a single cardiac cell depends on the stimulation profile of its neighboring cells.

Property C3. It has been reported that epicardial, endocardial, and midmyocardial cells have different AP morphologies [16,29]. In particular, a crucial "spike-and-dome" (i.e. a sharp peak followed by a blunt peak) AP morphology can only be observed in epicardial cells but not endocardial and midmyocardial cells (Fig. 3).

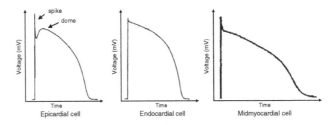

Fig. 3. The AP morphologies of epicardial [29], endocardial [29] and midmyocardial [16] cells.

We formulated the property for a spike-and-dome AP morphology as a quantitative property,

$$\mathbf{F}^{\leq 500}(\mathbf{G}^{\leq 1}([1.4 \leq u]) \wedge \mathbf{F}^{\leq 500}([0.8 \leq u] \wedge [u \leq 1.1] \wedge \mathbf{F}^{\leq 500}(\mathbf{G}^{\leq 50}([1.1 \leq u])))).$$

The property was verified to be *true* for epicardial cells and *false* for endocardial and midmyocardial cells under the healthy condition and transient stimulation. Among 26 model parameters, 20 of them have different values over different cell types. We then perturbed each epicardial parameter and checked if the above property still holds. Our results show that τ_{s2} is key to the AP morphology (i.e. the spike-and-dome AP morphology disappears when $\tau_{s2} = 2$), which highlights the importance of s gate to epicardial cells. This is consistent with [27] in that the model proposed in [17] (which does not include s gate) is unable to capture the dynamics of epicardial cells.

6.2 Circadian Rhythm Model

Mammalian cells follow a circadian rhythm with a 24 h period, which is generated and governed by a highly coupled transcription-translation network. The model diagram and the corresponding hybrid system dynamics proposed in [28,30] is described below.

The equations governing the dynamics of the circadian clock model are given in Fig. 4. The equations contain rate constants which are denoted k_1 to k_{28} and are set according to [30]. The combination of "mode indicator" binary variables θ_{CB} to $\theta_{RE}, \theta_{PC1}, \theta_{PC2}$ and θ_{PC3} define the mode of the dynamics, and each mode is defined by a unique value combination of the mode indicators. These value combinations are listed in Table 2. The guards associated with a source and target mode are constructed as follows. Each mode indicator corresponds to a guard component which is a threshold on a state variable. For instance, θ_{RE} has the corresponding guard component [REV-ERB]< 1.1. The guard to a target mode is enabled if all the mode indicators that are on in the mode are enabled according to their respective guard components. Finally, a transition between a source and a target mode only exists if there is only one difference in the combination of mode indicators. For instance, there is a transition from mode 1 to mode 2 but not from mode 1 to mode 9. The dynamics of the *Clock* mRNA is governed externally.

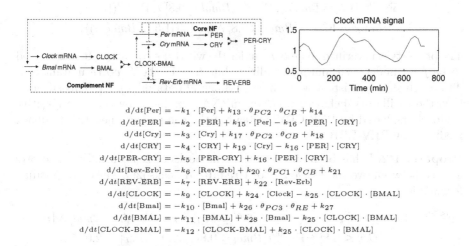

$$d/dt[\text{Per}] = -k_1 \cdot [\text{Per}] + k_{13} \cdot \theta_{PC2} \cdot \theta_{CB} + k_{14}$$
$$d/dt[\text{PER}] = -k_2 \cdot [\text{PER}] + k_{15} \cdot [\text{Per}] - k_{16} \cdot [\text{PER}] \cdot [\text{CRY}]$$
$$d/dt[\text{Cry}] = -k_3 \cdot [\text{Cry}] + k_{17} \cdot \theta_{PC2} \cdot \theta_{CB} + k_{18}$$
$$d/dt[\text{CRY}] = -k_4 \cdot [\text{CRY}] + k_{19} \cdot [\text{Cry}] - k_{16} \cdot [\text{PER}] \cdot [\text{CRY}]$$
$$d/dt[\text{PER-CRY}] = -k_5 \cdot [\text{PER-CRY}] + k_{16} \cdot [\text{PER}] \cdot [\text{CRY}]$$
$$d/dt[\text{Rev-Erb}] = -k_6 \cdot [\text{Rev-Erb}] + k_{20} \cdot \theta_{PC1} \cdot \theta_{CB} + k_{21}$$
$$d/dt[\text{REV-ERB}] = -k_7 \cdot [\text{REV-ERB}] + k_{22} \cdot [\text{Rev-Erb}]$$
$$d/dt[\text{CLOCK}] = -k_9 \cdot [\text{CLOCK}] + k_{24} \cdot [\text{Clock}] - k_{25} \cdot [\text{CLOCK}] \cdot [\text{BMAL}]$$
$$d/dt[\text{Bmal}] = -k_{10} \cdot [\text{Bmal}] + k_{26} \cdot \theta_{PC3} \cdot \theta_{RE} + k_{27}$$
$$d/dt[\text{BMAL}] = -k_{11} \cdot [\text{BMAL}] + k_{28} \cdot [\text{Bmal}] - k_{25} \cdot [\text{CLOCK}] \cdot [\text{BMAL}]$$
$$d/dt[\text{CLOCK-BMAL}] = -k_{12} \cdot [\text{CLOCK-BMAL}] + k_{25} \cdot [\text{CLOCK}] \cdot [\text{BMAL}]$$

Fig. 4. The model diagram, the Clock mRNA signal and the equations governing the circadian clock model.

The system comprises 16 modes, each of which contains 12 state variables and 29 parameters. Each mode corresponds to a particular combination of ON or OFF transcriptional states of genes *Per*, *Cry*, *Rev-Erb*, *Clock*, and *Bmal*. The switches between modes are guarded by the threshold levels of protein complexes PER-CRY, CLOCK-BMAL and REV-REB. The mRNA levels of *Per* and *Cry*

Table 2. The 5 mode indicator variables and their associated guard components (top). The 16 modes of the circadian clock model with the corresponding combination of binary mode indicator variables (bottom).

Mode indicator	Guard component
θ_{RE}	[REV-ERB]< 1.1
θ_{CB}	[CLOCK-BMAL]> 1.0
θ_{PC1}	[PER-CRY]< 1.4
θ_{PC2}	1.4 <[PER-CRY]< 1.5
θ_{PC3}	2.2 <[PER-CRY]

Mode	1	2	3	4
$(\theta_{PC1},\theta_{PC2},\theta_{PC3},\theta_{RE},\theta_{CB})$	(1,1,0,1,0)	(1,1,0,1,1)	(1,1,0,0,0)	(1,1,0,0,1)
Mode	5	6	7	8
$(\theta_{PC1},\theta_{PC2},\theta_{PC3},\theta_{RE},\theta_{CB})$	(0,1,0,1,0)	(0,1,0,1,1)	(0,1,0,0,0)	(0,1,0,0,1)
Mode	9	10	11	12
$(\theta_{PC1},\theta_{PC2},\theta_{PC3},\theta_{RE},\theta_{CB})$	(0,0,0,1,0)	(0,0,0,1,1)	(0,0,0,0,0)	(0,0,0,0,1)
Mode	13	14	15	16
$(\theta_{PC1},\theta_{PC2},\theta_{PC3},\theta_{RE},\theta_{CB})$	(0,0,1,1,0)	(0,0,1,1,1)	(0,0,1,0,0)	(0,0,1,0,1)

are known to be oscillating due to the negative feedback loops in the network. Specifically, there are two major negative feedback (NF) loops: (i) the core NF formed by PER-CRY, CLOCK-BMAL, PER, and CRY and (ii) a complement NF formed by REV-ERB, BMAL, and CLOCK-BMAL. The time constants appearing in the properties are in minute units.

Property R1. Similar to *Per* and *Cry*, the expression level of *Bmal* gene is also oscillating [32]. We formulated this property as

$$\mathbf{F}^{\leq 500}([1.5 \leq Bmal] \wedge \mathbf{F}^{\leq 500}([Bmal \leq 0.8] \wedge \mathbf{F}^{\leq 500}([1.5 \leq$$
$$Bmal] \wedge \mathbf{F}^{\leq 500}([Bmal \leq 0.8] \wedge \mathbf{F}^{\leq 500}([1.5 \leq Bmal])))))$$

The property was verified to be *true* under the wild type condition. It was verified to be *false* under *Cry* mutant condition but *true* in the *Rev-Erb* mutant condition, which is consistent with the experimental data in [26,32]. This suggests that the oscillatory behavior of *Bmal* mRNA is induced by the core negative feedback mediated by PER-CRY, instead of the complement negative feedback mediated by REV-ERB.

Property R2. It has been observed that the peaks of *Bmal* mRNA are always located between two successive *Per* or *Cry* mRNA peaks [26]. The corresponding formula is

$$\mathbf{F}^{\leq 500}([Bmal \leq 0.8] \wedge [2.0 \leq Per] \wedge [2.0 \leq Cry] \wedge \mathbf{F}^{\leq 500}([1.5 \leq Bmal] \wedge [Per \leq$$
$$0.8] \wedge [Cry \leq 0.8] \wedge \mathbf{F}^{\leq 500}([Bmal \leq 0.8] \wedge [2.0 \leq Per] \wedge [2.0 \leq$$
$$Cry] \wedge \mathbf{F}^{\leq 500}([1.5 \leq Bmal] \wedge [Per \leq 0.8] \wedge [Cry \leq 0.8]))))$$

The above query was verified to be *true* under wild type condition. If we remove the dependence between *Bmal* transcription and PER-CRY concentration, the property R2 was verified to be *false*, while the property R1 was verified to *true* (i.e. oscillating). Thus, our results suggest that the complement negative feedback mediated by REV-ERB is responsible for maintaining the oscillatory behavior of *Bmal* mRNA level while PER-CRY plays a role in delaying the *Bmal* expression responses.

Table 3 is a summary of the performance of the SMC procedure for the hybrid systems for the two case studies presented above.

Table 3. Results summary of SMC for hybrid systems

Property	Condition	Decision	# samples
C1	Epicardial, Healthy	True	459
C1	Endocardial, Healthy	True	459
C1	Midmyocardial, Healthy	True	459
C1	Epicardial, Diseased	False	1
C1	Endocardial, Diseased	False	1
C1	Midmyocardial, Diseased	False	1
C2	Epicardial, Transient	True	459
C2	Endocardial, Transient	True	459
C2	Midmyocardial, Transient	True	459
C2	Epicardial, Sustained	False	1
C2	Endocardial, Sustained	False	1
C2	Midmyocardial, Sustained	False	1
C3	Epicardial, $\tau_{s2} = 16$	True	459
C3	Epicardial, $\tau_{s2} = 2$	False	1
C3	Endocardial	False	1
C3	Midmyocardial	False	1
R1	Wild type	True	459
R1	Cry mutant	False	1
R1	Rev-Erb mutant	True	459
R2	Wild type	True	459
R2	Without PER-CRY dependence	False	1
R1	Without PER-CRY dependence	True	459

7 Conclusion

We have presented an approximate probabilistic verification method for analyzing the dynamics of a hybrid system H in terms of a Markov chain M. For bounded time properties, we have shown a strong correspondence between the behaviors of H and M. We have also extended this result to handle quantitative atomic propositions and shown a similar correspondence result for the sub-dynamics consisting of robust trajectories. Thus the intractable verification problem for H can be solved approximately using its Markov chain approximation. Accordingly, we have devised a statistical model checking procedure to verify that M almost certainly meets a BLTL specification and then applied this procedure to two examples to demonstrate the applicability of our approximation scheme. A hardware accelerated parallel implementation of the trajectory sampling procedure will considerably improve the performance and scalability of our method. Overall, we view our results as providing a mathematical basis for verifying if a hybrid system models satisfies a BLTL property with high probability.

As an extension, one could consider more sophisticated stochastic assumptions regarding the time points and value states at which the mode transitions take place. These assumptions will however have to be justified and motivated by the modeling problem at hand, especially in systems biology applications. Yet another valuable extension will be to study a network of hybrid systems. This will enable us to model the cross talk, feed-forward and feed-back loops involving multiple signaling pathways. A further discretization of the continuous component of the hybrid system could also be coupled with the proposed approach to reduce the complexity and increase the robustness of biological models [9].

References

1. Abate, A., Ames, A.D., Sastry, S.S.: Stochastic approximations of hybrid systems. In: ACC 2005. pp. 1557–1562 (2005)
2. Agrawal, M., Stephan, F., Thiagarajan, P.S., Yang, S.: Behavioural approximations for restricted linear differential hybrid automata. In: Hespanha, J.P., Tiwari, A. (eds.) HSCC 2006. LNCS, vol. 3927, pp. 4–18. Springer, Heidelberg (2006)
3. Alur, R., Henzinger, T.A., Lafferriere, G., Pappas, G.J.: Discrete abstractions of hybrid systems. Proc. IEEE **88**(7), 971–984 (2000)
4. Approximate probabilistic verification of hybrid systems (Technical report): http://arxiv.org/abs/1412.6953
5. Ballarini, P., Djafri, H., Duflot, M., Haddad, S., Pekergin, N.: COSMOS: a statistical model checker for the hybrid automata stochastic logic. In: QEST 2011. pp. 143–144 (2011)
6. Batt, G., Ropers, D., de Jong, H., Geiselmann, J., Page, M., Schneider, D.: Qualitative analysis and verification of hybrid models of genetic regulatory networks: nutritional stress response in *Escherichia coli*. In: Morari, M., Thiele, L. (eds.) Hybrid Systems: Computation and Control. LNCS, vol. 3414, pp. 134–150. Springer, Heidelberg (2005)
7. Biere, A., Cimatti, A., Clarke, E., Zhu, Y.: Symbolic model checking without BDDs. In: Cleaveland, W.R. (ed.) TACAS 1999. LNCS, vol. 1579, pp. 193–207. Springer, Heidelberg (1999)
8. Blom, H.A., Lygeros, J., Everdij, M., Loizou, S., Kyriakopoulos, K.: Stochastic Hybrid Systems: Theory and Safety Critical Applications. Springer, Heidelberg (2006)
9. Bortolussi, L., Policriti, A.: The importance of being (a little bit) discrete. ENTCS **229**(1), 75–92 (2009)
10. Bruce, D., Pathmanathan, P., Whiteley, J.P.: Modelling the effect of gap junctions on tissue-level cardiac electrophysiology. Bull. Math. Biol. **76**(2), 431–454 (2014)
11. Buckwar, E., Riedler, M.G.: An exact stochastic hybrid model of excitable membranes including spatio-temporal evolution. J. Math. Biol. **63**(6), 1051–1093 (2011)

12. Bueno-Orovio, A., Cherry, E.M., Fenton, F.H.: Minimal model for human ventricular action potentials in tissue. J. Theor. Biol. **253**, 544–560 (2008)
13. Cassandras, C.G., Lygeros, J.: Stochastic Hybrid Systems. CRC Press, Taylor & Francis Group (2006)
14. Clarke, E., Fehnker, A., Han, Z., Krogh, B.H., Stursberg, O., Theobald, M.: Verification of hybrid systems based on counterexample-guided abstraction refinement. In: Garavel, H., Hatcliff, J. (eds.) TACAS 2003. LNCS, vol. 2619, pp. 192–207. Springer, Heidelberg (2003)
15. Clarke, E.M., Grumberg, O., Peled, D.A.: Model Checking. MIT Press, Cambridge (1999)
16. Drouin, E., Charpentier, F., Gauthier, C., Laurent, K., Le Marec, H.: Electrophysiologic characteristics of cells spanning the left ventricular wall of human heart: evidence for presence of m cells. J Am Coll Cardiol **26**, 185–192 (1995)
17. Fenton, F., Karma, A.: Vortex dynamics in 3D continuous myocardium with fiber rotation: filament instability and fibrillation. Chaos **8**, 20–47 (1998)
18. Frehse, G.: PHAVer: algorithmic verification of hybrid systems past hytech. In: Morari, M., Thiele, L. (eds.) HSCC 2005. LNCS, vol. 3414, pp. 258–273. Springer, Heidelberg (2005)
19. Gao, S., Kong, S., Clarke, E.: Delta-complete reachability analysis (part i). In: Technical report, CMU SCS, CMU-CS-13-131 (2013)
20. Girard, A., Le Guernic, C., Maler, O.: Efficient computation of reachable sets of linear time-invariant systems with inputs. In: Hespanha, J.P., Tiwari, A. (eds.) HSCC 2006. LNCS, vol. 3927, pp. 257–271. Springer, Heidelberg (2006)
21. Grosu, R., Batt, G., Fenton, F.H., Glimm, J., Le Guernic, C., Smolka, S.A., Bartocci, E.: From cardiac cells to genetic regulatory networks. In: Gopalakrishnan, G., Qadeer, S. (eds.) CAV 2011. LNCS, vol. 6806, pp. 396–411. Springer, Heidelberg (2011)
22. Henzinger, T.: The theory of hybrid automata. In: LICS 1996. pp. 278–292 (1996)
23. Henzinger, T., Kopke, P.: Discrete-time control for rectangular hybrid automata. Theor. Comput. Sci. **221**(1), 369–392 (1999)
24. Hirsch, M., Smale, S., Devaney, R.: Differential Equations, Dynamical Systems, and an Introduction to Chaos. Academic Press, Waltham (2012)
25. Julius, A.A., Pappas, G.J.: Approximations of stochastic hybrid systems. IEEE Trans. Autom. Control **54**(6), 1193–1203 (2009)
26. Kim, J.K., Forger, D.B.: A mechanism for robust circadian timekeeping via stoichiometric balance. Mol Syst Biol **8**, 1–14 (2012)
27. Liu, B., Kong, S., Gao, S., Zuliani, P., Clarke, E.M.: Parameter synthesis for cardiac cell hybrid models using δ-decisions. In: Mendes, P., Dada, J.O., Smallbone, K. (eds.) CMSB 2014. LNCS, vol. 8859, pp. 99–113. Springer, Heidelberg (2014)
28. Matsuno, H., Inouye, S.T., Okitsu, Y., Fujii, Y., Miyano, S.: A new regulatory interaction suggested by simulations for circadian genetic control mechanism in mammals. J Bioinf. Comput Biol **4**(1), 139–153 (2006)
29. Nabauer, M., Beuckelmann, D.J., Uberfuhr, P., Steinbeck, G.: Regional differences in current density and rate-dependent properties of the transient outward current in subepicardial and subendocardial myocytes of human left ventricle. Circulation **93**, 169–177 (1996)
30. Nakamura, K., Yoshida, R., Nagasaki, M., Miyano, S., Higuchi, T.: Parameter estimation of in silico biological pathways with particle filtering towards a petascale computing. In: PSB 2009. pp. 227–238 (2009)

31. Palaniappan, S.K., Gyori, B.M., Liu, B., Hsu, D., Thiagarajan, P.S.: Statistical model checking based calibration and analysis of bio-pathway models. In: Gupta, A., Henzinger, T.A. (eds.) CMSB 2013. LNCS, vol. 8130, pp. 120–134. Springer, Heidelberg (2013)

32. Shearman, L., Sriram, S., Weaver, D., Maywood, E., Chaves, I., Zheng, B., Kume, K., Lee, C., van der Horst, G., Hastings, M., Reppert, S.: Interacting molecular loops in the mammalian circadian clock. Science **288**, 1013–1019 (2000)

33. Stephen, W.: General Topology. Dover Publications, Addison-Wesley Publishing Company, Reading, Massachusetts (1970)

34. Tanaka, K., Zlochiver, S., Vikstrom, K., Yamazaki, M., Moreno, J., Klos, M., Zaitsev, A., Vaidyanathan, R., Auerbach, D., Landas, S., Guiraudon, G., Jalife, J., Berenfeld, O., Kalifa, J.: Spatial distribution of fibrosis governs fibrillation wave dynamics in the posterior left atrium during heart failure. Circ. Res. **8**(101), 839–847 (2007)

Quantitative Analysis
of Biological Models

Synthesising Robust and Optimal Parameters for Cardiac Pacemakers Using Symbolic and Evolutionary Computation Techniques

Marta Kwiatkowska[1], Alexandru Mereacre[1], Nicola Paoletti[1]([⊠]), and Andrea Patanè[2]

[1] Department of Computer Science, University of Oxford, Oxford, UK
nclpltt@gmail.com
[2] Department of Mathematics and Computer Science, University of Catania, Catania, Italy

Abstract. We consider the problem of automatically finding safe and robust values of timing parameters of cardiac pacemaker models so that a quantitative objective, such as the pacemaker energy consumption or its cardiac output (a heamodynamic indicator of the human heart), is optimised in a finite path. The models are given as parametric networks of timed I/O automata with data, which extend timed I/O automata with priorities, real variables and real-valued functions, and specifications as Counting Metric Temporal Logic (CMTL) formulas. We formulate the parameter synthesis as a bilevel optimisation problem, where the quantitative objective (the outer problem) is optimised in the solution space obtained from optimising an inner problem that yields the maximal robustness for any parameter of the model. We develop an SMT-based method for solving the inner problem through a discrete encoding, and combine it with evolutionary algorithms and simulations to solve the outer optimisation task. We apply our approach to the composition of a (non-linear) multi-component heart model with the parametric dual chamber pacemaker model in order to find the values of multiple timing parameters of the pacemaker for different heart diseases.

1 Introduction

Motivation. The growing demand for wearable health monitoring devices, from fitness apps running on smart watches to implantable devices such as cardiac pacemakers and glucose monitoring, calls for design methodologies that can ensure their safety, effectiveness and energy efficiency. Model-based verification [9,21,32] has proved useful in establishing key correctness properties of cardiac pacemakers [19], but the approach has limitations, in that it is not clear how to redesign the model if it fails to satisfy a given property. Instead, the parameter synthesis problem aims to automatically find optimal values of parameters to guarantee that a given property is satisfied. Similarly to verification, this problem has prohibitive complexity and may suffer from undecidability, typically tackled through discretisation of the parameter space.

© Springer International Publishing Switzerland 2015
A. Abate and D. Šafránek (Eds.): HSB 2015, LNBI 9271, pp. 119–140, 2015.
DOI: 10.1007/978-3-319-26916-0_7

In [15], we presented a parameter synthesis method for timed I/O automata (TIOA) that optimises the choice of timing delays for a given objective function to guarantee that a property, expressed in Counting MTL, a generalisation of Metric Temporal Logic, is satisfied. The method is based on exploring finite discrete paths and the corresponding timing constraints. We have applied the techniques to cardiac pacemakers, deriving robust values for safety and energy efficiency, but could only guarantee partial coverage via sampling, as fully exhaustive exploration was not practical.

Contribution. In this paper, we tackle, for the first time, the problem of ensuring effectiveness for pacemakers defined in terms of cardiac output (a heamodynamic indicator of the human heart), as well as safety. To this end, we extend the models and logic of [15] with real-valued data variables, and provide a novel method for synthesising timing delays that are simultaneously safe *and* robust, whilst guaranteeing that a given quantitative objective is optimised. This is formulated as a bi-level optimisation problem, which we solve through a combination of symbolic, SMT-based, analysis of finite paths based on discrete encoding (for the inner problem), with evolutionary computation techniques (for the outer problem). We consider a novel multi-component heart model given as a network of TIOA with data [5] and extend it in order to compute the cardiac output. We apply the developed techniques to the synthesis of multiple pacemaker parameters for different heart conditions, in order to optimise, at the outer level, either energy consumption or cardiac output, on top of the solution space that yields a safe heart rhythm (formulated as a CMTL property) with maximum robustness.

Related Work. The undecidability of the parametric reachability problem is proved in [16]. The majority of work for timed systems concerns synthesis from logic formulas, e.g. [8], with the exception of [1,2] who consider a reference valuation. In [7,23], the authors show PSPACE-completeness of the emptiness problem and TCTL, respectively. Robustness under a given timed perturbation is considered in [38] and parameter synthesis for reachability for probabilistic timed automata in [22]. SMT-based verification of timed and hybrid systems has received a lot of attention recently, see e.g. [10]. In [26], the authors present an SMT-based timed system extension to the IC3 algorithm. [25] and [27] respectively develop real-time bounded model checking (BMC) approaches for LTL and CTL. [20] presents an SMT technique to generate inductive invariants for hybrid systems. Sturm et al. [37] applies real quantifier elimination tools to synthesise continuous and switched dynamical systems. The dReal solver [18] uses a relaxed notion of satisfiability in order to provide decision procedures for non-linear hybrid systems.

In this paper, we extend the model and logic of [15], and replace the path exploration with a fully symbolic BMC-based algorithm. We adopt the pacemaker model from [21] but consider a different heart model [5], which we enhance with the blood pressure component.

2 Background

2.1 Timed I/O Automata with Priorities and Data

We extend the timed I/O automata model with priorities of [15] with data variables. Let $\mathcal{X}$ be a set of *non-negative* real-valued variables, called *clocks*. Let $\mathcal{D}$ be a set of real-valued variables, called *data*. A variable valuation is a function $\eta = \eta_{|\mathcal{X}} \cup \eta_{|\mathcal{D}}$ where $\eta_{|\mathcal{X}} : \mathcal{X} \to \mathbb{R}_{\geq 0}$ and $\eta_{|\mathcal{D}} : \mathcal{D} \to \mathbb{R}$. We denote the set of variables with $V = \mathcal{X} \cup \mathcal{D}$. Let Γ be a set of real-valued *parameters*. A parameter valuation is a function $\gamma : \Gamma \to \mathbb{R}$ mapping each parameter p to a value in its domain $\mathrm{dom}(p) \subseteq \mathbb{R}$.

Let $\mathcal{Y}$ be a set and $\mathcal{V}(\mathcal{Y})$ denote the set of all valuations over $\mathcal{Y}$. We consider guard constraints of the form $\bigwedge_i v_i \bowtie_i f_i$, where $v_i \in \mathcal{X}$ is a clock, $\bowtie_i \in \{<, \leqslant, >, \geqslant\}$ and $f_i : \mathcal{V}(\mathcal{D}) \times \mathcal{V}(\Gamma) \to \mathbb{R}$ is a real-valued function over data variable and parameter valuations. A variable valuation η and a parameter valuation γ satisfy the above constraint iff $\bigwedge_i \eta(v_i) \bowtie_i f_i(\eta_{|\mathcal{D}}, \gamma)$ holds. We denote with $\mathcal{B}(V)$ the set of guard constraints over V. The reset of a set of variables $V' \subseteq V$ is an arbitrary function $r : V' \times \mathcal{V}(V) \times \mathcal{V}(\Gamma) \to \mathbb{R}$. Given valuations η and γ, η is updated by reset r to the valuation $\eta[r] = \{v \mapsto r(v, \eta, \gamma) \mid v \in V'\} \cup \{v \mapsto \eta(v) \mid v \notin V'\}$ that applies the reset r to the variables in V' and leaves unchanged the others. We denote with $\mathcal{R}$ the set of reset functions. The valuation η after time $\delta \in \mathbb{R}_{\geq 0}$ has elapsed is denoted with $\eta + \delta$ and is such that $\eta + \delta(v) = \eta(v) + \delta$ if $v \in \mathcal{X}$ and $\eta + \delta(v) = \eta(v)$ otherwise. This implies that all clocks proceed at the same speed and data variables are not affected by the passage of time.

Definition 1 (Deterministic Timed I/O Automaton with Priority and Data). *A deterministic timed I/O automaton (TIOA) with priority and data $\mathcal{A} = (\mathcal{X}, \Gamma, \mathcal{D}, Q, q_0, \Sigma_{\mathrm{in}}, \Sigma_{\mathrm{out}}, \to)$ consists of:*

- *A finite set of clocks $\mathcal{X}$, data variables $\mathcal{D}$ and parameters Γ.*
- *A finite set of locations Q, with the initial location $q_0 \in Q$.*
- *A finite set of input actions Σ_{in} and a finite set of output actions Σ_{out}.*
- *A finite set of edges $\to \subseteq Q \times (\Sigma_{\mathrm{in}} \cup \Sigma_{\mathrm{out}}) \times \mathbb{N} \times \mathcal{B}(V) \times \mathcal{R} \times Q$. Each edge $e = (q, a, pr, g, r, q')$ is described by a source location q, an action a, a priority pr, a guard g, a reset r and a target q'.*

We require that priorities define a total ordering of the edges out of any location, and that output actions have higher priority than input actions. The TIOAs as defined above are able to synchronise on matching input and output actions, thus forming *networks of communicating automata*. We say that an output edge is *enabled* when the associated guard holds. On the other hand, an input edge is enabled when both its guard holds and it can synchronise with a matching output action fired by another component of the network. A component of a network of TIOAs is enabled if, from its current location, there is at least one outgoing edge enabled. Also, we assume that output edges are *urgent*, meaning that they are taken as soon as they become enabled. As shown in [15], priority and urgency imply that *the TIOA is deterministic*.

Definition 2 (Network of TIOAs). *A network of TIOAs with m components is a tuple $\mathcal{N} = (\{\mathcal{A}^1, \ldots, \mathcal{A}^m\}, \mathcal{X}, \Gamma, \mathcal{D}, \Sigma_{\text{in}}, \Sigma_{\text{out}})$ of TIOAs, where*

- *for $j = 1, \ldots, m$, $\mathcal{A}^j = (\mathcal{X}, \Gamma, \mathcal{D}, Q^j, q_0^j, \Sigma_{\text{in}}, \Sigma_{\text{out}}, \rightarrow^j)$ is a TIOA,*
- *$\mathcal{X}, \Gamma, \mathcal{D}, \Sigma_{\text{in}}, \Sigma_{\text{out}}$ are the common sets of clocks, parameters, data variables, input and output actions, respectively,*

We define the set of network modes by $Q = Q^1 \times \cdots \times Q^m$, with initial mode $q_0 = (q_0^1, \ldots, q_0^m)$ and the initial variable valuation η_0. A state of the network is a pair (q, η) where $q \in Q$ and $\eta \in \mathcal{V}(V)$.

A *parametric network of TIOAs* is a network where the parameter valuation is unknown, and is denoted by $\mathcal{N}(\cdot)$. $\mathcal{N}(\gamma)$ denotes the network obtained by instantiating valuation γ. We describe the formal semantics of a network of TIOAs in terms of timed paths. In the following, we use the predicate $\mathsf{enabled}(\mathcal{N}, j, q, \eta, \gamma)$ (see [30] for its formal encoding) to indicate whether the j-th component of network $\mathcal{N}$ is enabled from the network mode q under variable valuation η and parameter valuation γ.

Definition 3 (Path of a TIOA Network). *Let $\mathcal{N}$ be a network of TIOAs and $n \in \mathbb{N}^+$. Let $\rho = (q_0, \eta_0) \xrightarrow{t_0} (q_1, \eta_1) \xrightarrow{t_1} \cdots \xrightarrow{t_{n-2}} (q_{n-1}, \eta_{n-1})$ be a timed sequence of length n where, for $i = 1, \ldots, n-1$, $t_{i-1} \geq 0$, $q_i \in Q$ and η_i is a variable valuation. Then, ρ is the timed path of network $\mathcal{N}$ if for any position $i = 0, \ldots, n-2$:*

(I) there exists at least one component enabled: $\exists j.\ \mathsf{enabled}(\mathcal{N}, j, q_i^j, \eta_i + t_i, \gamma)$; let E_{i, t_i} be the set of such components;

(II) each component $j \in E_{i, t_i}$ fires the edge $e_i^j = (q_i^j, a_i^j, pr_i^j, g_i^j, r_i^j, q_{i+1}^j) \in \rightarrow^j$ that is enabled and with maximum priority among the enabled edges;

(III) the variable valuation is updated according to the elapsed time and the resets of enabled components[1]: $\eta_{i+1} := \eta_i + t_i[\bigcup_{j \in E_{i, t_i}} r_i^j]$; and

(IV) t_i is the least time for which there are enabled components $\forall t' < t_i$. $E_{i, t'} = \varnothing$:

For $k, m \in \mathbb{N}$, $\rho[k] = (q_k, \eta_k)$ is the k-th state of the path, $\rho^{[k,m]}$ is the subpath of length $m - k + 1$ starting at position k, $\rho\langle k \rangle = t_k$ is the k-th delay and $\rho\langle k, m \rangle = \sum_{k'=k}^{m} t_{k'}$ is the total time spent in the subpath $\rho^{[k,m]}$. For $t \in \mathbb{R}_{\geqslant 0}$, we denote with $\rho@t$ the smallest index o such that $\sum_{k=0}^{o} \rho\langle k \rangle > t$. If no such index exists, then $\rho@t = n - 1$. When the network is parametric, i.e. of the form $\mathcal{N}(\cdot)$, the corresponding parametric path is denoted with $\rho(\cdot)$.

In the following, we denote with Π the set of finite timed paths. Given $t \in \mathbb{R}^{\geq 0}$, we also define the *path up to time t* as the path ρ with length $|\rho| = \rho@t$, that is, such that: (a) $\rho\langle 0, |\rho| - 1 \rangle \leq t$; and (b) let ρ' be the 1-step extension of ρ, then $\rho'\langle 0, |\rho'| - 1 \rangle > t$. We say that a parametric path $\rho(\cdot)$ is up to time t if, for each $\gamma \in \mathcal{V}(\Gamma)$, $\rho(\gamma)$ is up to t.

---

[1] In order to have consistent resets, we assume that different components cannot update the same variable with different values during the same transition.

Example 1. Consider the TIOAs $\mathcal{A}_1$ and $\mathcal{A}_2$ in Fig. 1. The automata $\mathcal{A}_1$ and $\mathcal{A}_2$ form a network. They communicate with each other by means of actions $\{\mathsf{VP}, \mathsf{AP}, \mathsf{AS}\} \in \Sigma_{\mathrm{in}} \cup \Sigma_{\mathrm{out}}$. We distinguish input (marked with ?) and output actions (marked with !). For instance, when automaton $\mathcal{A}_2$ takes a transition and outputs the action VP!, the automaton $\mathcal{A}_1$ synchronises by taking the corresponding transition with the input action VP?. We use Roman numbers to denote priorities, with the lowest number denoting the highest priority. The network $\mathcal{N}$ has three clocks t, x and y, and two variables α, and β. The initial mode of the network is (q, z) and the initial values for the α and β variables are zero. Each edge of the automaton is labelled with an action, a guard over the set of clocks and a reset over the set of clocks and variables. For instance, one of the edges from q' to q' is labelled with the guard $t \geq T - \beta$, action AP and clock reset $t := 0$. The network $\mathcal{N}$ has also three parameters T, P and J.

There are two ways to take an edge. First, when an input action is enabled. Second, when the clock satisfies a given guard. For example, automaton $\mathcal{A}_2$ has two transitions labelled with the conditions $x \geq P - \alpha$ and $y \geq J$. As soon as the clock y satisfies the guard $y \geq J$, the automaton takes the corresponding transition and outputs the action VP!, resetting to zero the value of the clock y and assigning the value of five to the variable β. When multiple transitions are enabled in a location, then the one with the highest priority will be taken. Consider the finite path below, where transitions are labelled with enabled output actions:

$$((q, z), (\alpha = 0, \beta = 0, t = 0, x = 0, y = 0))$$
$$\downarrow J, \mathsf{VP}$$
$$((q', z), (\alpha = 0, \beta = 5, t = 0, x = J, y = 0))$$
$$\downarrow T{-}5, \mathsf{AP}$$
$$((q', z), (\alpha = 0, \beta = 5, t = 0, x = J{+}T{-}5, y = T{-}5))$$
$$\downarrow P{-}J{-}T{+}5, \mathsf{AS}$$
$$((q, z), (\alpha = 10, \beta = 5, t = P{-}J{-}T{+}5, x = 0, y = P{-}J)).$$

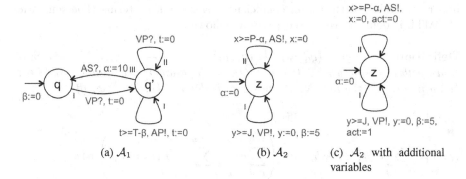

(a) $\mathcal{A}_1$ (b) $\mathcal{A}_2$ (c) $\mathcal{A}_2$ with additional variables

Fig. 1. Example network $\mathcal{N}$ with two components, $\mathcal{A}_1$ and $\mathcal{A}_2$.

Each element of the tuple represents the state of the network and the values of the variables. The network starts in the initial state (q, z) with the values of the variables $(\alpha = 0, \beta = 0, t = 0, x = 0, y = 0)$. In the automaton $\mathcal{A}_2$, after J time units have passed, the guard $y \geq J$ becomes true and the corresponding transition is triggered at this point, outputting the action VP and resetting the clock t to 0 and the variable β to 5. The automaton $\mathcal{A}_1$ then synchronises with $\mathcal{A}_2$ via the matching input, VP, which moves the automaton to q'. Then $\mathcal{A}_1$ takes transition labelled with $T - \beta$ and $\mathcal{A}_2$ does no transition. Then the automaton takes the transitions labelled with $P - \alpha$ outputting the action AS and the state of the network becomes (q, z). Note that, in order to take the transition labelled with action VP in $\mathcal{N}$, the parameters P and J have to satisfy the urgency constraint $P - \alpha < J$. Similar relations can be derived for the remaining transitions of the path.

2.2   Counting MTL

We work with the *Counting Metric Temporal Logic* (CMTL), which is an extension of MTL with the counting operator ($\#$) [15,34], now interpreted over TIOAs with data.

Let $\mathcal{E}(V)$ be the set of constraints $\bigwedge_i c_i \bowtie_i g_i$ over variables in $V = \mathcal{X} \cup \mathcal{D}$, where $c_i \in \mathbb{R}$, $\bowtie_i \in \{<, \leq, =, \geq, >\}$ and $g_i : \mathcal{V}(V) \to \mathbb{R}$ is a real-valued function over V. To this end, we replace CMTL atomic propositions by predicates from $\mathcal{E}(V)$. For instance, given two variables $x, y \in V$, a constraint from $\mathcal{E}(V)$ is $x = 1 \wedge y \geq 10$. The syntax of CMTL is defined by

$$\varphi ::= e \mid \sum_{j \in J} c_j \#_{\ell_j}^{u_j} e_j \bowtie b \mid \varphi \wedge \varphi \mid \neg \varphi \mid \varphi \, \mathcal{U}^{[\ell, u]} \varphi,$$

where J is a finite set of indices, $\bowtie \in \{>, \geqslant, <, \leqslant\}$, $b \in \mathbb{Z}$, $c_j \in \mathbb{Z}$, $\ell \in \mathbb{R}_{\geqslant 0}$, $\ell_j \in \mathbb{R}_{\geqslant 0}$, $u \in \mathbb{R}_{\geqslant 0} \cup \{\infty\}$, $u_j \in \mathbb{R}_{\geqslant 0} \cup \{\infty\}$ are time points such that $\ell \leqslant u$ and $\ell_j \leqslant u_j$, and $e, e_j \in \mathcal{E}(V)$ for all $j \in J$. The counting term $\#_{\ell_j}^{u_j} e_j$ counts how many times, in the interval of time $[\ell_j, u_j]$, e_j holds true. Such terms can be combined to form a so-called basic counting formula $\sum_{j \in J} c_j \#_{\ell_j}^{u_j} e_j \bowtie b$, i.e. a linear constraint (with integer coefficients) over counting terms. The semantics of CMTL is defined over timed paths as follows.

Definition 4. *Let* $\rho = (q_0, \eta_0) \xrightarrow{t_0} (q_1, \eta_1) \xrightarrow{t_1} \cdots \xrightarrow{t_{n-1}} (q_n, \eta_n)$ *be the finite timed path of the network* $\mathcal{N}(\gamma)$ *of TIOAs with parameter valuation* γ *and* $i \in \mathbb{N}$ *be an index. We say that* $\mathcal{N}$ *satisfies* φ *at* i, *denoted* $(\rho, i) \models^{\mathcal{N}} \varphi$, *iff*

$(\rho, i) \models^{\mathcal{N}} e$ iff $\eta_i \models e$

$(\rho, i) \models^{\mathcal{N}} \sum_{j \in J} c_j \#_{\ell_j}^{u_j} e_j \bowtie b$ iff $\left(\sum_{j \in J} c_j \sum_{k = \rho^{[i, |\rho|]} @ \ell_j}^{\rho^{[i, |\rho|]} @ u_j - 1} \mathbf{1} \left(\eta_k \models e_j \right) \right) \bowtie b$

$(\rho, i) \models^{\mathcal{N}} \varphi_1 \wedge \varphi_2$ iff $(\rho, i) \models^{\mathcal{N}} \varphi_1 \wedge (\rho, i) \models^{\mathcal{N}} \varphi_2$

$(\rho, i) \models^{\mathcal{N}} \neg\varphi_1$ $\qquad\qquad$ iff $(\rho, i) \not\models^{\mathcal{N}} \varphi_1$

$(\rho, i) \models^{\mathcal{N}} \varphi_1 \, \mathcal{U}^{[\ell, u]} \varphi_2$ $\qquad$ iff $\exists i'. \, i \leqslant i'$ s.t. $\displaystyle\sum_{k=i}^{i'} \rho\langle k \rangle \in [\ell, u] \wedge (\rho, i') \models^{\mathcal{N}} \varphi_2 \wedge$

$$\forall i''. \, i \leqslant i'' < i' \wedge (\rho, i'') \models^{\mathcal{N}} \varphi_1,$$

where φ_1, φ_2 are CMTL formulas, $i', i'' \in \mathbb{N}$, $e \in \mathcal{E}(V)$, $\ell_j \in \mathbb{R}_{\geqslant 0}$ and $\mathbf{1}\,(\eta_k \models e_j)$ is the characteristic function that returns 1 if $\eta_k \models e_j$ and 0 otherwise.

We define $\Diamond^{[\ell, u]} \varphi := \mathbf{true} \, \mathcal{U}^{[\ell, u]} \varphi$ and $\Box^{[\ell, u]} \varphi := \neg\Diamond^{[\ell, u]} \neg\varphi$. Details on the decidability and complexity of the logic can be found in [33, 34].

Example 2. Let $\mathcal{A}_2$ from Fig. 1(c) be the modified version of the TIOAs $\mathcal{A}_2$ from Fig. 1, where we add a new variable *act*. The variable *act* identifies the presence of the action VP or AS through expression $act = 1$ or $act = 0$, respectively. We also modify automaton $\mathcal{A}_1$ by adding the update $act := 2$ to the edge labelled with the action AP!. We set the initial valuation to $act := -1$. We consider the following CMTL formula which states that, starting from any time in the interval $[0, 100]$, the number of performed VP actions in the interval of time $[0, 7]$ has to be no lower than 1 and at most 4:

$$\Box^{[0,100]} \left(\#_0^7 (act = 1) \geq 1 \wedge \#_0^7 (act = 1) \leq 4 \right) \tag{1}$$

3 Robust Optimal Synthesis Problem

We introduce a parameter synthesis problem for networks of TIOAs that asks for the parameter valuation that, first, maximises parameter robustness and, second, minimises some cost function, e.g. energy consumption. This problem is motivated by the fact that, in the design of medical devices, safety is of paramount importance and it is desirable to maintain patient's physiological properties in a robust way w.r.t. perturbations of parameter values. We express such properties in CMTL, which we use to formulate the requirement of a safe heart rhythm (see Sect. 4). We also assume a cost function $f : \Pi \to \mathbb{R}$ that maps timed paths to reals.

Thus, we aim at finding a valuation γ with maximum *robustness radius*, i.e. a quantity $\epsilon \in \mathbb{R}^+$ such that a CMTL formula ϕ is guaranteed to hold for any perturbation of γ bounded by ϵ. Then, we synthesise parameters that yield the minimum cost on top of the solution space with maximum ϵ. This problem can be effectively formulated as a *bi-level optimisation* problem (see e.g. [12]), where robustness maximisation and cost optimisation represent the so-called *inner and outer problems*, respectively.

Let $\epsilon \in \mathbb{R}^+$ and $\gamma \in \mathcal{V}(\Gamma)$. The ϵ-bounded perturbations of γ are denoted by the set $B_\epsilon(\gamma) = \{\gamma' \mid \forall p \in \Gamma. \, |\gamma'(p) - \gamma(p)| \leq \epsilon\}$. Given property ϕ and path $\rho(\cdot)$ of network $\mathcal{N}(\cdot)$, we say that a parameter valuation γ is ϵ-*robust* w.r.t. ϕ if it holds that $\forall \gamma' \in B_\epsilon(\gamma). \, \rho(\gamma') \models^{\mathcal{N}(\gamma')} \phi$. Note that, for arbitrary ϵ, there can exist perturbed valuations outside the domain of parameters, i.e. $B_\epsilon(\gamma) \not\subseteq \mathcal{V}(\Gamma)$. In the following, we only admit the case when $B_\epsilon(\gamma) \subseteq \mathcal{V}(\Gamma)$.

Problem 1 (Robust Optimal Synthesis). Let $\mathcal{N}(\cdot)$ be a parametric network of TIOAs, $t \in \mathbb{R}^{\geq 0}$, $k \in \mathbb{N}^+$, ϕ be a CMTL property and f be a cost function. Let $\rho(\cdot)$ be the parametric path of $\mathcal{N}(\cdot)$ of length k and $\rho'(\cdot)$ be the parametric path up to time t. The *robust optimal synthesis problem* is finding a parameter valuation γ that solves the following bi-level optimisation problem:

$$\min_{\gamma_o \in V(\Gamma)} f(\rho'(\gamma_o)) \quad \text{subject to } \gamma_o \in \arg\max_{\gamma_i \in V(\Gamma)} \in$$

$$\text{subject to } B_\in(\gamma_i) \subseteq V(\Gamma) \text{ and } \forall \gamma' \in B_\in(\gamma_i).\ \rho(\gamma') \vDash^{N(\gamma')} \phi.$$

Note that in the above problem the path lengths for the inner and outer problem are arbitrary and in general not interrelated. In practice, as explained in Sect. 5.3, f is evaluated through simulation and thus we can support longer lengths for ρ'.

Running Example. We formulate an instance of Problem 1 by taking the CMTL property in Example 2 and the modified network defined therein. In the inner problem, we take the parametric path ρ of length $k = 15$. In the outer problem, we consider the path ρ' up to time $t = 100$ and aim to minimise the number of AS actions performed along ρ', leading to the objective $f(\rho') = \#_0^{100}\,(act = 0)$.

4   Heart and Pacemaker Models

In this section we describe the heart and the pacemaker models, and provide the properties and functions for the synthesis problem. The pacemaker has the role of maintaining the synchronisation between the atrium and the ventricle. In particular, we consider a basic DDD pacemaker specification [36], that is, pacing and sensing both the atrium and the ventricle, and provide a TIOA network adapted from [21].

The heart model is used to reproduce the propagation of the cardiac action potential, and is a TIOA translation of the model by Lian et al. [31] (see [5] for details). In [5], the authors provide a probabilistic model where parameters are drawn from probability distributions derived from patients data. Here, parameters are set to a given fixed value, thus resulting in a fully deterministic model. The composed heart-pacemaker model consists of 11 TIOA components, with 11 clock variables, 7 data variables and 18 action labels.

Heart Model. The high-level structure of the model is depicted in Fig. 2(a). It comprises five main conduction nodes and two main conduction paths: from the atrium to the ventricle (*antegrade* conduction) or vice-versa (*retrograde*). The *Atrium* component is responsible for modelling the sinoatrial (SA) node, i.e. the natural pacemaker of the heart. This has a predefined firing rate, given by parameter SA_d. It can also produce ectopic beats with rate SA_ectopD. In Fig. 3 we depict the sub-network that models the atrium. In Fig. 3(a), we illustrate the automata for the SA node and the ectopic beat generator, respectively.

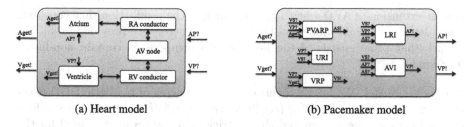

(a) Heart model   (b) Pacemaker model

Fig. 2. Heart and pacemaker models.

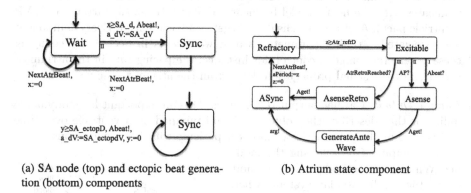

(a) SA node (top) and ectopic beat genera-   (b) Atrium state component
tion (bottom) components

Fig. 3. The *Atrium* component.

When their clocks (x and y) satisfy the corresponding guards, the output action Abeat is produced, which indicates the generation of an atrial impulse. a_dV is a variable storing the action potential in the atrium, which might vary depending on whether the beat is regular or ectopic. The TIOA in Fig. 3(b) models the current state of the atrium. In the *Refractory* mode the atrium component starts a timer, modelled by clock z. After the atrial refractory period has elapsed ($z \geq$ Atr_refrD), the atrium changes its mode to *Excitable*. In this mode, it can receive three types of actions: an SA node signal, Abeat; a pacing signal from the pacemaker, AP; or a retrograde signal from the ventricle, AtrRetroReached. Finally, the atrium generates the output action Aget to notify the pacemaker of the atrial impulse, and returns to the *Refractory* mode. The *aPeriod* variable is used to store the duration of the last atrial cycle.

The *Ventricle* component is similar to the *Atrium* component, i.e., it has an intrinsic beat generator and an ectopic beat generator. In addition, it has a variable *vPeriod* to store the ventricular period. The *RA conductor* and *RV conductor* are used to model the propagation delay of the action potential from the atrium to the ventricle and back (antegrade or retrograde conduction).

The *AV node* component is responsible for delaying the entrance of the action potential from the atrium into the ventricle. The AV conduction delay (AVD) is given by an exponential function AVD $:=$ AVD$_{min} + \alpha \exp(\frac{-T_{rec}}{\tau_c})$, where T_{rec} is the AV recovery time, AVD$_{min}$ is the shortest AVD when $T_{rec} \to \infty$, α is the longest extension of AVD when $T_{rec} = 0$ and τ_c is the conduction time constant. More details of the AV node constants and parameters are provided in [5]. We remark that some guards of the heart model components contain non-linear functions.

Pacemaker Model. We briefly describe the components of the basic DDD pacemaker model shown in Fig. 2(b) (see [9] for details): *AVI* maintains the synchronisation between the atrium and the ventricle, *LRI* sets a lower bound for the heart rate, *URI* sets an upper bound for the heart rate, *PVARP* detects intrinsic atrial events, and *VRP* detects intrinsic ventricle events. The pacemaker communicates with the heart model by means of four actions: AP (atrial pace), VP (ventricle pace), Aget (atrial sense) and Vget (ventricle sense). Every component has associated a timing parameter, which we discuss in Sect. 6. By changing these parameters one can control, for instance, the pacing rate in the atrium or ventricle, or the signal propagation delay from the atrium to the ventricle.

Cardiac Output. Cardiac output (CO, cm³·s⁻¹) is an important heamodynamic indicator that describes the volume of blood pumped by a ventricle over time and is used in clinical practice to monitor patients with heart conditions.

We compute CO following the modified Windkessel method in [17] for modelling the cardiovascular system, where the aortic flow is modelled as a square wave, which is more realistic than the standard Windkessel model (see e.g. [3]) where it is approximated as a series of impulses.

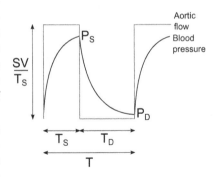

Fig. 4. Blood pressure (black) computed considering a square wave aortic flow (red) (Colour figure online).

The ventricular period alternates in two parts: the systole (T_S) and the diastole (T_D). During systole the ventricles first contract and reach a maximum pressure giving rise to a heart beat. Then, they drive the blood flow to the pulmonary and aortic valves. During diastole, ventricular pressure drops to its minimum and blood starts flowing from the atria to the ventricles until it is ejected in the next systole. Therefore, each wave in the the aortic flow signal has amplitude $\frac{SV}{T_S}$ and period T_S (see Fig. 4), where SV (cm³) is the stroke volume, i.e. the difference between the volume at diastole and that at systole.

Following [17], the maximum arterial pressure at systole P_S (mmHg) is computed as $P_S = P_D \cdot \exp\left(\frac{-T_S}{R \cdot C}\right) + R \cdot \frac{SV}{T_S}\left(1 - \exp\left(\frac{-T_S}{R \cdot C}\right)\right)$, where R (mmHg·s·cm⁻³) and C (cm³·mmHg⁻¹) are the aortic resistance and compliance parameters, respectively. The first term of the equation describes the decay of the diastolic pressure at the previous cycle with rate $\frac{1}{R \cdot C}$ during T_S, while the second term accounts for the pressure given by the aortic flow. The equation for the minimum pressure at diastole, P_D (mmHg), is $P_D = P_S \cdot \exp\left(\frac{-T_D}{R \cdot C}\right)$, which describes the decay of the pressure at systole during T_D. Finally, $CO = C \cdot (P_S - P_D)/T$ depends on the difference between P_S and P_D over the heart period $T = T_S + T_D$.

At each ventricular event (actions Vget or VP), we compute the CO of the previous cycle, assuming $T_S = 0.25 \cdot T$ and $T_D = 0.75 \cdot T$, where T is the time elapsed from the previous ventricular event. We consider the parameters of a healthy patient, namely, $C = 1.3$, $R = 0.9$ and SV = 90 [24]. The initial diastolic pressure is set to 80.

Properties. We consider two variants of the robust optimal synthesis problems (Problem 1) where, in the outer problem, we minimise the energy consumption of the pacemaker or optimise the cardiac output, respectively. In the first variant, we take a simplified model assuming that only atrial and ventricular pacing contribute to energy consumption, with weights 2 and 3 respectively. Thus, the cost function is given by $f(\rho') = 2 \cdot \#_0^{60000} (act = AP) + 3 \cdot \#_0^{60000} (act = VP)$, where act is a variable storing the last performed action, and $AP, VP \in \mathbb{R}$ identify an AP or a VP action, respectively. By abuse of notation, the operator $\#$ denotes a function of ρ', i.e., it counts how often an expression from $\mathcal{E}(V)$ holds true in the path ρ'. In the second variant, we seek to find parameters that make the computed cardiac output as close as possible to a given reference value $\overline{CO}$ (set to 80 cm$^3 \cdot$s^{-1}). Let $Vbeat(\rho')$ be the set of states along path ρ' where a ventricular beat happens. Then, $f(\rho') = \left(\sum_{(q,\eta) \in Vbeat(\rho')} |\eta(CO) - \overline{CO}|\right)/(|Vbeat(\rho')|)$ is the cost function, i.e. the mean difference between the valuations of CO in $Vbeat(\rho')$ and the reference $\overline{CO}$.

For both variants, in the inner problem we find parameters that guarantee a safe heart rhythm with maximum robustness. We express this requirement by imposing that the ventricular period (the time distance between two consecutive ventricular beats) is always within the interval $[500, 1000]$ ms, i.e. between 60 and 120 BPM. The corresponding CMTL property is $\phi = \Box^{[0,T]} (vPeriod \in [500, 1000])$, where the time bound $T = \rho\langle 0, |\rho| - 1\rangle$ is chosen to cover the whole length of path ρ.

5 Parameter Synthesis Algorithms

We now present methods to solve the bi-level optimisation problem introduced in Sect. 3. For space reasons, we restrict ourselves to *safety properties*, i.e. formulas of the form $\Box^{[\ell,u]}\phi$, where ϕ is a CMTL formula without temporal operators. In [30], we also provide algorithms for the reachability fragment. First, we describe the algorithm for the inner problem, namely maximising the robustness radius w.r.t. a given safety property. Second, we devise a method for optimising the outer objective by exploiting the solution of the inner problem. Both methods are based on encoding the TIOA model and the synthesis problem as a Satisfiability Modulo Theories (SMT) problem, which we describe below and, in more detail, in [30]. In our implementation, we use the Z3 theorem prover [13].

5.1   SMT Encoding

We now introduce the notion of TIOAs extended with *non-deterministic variables*, which are necessary to provide a sound encoding of the problem. Intuitively, such variables can be updated in a non-deterministic way to a number of possible values.

Let $\bar{V}$ be the set of non-deterministic variables; let $\bar{v} \in \bar{V}$ and η and γ be valuations of variables and parameters, respectively. Then, the reset r of $\bar{v}$ induces a set of admissible values $r(\bar{v}, \eta, \gamma)$ and a set of admissible updated valuations $\eta[r] = \{\eta' \mid \eta'(\bar{v}) \in r(\bar{v}, \eta, \gamma)$ and $\eta'(v') = \eta(v')$ for $v' \neq \bar{v}\}$. This implies that a network $\mathcal{N}$ of thus extended TIOAs can have multiple admissible paths. We denote with $\Pi(\mathcal{N})$ the set of such paths. Clearly, fixing a valuation for the non-deterministic variables at each state of the path induces a deterministic path according to Definition 3. We provide a discrete encoding of the problem in the theory of bit-vectors (SMT UF_BV). Clocks, parameters and variables are expressed as bit-vectors and therefore have finite domains. Non-deterministic variables are used to provide an interval-based abstraction for non-integer values and non-linear functions as follows. Consider a generic update $y := f(x)$ with $f : \mathbb{D} \rightarrow \mathbb{R}$, where $y \in V$ and $\mathbb{D}$ is the discrete and finite domain of f. For each $x \in \mathbb{D}$, we pre-compute the discrete bounds of f on x, $[f(x)^{\perp}, f(x)^{\top}]$ as

$$f(x)^{\perp} \leq \left\lfloor \min_{x' \in [x, x+1)} f(x') \right\rfloor \text{ and } f(x)^{\top} \geq \left\lceil \max_{x' \in [x, x+1)} f(x') \right\rceil$$

Then, we encode the update $y := f(x)$ as $y' \in [f(x)^{\perp}, f(x)^{\top}]$ using a non-deterministic variable $y' \in \bar{V}$. In this way, we provide a *conservative over-approximation* of the original system, since the above interval-based abstraction induces additional behaviours but preserves the original ones. As discussed later, this potentially leads to spurious counter-examples, i.e. valuations that violate the given property in the abstracted system but not in the original one. In this work, we did not implement a refinement step for excluding spurious counter-examples [11], but we experimentally evaluated their number (see Sect. 6). In general, the quality of the abstraction is affected by the dynamics of the functions involved, e.g. the presence of large variations in small intervals.

Through this discrete SMT encoding, the verification problem for TIOAs with data and CMTL formulas is NExpTime [28].

5.2   The Inner Problem

The main algorithm for solving the inner problem is given in Algorithm 1, which extends the SMT-based method for *bounded model checking (BMC)* [4] in order to synthesise the space of parameters that yields maximum robustness. Given a safety property φ, the algorithm returns the maximum robustness radius ϵ, a parameter valuation $\bar{\gamma}$ that is ϵ-robust w.r.t. φ, and an *under-approximation* Unsafe of the true unsafe parameter valuations. We encode an SMT problem where the Unsafe region is built by searching for counter-examples (CEs) to safety, which amounts to finding valuations s.t. $\neg\varphi$ holds at some point in the path, up to a fixed length n. Enumerating all possible counter-examples up to n, especially when n is large, is clearly infeasible. Here, we implement several solutions to overcome this problem.

First, we exploit *incremental solving*, so that CEs are computed step-wise, for increasing path lengths, exploiting the fact that SMT solvers can use the clauses learned in the previous steps to improve the solution time of the current step. Second, we include an algorithm for *counter-example generalization* (procedure GeneralizeCE, Algorithm 2), that, given a CE, attempts to derive an unsafe region that contains the CE. Third, we *restrict the search space for counter-examples* to the extent necessary to prove the actual maximum robustness radius ϵ, thus avoiding the computation of irrelevant CEs.

Counter-Example Generation. In Algorithm 1, we first initialize ϵ, Unsafe and $\bar{\gamma}$ (lines 2–4). The *Init* predicate (line 5) is used to constrain the initial automata locations and variable valuation. Command Assert adds in the SMT solver a formula that must hold true. At a generic step k of the path, we first assert the safety property up to the current total time $\rho\langle 0, k\rangle$ if this is lower than the time bound u (line 7). In this case, the assertion is named with a literal p_k, meaning that the satisfaction value of $\Box^{[\ell,\min(u,\rho\langle 0,k\rangle)]}\phi$ is the same as p_k. During the counter-example generation cycle (lines 8–17), MaxRadius procedure is called to update the maximum ϵ and ϵ-robust valuation $\bar{\gamma}$ according to the current Unsafe region. This information is used to temporarily restrict the search space for CEs to the region $B_\epsilon(\bar{\gamma})$ (line 11). Solve $\neg p_k$ checks if the negated safety is satisfied under the current assertions. If so, the solver returns a model, in our case a counter-example γ_{CE}, which we generalize to γ'_{CE} by calling GeneralizeCE. γ'_{CE} is then excluded from the search space (line 15) and added to Unsafe (line 16). The Pop command removes from the solver all the constraints asserted after the last Push (in this case, only $B_\epsilon(\bar{\gamma})$). If instead no CEs can be found in $B_\epsilon(\bar{\gamma})$, we can conclude that, up to step k, ϵ is the actual max radius and $\bar{\gamma}$ is ϵ-robust. Thus, we can exit the CE generation loop, assert the transition constraints (line 21) and increase the step to $k + 1$. When $\epsilon < 1$, the algorithm terminates since it implies that no robust parameters exist (lines 18–19). In the algorithm, T indicates the transition predicate between states of the path, i.e. $T(s, s') = \exists t.\ s \xrightarrow{t} s'$. We remark that, by bounding and discretising the parameter space, we can ensure that the cycle at lines 8–17 always terminates. Note that the bound n on the path length is given in input to the algorithm. Other stopping criteria could be considered based on, for instance, the size of Unsafe or the worst-case time bound.

Spurious Counter-Examples. Due to the abstraction induced by the non-deterministic variables, a CE γ_{CE} can be spurious, i.e. it does not violate the property in the original system. Let $\boldsymbol{\eta} = \bar{\eta}_1, \ldots, \bar{\eta}_k$ be a sequence of valuations over $\bar{V}$, $\gamma \in \mathcal{V}(\Gamma)$, and $\rho(\gamma, \boldsymbol{\eta})$ be the path of $\mathcal{N}(\gamma)$ where the non-deterministic variables at i-th state are set to $\bar{\eta}_i$. Let $\boldsymbol{\eta}^*$ be the sequence of valuations describing the evolution of the original system. For a safety property φ, any CE γ_{CE} generated by Algorithm 1 is such that $\exists \boldsymbol{\eta}.\ \rho(\gamma_{CE}, \boldsymbol{\eta}) \in$

$\Pi(\mathcal{N}(\gamma_{CE})) \wedge \rho(\gamma_{CE}, \boldsymbol{\eta}) \not\models^{\mathcal{N}(\gamma_{CE})} \varphi$, i.e. γ_{CE} violates φ for some valuations $\boldsymbol{\eta}$ of $\bar{V}$. The first term of the conjunction expresses that $\boldsymbol{\eta}$ is admissible, that is, $\rho(\gamma_{CE}, \boldsymbol{\eta})$ is a path of $\mathcal{N}(\gamma_{CE})$. Then, γ_{CE} is spurious if $\rho(\gamma_{CE}, \boldsymbol{\eta}^*) \models^{\mathcal{N}(\gamma_{CE})} \varphi$.

Counter-Example Generalisation. The GeneralizeCE procedure is executed on top of the solver used in Algorithm 1 and exploits the ability of SMT solvers to generate unsatisfiable cores, i.e., when a formula is unsatisfiable under the current assertions, produce a subset of its clauses whose conjunction is still unsatisfiable. Given a CE γ_{CE}, the idea is to derive a larger unsafe region γ'_{CE} that contains γ_{CE}. This is achieved by asserting the safety property (line 3) and the valuation γ_{CE} (line 4). In particular, we associate each assertion $p = \gamma_{CE}(p)$ for $p \in \Gamma$ (used to assert γ_{CE}) with a literal g_p. If formula $\bigwedge_{p \in \Gamma} g_p$ (line 5) is unsatisfiable, the solver returns an unsat core, i.e. a set UnsatCore $\subseteq \{g_p \mid p \in \Gamma\}$. If UnsatCore is a strict subset of the g_p literals, we say that the generalization is successful since we obtain a larger region: $\gamma'_{CE} = \{\gamma \mid \gamma(p) = \gamma_{CE}(p) \text{ if } g_p \in \text{UnsatCore}\}$. Otherwise, $\gamma'_{CE} = \gamma_{CE}$. As an example, let $\gamma_{CE} = (p_1 = 3 \wedge p_2 = 5)$, and let g_{p_1} and g_{p_2} be the corresponding literals. If UnsatCore $= \{g_{p_2}\}$, then the generalization $\gamma'_{CE} = (p_2 = 5)$ strictly contains γ_{CE}. Importantly, $\bigwedge_{p \in \Gamma} g_p$ being unsatisfiable means that γ_{CE} violates safety for any valuation of the non-deterministic variables and, therefore, it is a CE also for the original system, i.e., it holds that

$$\forall \boldsymbol{\eta}. \; \rho(\gamma_{CE}, \boldsymbol{\eta}) \in \Pi(\mathcal{N}(\gamma_{CE})) \implies \rho(\gamma_{CE}, \boldsymbol{\eta}) \not\models^{\mathcal{N}(\gamma_{CE})} \varphi. \tag{2}$$

This applies also to its generalization γ'_{CE}. In the implementation, we use a more advanced algorithm that can rule out even larger unsafe regions, which is reported in [30].

Maximum Robustness Radius. Procedure MaxRadius (Algorithm 3) takes the current Unsafe region and the previous maximum radius ϵ, and returns the updated maximum ϵ together with an ϵ-robust valuation $\bar{\gamma}$, such that $B_\epsilon(\bar{\gamma}) \wedge$ Unsafe is unsatisfiable. To this aim, we just need to find a valuation $\bar{\gamma}$ such that

$$B_\epsilon(\bar{\gamma}) \subseteq \mathcal{V}(\Gamma) \wedge \forall \gamma' \in B_\epsilon(\bar{\gamma}). \neg(\gamma' \wedge \text{Unsafe}). \tag{3}$$

Procedure FindRobustParam (not shown) performs this check and returns such $\bar{\gamma}$ if any exists. In this case, we increment ϵ and repeat the procedure as long as an ϵ-robust valuation is found. Otherwise, we decrement it and repeat the procedure until Eq. 3 is met. In our implementation, ϵ is discretized too. We remark that this procedure uses a separate SMT solver (SMT QBVF) so it can be efficiently parallelized. Since we consider parameters with bounded domains, $B_\epsilon(\bar{\gamma}) \subseteq \mathcal{V}(\Gamma)$ in Eq. 3 implies that ϵ is bounded too, and thus, that the algorithm terminates.

Spurious Robust Valuations. Since we do not exhaustively search for CEs, Unsafe is an under-approximation of the true unsafe set. For the same reason, there could be[2] *spurious ϵ-robust valuations* γ s.t. they meet Eq. 3, but CEs exist in $B_\epsilon(\gamma)$. This happens when Algorithm 1 terminates without inspecting region $B_\epsilon(\gamma)$. The

---

[2] Not to be confused with the spurious counter-examples discussed before.

following proposition characterises when a valuation is in the solution space of the inner problem.

Proposition 1 (Inner Problem Solution). *Let $\gamma \in \mathcal{V}(\Gamma)$, Unsafe and ϵ be as returned by Algorithm 1. Then, γ is a solution of the inner problem in Problem 1 iff it holds that:*

(i) γ is ϵ-robust w.r.t. Unsafe, i.e. it satisfies Eq. 3; and
(ii) no counter-examples can be found in $B_\epsilon(\gamma)$.

Note that (ii) can be decided with one iteration of the CE generation loop in Algorithm 1 within region $B_\epsilon(\gamma)$. Nevertheless, the algorithm guarantees that the returned ϵ is the maximum robust radius and that $\bar{\gamma}$ is a solution of the inner problem. Indeed, $\bar{\gamma}$ is computed by Algorithm 3 and therefore meets Eq. 3. Further, no CEs exist in $B_\epsilon(\bar{\gamma})$ ($\bar{\gamma}$ is not spurious), otherwise the incremental synthesis algorithm could not exit the loop at lines 8–17 and would proceed by updating ϵ and $\bar{\gamma}$.

Algorithm 1. Incremental Synthesis

Require: Parametric network $\mathcal{N}(\cdot)$, CMTL property $\Box^{[\ell,u]}\phi$, path length $n \in \mathbb{N}^+$
Ensure: Maximum robust radius ϵ, ϵ-robust valuation $\bar{\gamma}$ and Unsafe region
1: **function** IncrementalSynth($\mathcal{N}(\cdot), \phi, n$)
2:      $\epsilon := 1$
3:      Unsafe $:= \bot$
4:      $\bar{\gamma} := \bot$
5:      Assert $Init(\rho[0])$
6:      **for** $k = 0, \ldots, n-1$ **do**
7:          Assert $p_k : \Box^{[\ell,\min(u,\rho\langle 0,k\rangle)]}\phi$
8:          **repeat** $\triangleright$ CE generation cycle
9:              $(\epsilon, \bar{\gamma}) :=$ MaxRadius(Unsafe, ϵ)
10:             Push
11:             Assert $B_\epsilon(\bar{\gamma})$
12:             $(SAT, \gamma_{CE}) :=$ Solve $\neg p_k$
13:             Pop
14:             $\gamma'_{CE} :=$ GeneralizeCE(γ_{CE})
15:             Assert $\neg\gamma'_{CE}$
16:             Unsafe $:=$ Unsafe $\vee \gamma'_{CE}$
17:         **until** SAT
18:         **if** $\epsilon < 1$ **then**
19:             **return** $(0, \bot, \top)$
20:         **if** $k < n-1$ **then**
21:             Assert $T(\rho[k], \rho[k+1])$
22:      **return** $(\epsilon, \bar{\gamma}, $Unsafe$)$

Algorithm 2. CE generalization

Require: Counter-example γ_{CE}
Ensure: Generalization γ'_{CE} s.t. $\gamma_{CE} \implies \gamma'_{CE}$
1: **function** GeneralizeCE(γ_{CE})
2:      Push
3:      Assert p_k
4:      **for all** $p \in \Gamma$ **do** Assert $g_p : p = \gamma_{CE}(p)$
5:      $(SAT, \gamma) :=$ Solve $\bigwedge_{p \in \Gamma} g_p$
6:      **if** SAT **then** $\gamma'_{CE} := \gamma_{CE}$
7:      **else** $\gamma'_{CE} := \bigwedge_{p.g_p \in \mathsf{UnsatCore}} p = \gamma_{CE}(p)$
8:      Pop
9:      **return** γ'_{CE}

Algorithm 3. Computation of maximum robust radius

Require: Unsafe region, starting radius ϵ
Ensure: Maximum robust radius ϵ and valuation $\bar{\gamma}$ that is ϵ-robust
1: **function** MaxRadius(Unsafe, ϵ)
2:      $\bar{\gamma} :=$ FindRobustParam(Unsafe, $\epsilon, \bot$)
3:      **if** $\bar{\gamma} = \bot$ **then** $inc := -1$
4:      **else** $inc := 1$
5:      **repeat**
6:          $\epsilon := \epsilon + inc$
7:          $\bar{\gamma} :=$ FindRobustParam(Unsafe, ϵ)
8:      **until** $(inc < 0 \iff \bar{\gamma} = \bot) \wedge \epsilon > 0$
9:      **if** $inc > 0$ **then** $\epsilon := \epsilon - inc$
10:     **return** $(\epsilon, \bar{\gamma})$

The incremental synthesis algorithm for reachability formulas (explained in [30]) follows the same structure as Algorithm 1 and proceeds by finding CEs to reachability, i.e. valuations such that the property never holds.

Running Example. To simplify the presentation, we fix $T = 10$ and consider only parameters $J \in [1, 41]$ and $P \in [11, 51]$. Figure 5 shows the incremental synthesis algorithm run on our example. The counter-example $J = 33$ and $P = 49$ indicated in plot (b) clearly violates the property (Eq. 1), since it gives the following path:

$$((q,z),(\alpha = 0, \beta = 0, t = 0, x = 0, y = 0, act = -1)) \xrightarrow{33}$$
$$((q',z),(\alpha = 0, \beta = 5, t = 0, x = 33, y = 0, act = 1))\dots$$

where, starting from position 0, no VP action is fired in the time interval $[0, 7]$. At the final step, we obtain $\epsilon = 2$ and $\bar{\gamma} = \{J \mapsto 4, P \mapsto 32\}$. Such parameters lead to the following path which can be shown to meet our CMTL property:

$$((q,z),(0,0,0,0,0,-1)) \xrightarrow{4} ((q',z),(0,5,0,4,0,1)) \xrightarrow{4} ((q',z),(0,5,0,8,0,1)) \xrightarrow{4}$$
$$((q',z),(0,5,0,12,0,1)) \xrightarrow{4} ((q',z),(0,5,0,16,0,1)) \xrightarrow{4} ((q',z),(0,5,0,20,0,1)) \xrightarrow{4}$$
$$((q',z),(0,5,0,24,0,1)) \xrightarrow{4} ((q',z),(0,5,0,28,0,1)) \xrightarrow{4} ((q',z),(0,5,0,32,0,1)) \xrightarrow{0}$$
$$((q,z),(10,5,0,0,0,0)) \xrightarrow{4} ((q',z),(10,5,0,4,0,1)) \xrightarrow{4} ((q',z),(10,5,0,8,0,1)) \xrightarrow{4}$$
$$((q',z),(10,5,0,12,0,1)) \xrightarrow{4} ((q',z),(10,5,0,16,0,1)) \xrightarrow{4} ((q',z),(10,5,0,20,0,1))$$

In the above, variable names are omitted and their ordering is as in the previous path. The validity of Eq. 1 can be shown for every valuation in $B_\epsilon(\bar{\gamma})$ in a similar way.

5.3   The Outer Problem

We present two methods for solving the outer problem. The former is based on the enumeration of ϵ-robust valuations, thus providing an exact solution to the outer problem, but is infeasible with high-dimensional parameter spaces.

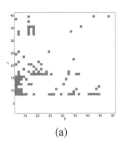
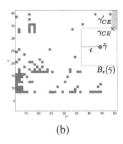
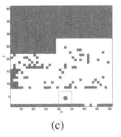

(a)   (b)   (c)

Fig. 5. Counter-examples generation cycle for the running example. Plot (a) shows the Unsafe region (red points) during step $k = 1$. Procedure MaxRadius computes the maximum ϵ (here, 8) and the ϵ-robust valuation $\bar{\gamma}$ ($J = 25, P = 43$, blue dot in plot b). The search for further CEs is restricted to $B_\epsilon(\bar{\gamma})$ (grey-bordered). Then, a CE γ_{CE} is found ($J = 33, P = 49$, red cross). The GeneralizeCE procedure manages to find the larger unsafe region $\gamma'_{CE} = J \geq 33 \wedge P \geq 49$ (light red). Plot (c) shows the results at the final step ($k = 15$) with $\epsilon = 2$ and $\bar{\gamma} = \{J \mapsto 4, P \mapsto 32\}$ (Colour figure online).

Note that enumeration is possible because we discretise the parameter space. The latter method provides an approximate solution by exploiting evolutionary strategies (ES). In both methods, the outer objective is evaluated through simulation. Importantly, simulation can cover path lengths that are prohibitive for BMC, which allows us to consider objective functions over large time bounds.

Exact Solution. The method consists of the following three steps:

1. Enumerate all valuations that meet Eq. 3. Let Γ' be the set of such valuations.
2. Simulate all $\gamma \in \Gamma'$ and compute the outer objective $f(\gamma)$.
3. Following the ordering by the cost function $f(\gamma)$, return the first valuation that meets condition (ii) of Proposition 1.

Optimisation with Evolutionary Strategies. ES are a class of stochastic optimisation methods which mimic the principles of Darwinian evolution in order to optimise a given objective. They work on a set of candidate solutions, the *population*, which at each iteration of the algorithm (*generation*) is subjected to various *natural operators*, until a pre-defined termination criterion is satisfied (e.g. max number of generations).

We implement a *non-isotropic self-adaptive* $(\mu/\rho + \lambda)$-ES, i.e. μ parents are used to generate λ offspring candidates through a ρ-parents recombination, and only the μ best solutions of the combined parents together with the offspring set are used in the next generation. In particular, we consider a 2-*parents dominant recombination*, which randomly takes two candidates from the parents set and generates a child by a parameter-wise random selection of the two parents' valuations. Since we deal with discrete parameters, we use the mutation operator in [35], which extends the principle of *maximum entropy* used in real ES problems to the integer case. We also include a *self-adaptation* mechanism [6] that changes the parameters of the mutation operator at each iteration.

In order to determine the best valuations at each generation, we define an order $\preceq$ that takes into account the outer objective and, following the *feasible-over-infeasible* principle [14], *penalizes valuations outside the solution space of the inner problem*. Let γ_i and γ_j be two valuations, $f(\gamma_i)$ and $f(\gamma_j)$ be their objective function values. Then, $\gamma_i \preceq \gamma_j$ if either:

1. γ_i meets condition (i) of Proposition 1, and γ_j does not; or
2. γ_i meets (i) and (ii), and γ_j meets only (i); or
3. both γ_i and γ_j meet (i) and (ii), and $f(\gamma_i) \leq f(\gamma_j)$.

We say that a solution is feasible for the outer problem if it solves the inner problem as per Proposition 1. Note that, if the population at a generic iteration i, $\mathcal{P}^i$, contains feasible solutions, then, for any $j > i$, $\mathcal{P}^j$ will contain feasible solutions too. Indeed, by the order defined, if $\mathcal{P}^i$ has at least one feasible point, then the best solution in $\mathcal{P}^i$ is also feasible. Since, for any k, the best solutions of $\mathcal{P}^k$ are kept in $\mathcal{P}^{k+1}$, we conclude that, for $j > i$, $\mathcal{P}^j$ will contain feasible solutions too. For the full ES algorithm, see [30].

Running Example. We obtain the exact solution $J = 4, P = 48$, which gives an outer objective of 2 (the number of AS actions fired within time 100). Due to the size of the problem, this required enumerating and simulating only 133 valuations at step 1 of the exact method. For the same reason, the ES algorithm is also able to achieve the optimal solution, being in this case $J = 4, P = 45$. In particular, this was obtained at the *first* iteration of the algorithm, run with $\lambda = 100$, $\mu = 50$ and $\rho = 2$.

6   Results

We apply our method to synthesise pacemaker parameters that ensure a safe heart rhythm and optimise either energy consumption or cardiac output (see Sect. 4 for the formulation of the problem and properties). In [30], we provide a more detailed experimental evaluation of our methods, with different numbers of parameters and path lengths. Here we consider two parameters that are critical for the correct functioning of the pacemaker device. The first parameter, TLRI, regulates the frequency of atrial impulses: TLRI − TAVI is the amount of time that the pacemaker waits before delivering an atrial pace when no atrial or ventricular events are detected, where TAVI is the pacemaker atrioventricular delay (default value: 150 ms). The second parameter, TURI, sets an upper bound on the heart rate. In particular, it is the amount of time that the pacemaker waits before pacing the ventricle, after an atrial stimulus has occurred and TAVI elapsed. We set the domain of both parameters to [10, 2000] ms, and add constraints to exclude from the search pacemaker parameters that are not physiologically meaningful: we require TLRI ≥ TURI, TLRI > TAVI and TURI ≥ TAVI. Note that the approach can be applied also to other pacemaker parameters: TAVI, TVRP (ventricular refractory period), TPVARP (post-ventricular atrial refractory period) and TPVABP (post-ventricular atrial blanking period).

Figure 6 summarizes the synthesis results obtained with the following heart conditions: *bradycardia*, i.e. slow heart rate, reproduced through an increased SA node firing rate (SA_d = 1500 ms, i.e. 40 BPM), and the *AV conduction defect* obtained by increasing the AV delay (AVD_{min} = 150 ms, default: 50 ms). In the experiments we consider a path length of 20 for solving the inner problem and solve the outer problem with both exact and ES methods.

The two experiments return similar robustness radii: $\epsilon = 240$ for bradycardia and $\epsilon = 250$ for AV defect. In the bradycardia case, we obtain TLRI = 770 ms, i.e. a pacing rate in the atrium of 77.92 BPM, for all objectives and solution methods for the outer problem. In the AV defect case, the synthesis experiments yield a similar TLRI value (750 ms, i.e. 80 BPM) and optimal cardiac output. However, the energy consumption of the pacemaker is much higher since, with this heart condition, impulses from the atrium are not correctly propagated, and thus a higher number of paces is required in the ventricle. We remark that, with our method, we are able to find the parameters that guarantee a safe heart rhythm despite large perturbations. For instance, the exact solution γ_o to the AV defect and energy experiment is TLRI = 750 ms and TURI = 480 ms, which

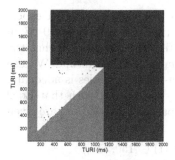

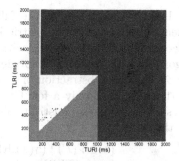

(a) Bradycardia. Inner problem solution time: 7354 s, $\epsilon = 250$ ms.

(b) AV defect. Inner problem solution time: 6601 s, $\epsilon = 240$ ms.

Outer objective:	Energy				Cardiac Output			
Condition:	Bradycardia		AV defect		Bradycardia		AV defect	
Method:	Exact	ES	Exact	ES	Exact	ES	Exact	ES
Best:	770,300	770,640	750,480	750,320	770,320	770,320	750,630	750,350
Cost:	158	158	400	400	9.14	9.14	9.37	9.37
Runtime:	2369	1101	913	1268	1547	118	848	111

Fig. 6. Unsafe regions (red areas and dots) returned by Algorithm 1 in the two experiments. Grey areas indicate pacemaker parameters that are not physiologically relevant and thus are excluded from the search space. The table shows the results of the outer optimisation for the energy and cardiac output objectives (see Sect. 4), comparing the exact and the ES-based methods. The best solutions are in the format TLRI, TURI. Runtimes are in seconds. ES parameters are $\lambda = 100$, $\mu = 50$, $\rho = 2$ and 50 generations (Color figure online).

implies that safety holds for all the parameters in $B_\epsilon(\gamma_o)$, i.e., with $\epsilon = 250$, for all TLRI $\in [500, 1000]$ ms and TURI $\in [230, 730]$ ms. For all our results, we observe that the nominal parameter values (TLRI $= 1000$ ms and TURI $= 500$ ms [36]) are included in $B_\epsilon(\gamma_o)$, meaning that the default pacemaker settings are safe but have a smaller tolerance.

Notably, the evolutionary approach is able to yield the same optimal objective value as the exact method. This is due to the fact that, with two parameters, the solution space of the inner problem (which corresponds to the domain of the outer problem) is quite small. Indeed, we obtain only 107 ϵ-robust valuations for bradycardia, and 52 for the AV defect. With the ES algorithm, we also achieve better performance in most cases, and the runtime improvement becomes even more marked with higher-dimensional parameter spaces, as reported in [30]. The only exception is the runtime obtained for the energy objective in the AV

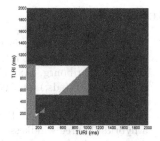

Fig. 7. Full Unsafe region for bradycardia (see Fig. 6(a)).

defect experiment, where the exact method performs slightly better than ES, which is explained by the small number of feasible points in the outer problem.

In Fig. 7, we illustrate the full synthesis region for the bradycardia experiment, obtained without restricting the search space for CEs (lines 9–13 of Algorithm 1). By comparing with the region in Fig. 6(a), we observe that the algorithm explores considerably fewer CEs, thus improving the runtime. We also report that the abstraction of real-valued and non-linear variables is adequate, in the sense that only a few CEs are *potentially* spurious, i.e. such that Eq. 2 does not hold. These constitute only 0.31 % of the parameter space and are given by the set

$$\{\gamma \in \mathcal{V}(\varGamma) \mid (\gamma(\text{TLRI}) = 510 \land \gamma(\text{TURI}) \in [150, 510])$$
$$\lor (\gamma(\text{TLRI}) = 1020 \land \gamma(\text{TURI}) \in [150, 1010])\}$$

With our approach, we can also synthesise parameters that are safe for a range of possible heart conditions, thus taking into account uncertainty in the heart dynamics [30].

7   Conclusion and Future Work

We have studied the problem of robust optimal parameter synthesis for networks of TIOAs with priorities and data and proposed a solution based on SMT solving and evolutionary strategies. We have applied the method to synthesise pacemaker parameters that are both safe and robust, while optimising energy consumption or cardiac output. As the main property specification language, we have considered the safety and reachability fragments of CMTL, which are sufficient to express relevant properties for cardiac pacemakers.

As future work we plan to apply the approach to additional safety properties, validate synthesis results with cardiologists and include advanced pacemaker features like rate-modulation [29], hysteresis and a battery model for optimising the expected lifetime of the device.

Acknowledgments. This work is supported by the ERC AdG VERIWARE and ERC PoC VERIPACE.

References

1. André, É., Chatain, T., Fribourg, L., Encrenaz, E.: An inverse method for parametric timed automata. Int. J. Found. Comput. Sci. **20**(05), 819–836 (2009)
2. André, É., Fribourg, L.: Behavioral cartography of timed automata. In: Kučera, A., Potapov, I. (eds.) Reachability Problems. LNCS, vol. 6227, pp. 76–90. Springer, Heidelberg (2010)
3. Arai, T., Lee, K., Cohen, R.J.: Cardiac output and stroke volume estimation using a hybrid of three windkessel models. In: EMBC, pp. 4971–4974. IEEE (2010)
4. Armando, A., Mantovani, J., Platania, L.: Bounded model checking of software using SMT solvers instead of SAT solvers. STTT **11**(1), 69–83 (2009)
5. Barbot, B., Kwiatkowska, M., Mereacre, A., Paoletti, N.: Estimation and verification of hybrid heart models for personalised medical and wearable devices. In: Roux, O., Bourdon, J. (eds.) CMSB 2015. LNCS, vol. 9308, pp. 3–7. Springer, Heidelberg (2015)

6. Beyer, H.-G., Schwefel, H.-P.: Evolution strategies-a comprehensive introduction. Nat. Comput. **1**(1), 3–52 (2002)
7. Bozzelli, L., La Torre, S.: Decision problems for lower/upper bound parametric timed automata. FMSD **35**(2), 121–151 (2009)
8. Bruyère, V., Raskin, J.-F.: Real-time model-checking: parameters everywhere. In: Pandya, P.K., Radhakrishnan, J. (eds.) FSTTCS 2003. LNCS, vol. 2914, pp. 100–111. Springer, Heidelberg (2003)
9. Chen, T., Diciolla, M., Kwiatkowska, M., Mereacre, A.: Quantitative verification of implantable cardiac pacemakers over hybrid heart models. Inf. Comput. **236**, 87–101 (2014)
10. Cimatti, A., Mover, S., Tonetta, S.: SMT-based verification of hybrid systems. In: Hoffmann, J., Selman, B. (eds.) Proceedings of the Twenty-Sixth AAAI Conference on Artificial Intelligence, 22–26 July 2012, Toronto, Ontario, Canada. AAAI Press (2012)
11. Clarke, E., Grumberg, O., Jha, S., Lu, Y., Veith, H.: Counterexample-guided abstraction refinement. In: Emerson, E.A., Sistla, A.P. (eds.) CAV 2000. LNCS, vol. 1855. Springer, Heidelberg (2000)
12. Colson, B., Marcotte, P., Savard, G.: An overview of bilevel optimization. Ann. Oper. Res. **153**(1), 235–256 (2007)
13. de Moura, L., Bjørner, N.S.: Z3: an efficient SMT solver. In: Ramakrishnan, C.R., Rehof, J. (eds.) TACAS 2008. LNCS, vol. 4963, pp. 337–340. Springer, Heidelberg (2008)
14. Deb, K.: An efficient constraint handling method for genetic algorithms. Comput. Methods Appl. Mech. Eng. **186**(2), 311–338 (2000)
15. Diciolla, M., Kim, C.H.P., Kwiatkowska, M., Mereacre, A.: Synthesising optimal timing delays for timed I/O automata. In: EMSOFT 2014. ACM (2014)
16. Doyen, L.: Robust parametric reachability for timed automata. Inf. Process. Lett. **102**(5), 208–213 (2007)
17. Fazeli, N., Hahn, J.-O.: Estimation of cardiac output and peripheral resistance using square-wave-approximated aortic flow signal. Front. Physiol **3**, 736–743 (2012)
18. Gao, S., Kong, S., Clarke, E.M.: Satisfiability modulo ODEs. In: Formal Methods in Computer-Aided Design (FMCAD), pp. 105–112. IEEE (2013)
19. Gomes, A.O., Oliveira, M.V.M.: Formal specification of a cardiac pacing system. In: Cavalcanti, A., Dams, D.R. (eds.) FM 2009. LNCS, vol. 5850, pp. 692–707. Springer, Heidelberg (2009)
20. Gulwani, S., Tiwari, A.: Constraint-based approach for analysis of hybrid systems. In: Gupta, A., Malik, S. (eds.) CAV 2008. LNCS, vol. 5123, pp. 190–203. Springer, Heidelberg (2008)
21. Jiang, Z., Pajic, M., Moarref, S., Alur, R., Mangharam, R.: Modeling and verification of a dual chamber implantable pacemaker. In: Flanagan, C., König, B. (eds.) TACAS 2012. LNCS, vol. 7214, pp. 188–203. Springer, Heidelberg (2012)
22. Jovanović, A., Kwiatkowska, M.: Parameter synthesis for probabilistic timed automata using stochastic game abstractions. In: Ouaknine, J., Potapov, I., Worrell, J. (eds.) RP 2014. LNCS, vol. 8762, pp. 176–189. Springer, Heidelberg (2014)
23. Jovanović, A., Lime, D., Roux, O.H.: Integer parameter synthesis for timed automata. In: Piterman, N., Smolka, S.A. (eds.) TACAS 2013. LNCS, vol. 7795, pp. 401–415. Springer, Heidelberg (2013)
24. Kerner, D.R.: Solving windkessel models with mlab (2007). http://www.civilized.com/mlabexamples/windkesmodel.htmld

25. Kindermann, R., Junttila, T., Niemelä, I.: Beyond lassos: complete SMT-based bounded model checking for timed automata. In: Giese, H., Rosu, G. (eds.) FORTE 2012 and FMOODS 2012. LNCS, vol. 7273, pp. 84–100. Springer, Heidelberg (2012)
26. Kindermann, R., Junttila, T., Niemelä, I.: SMT-based induction methods for timed systems. In: Jurdziński, M., Ničković, D. (eds.) FORMATS 2012. LNCS, vol. 7595, pp. 171–187. Springer, Heidelberg (2012)
27. Knapik, M., Penczek, W.: Bounded model checking for parametric timed automata. In: Jensen, K., Donatelli, S., Kleijn, J. (eds.) ToPNoC V. LNCS, vol. 6900, pp. 141–159. Springer, Heidelberg (2012)
28. Kovásznai, G., Fröhlich, A., Biere, A.: On the complexity of fixed-size bit-vector logics with binary encoded bit-width. In: SMT, pp. 44–56 (2012)
29. Kwiatkowska, M., Lea-Banks, H., Mereacre, A., Paoletti, N.: Formal modelling and validation of rate-adaptive pacemakers. In: ICHI, pp. 23–32. IEEE (2014)
30. Kwiatkowska, M., Mereacre, A., Paoletti, N., Patanè, A.: Synthesising robust and optimal parameters for cardiac pacemakers using symbolic and evolutionary computation techniques. Technical Report RR-15-09, Department of Computer Science, University of Oxford (2015)
31. Lian, J., Krätschmer, H., Müssig, D., Stotts, L.: Open source modeling of heart rhythm and cardiac pacing. Open Pacing Electrophysiol. Ther. J. 3, 4 (2010)
32. Méry, D., Singh, N.K.: Closed-loop modeling of cardiac pacemaker and heart. In: Weber, J., Perseil, I. (eds.) FHIES 2012. LNCS, vol. 7789, pp. 151–166. Springer, Heidelberg (2013)
33. Ouaknine, J., Worrell, J.: On the decidability of metric temporal logic. In: Proceedings of the 20th Annual IEEE Symposium on Logic in Computer Science, LICS 2005, pp. 188–197. IEEE (2005)
34. Rabinovich, A.: Complexity of metric temporal logics with counting and the pnueli modalities. Theor. Comput. Sci. 411(22), 2331–2342 (2010)
35. Rudolph, G.: An evolutionary algorithm for integer programming. In: Davidor, Y., Schwefel, H.-P., Männer, R. (eds.) PPSN III. LNCS, vol. 866, pp. 139–148. Springer, Heidelberg (1994)
36. Boston Scientific: Pacemaker System Specification. Boston Scientific, Boston (2007)
37. Sturm, T., Tiwari, A.: Verification and synthesis using real quantifier elimination. In: Proceedings of the 36th International Symposium on Symbolic and Algebraic Computation, pp. 329–336. ACM (2011)
38. Traonouez, L.-M.: A parametric counterexample refinement approach for robust timed specifications. arXiv preprint arXiv:1207.4269 (2012)

Model-Based Whole-Genome Analysis of DNA Methylation Fidelity

Christoph Bock[3,4,5], Luca Bortolussi[1,2(✉)], Thilo Krüger[1],
Linar Mikeev[1], and Verena Wolf[1]

[1] Modelling and Simulation Group, University of Saarland, Saarbrücken, Germany
{thilo.krueger,linar.mikeev,verena.wolf}@uni-saarland.de
[2] DMG, University of Trieste, Trieste, Italy
luca@dmi.units.it
[3] CeMM Research Center for Molecular Medicine of the Austrian
Academy of Sciences, Vienna, Austria
cbock@cemm.oeaw.ac.at
[4] Department of Laboratory Medicine, Medical University of Vienna, Vienna, Austria
[5] Max Planck Institute for Informatics, Saarbrücken, Germany

Abstract. We consider the problem of understanding how DNA methylation fidelity, i.e. the preservation of methylated sites in the genome, varies across the genome and across different cell types. Our approach uses a stochastic model of DNA methylation across generations and trains it using data obtained through next generation sequencing. By training the model locally, i.e. learning its parameters based on observations in a specific genomic region, we can compare how DNA methylation fidelity varies genome-wide. In the paper, we focus on the computational challenges to scale parameter estimation to the whole-genome level, and present two methods to achieve this goal, one based on moment-based approximation and one based on simulation. We extensively tested our methods on synthetic data and on a first batch of experimental data.

Keywords: DNA methylation · Epigenomics · Branching processes · Parameter estimation · Next generation sequencing

1 Introduction

Epigenetic marks such as DNA methylation provide a mechanism by which cells can control gene activity in a manner that is heritable between cell generations and adaptive to external stimuli [3]. Biochemically, DNA methylation is a covalent modification of the DNA. In vertebrates, DNA methylation occurs

---

L.B., T.K., L.M., and V.W. are partially funded by the German Research Council (DFG) as part of the Cluster of Excellence on Multimodal Computing and Interaction at Saarland University and the Collaborative Research Center SFB 1027. C.B. was supported by a New Frontiers Group award of the Austrian Academy of Sciences.

© Springer International Publishing Switzerland 2015
A. Abate and D. Šafránek (Eds.): HSB 2015, LNBI 9271, pp. 141–155, 2015.
DOI: 10.1007/978-3-319-26916-0_8

almost exclusively in the context of a cytosine (C) followed by a guanine (G). These so-called CpG sites are palindromic and can carry one DNA methylation group on each strand. As the result, a single CpG site can be symmetrically unmethylated, asymmetrically methylated on either the forward or the reverse strand of the DNA (hemimethylated), or symmetrically methylated on both strands. In most cases, DNA methylation is symmetrical, which provides redundant information on both strands. When cells divide and the DNA is copied in a semi-conservative manner (i.e., each daughter cell receives one strand of the double-stranded DNA molecule), the DNA methylation on the newly synthesized strand can be reconstructed from the DNA methylation on the conserved DNA strand. The process of copying DNA methylation patterns is called maintenance methylation.

Compared to the very high efficiency with which the DNA sequence is copied and maintained during cell division (typically with error rates in the order of 10^{-8}), the fidelity of DNA methylation maintenance is much lower. Based on single-locus studies, error rates have been estimated to be in the order of 10^{-2} to 10^{-3}. To maintain high DNA methylation levels in specific regions of the genome despite the relatively low fidelity of maintenance DNA methylation, cells utilize a second mechanism called de novo methylation to methylate previously unmethylated cytosines independent of the DNA methylation status of the second cytosine within a CpG site. The rates of de novo methylation during normal cell growth are relatively low, but they appear to be sufficient to compensate for the gradual loss of DNA methylation that would normally result from the limited fidelity of DNA methylation maintenance.

Comprehensive genome-wide assessments of the fidelity of DNA methylation and of the de novo DNA methylation rate and their comparison between different cell types have been lacking, and prior work has focused on small parts of the genome. With genome-wide methods for DNA methylation mapping and analysis [4,5], even in single cells [7], it is now possible to collect comprehensive datasets to estimate these important biological parameters in a genome-wide manner and to systematically search for differences between cell types. In this study, we address the computational challenges of inferring these parameters in a manner that is sufficiently high-throughput and scalable to support the genome-wide analysis of large numbers of samples.

The assumed experimental design is as follows: A single cell is isolated and left to grow exponentially over n generations, typically $n = 20$. The resulting cell population (approximately $2^{20} \approx 1$ million cells) is subjected to genome-scale bisulfite sequencing, the reads are aligned to the reference genome, and for each CpG site the number of methylated and unmethylated reads are counted for a subsample of the population (typically 10 to 100 measurements per CpG, with several million assayed CpGs). In these experiments, we are interested in the dynamic nature of the methylation process, in particular how it is propagated through the n cell generations. In order to understand this behaviour, we need a mathematical model of the methylation fidelity and of the de novo methylation rate, which must be trained using the experimental data available. In building the model, our main goal was to predict its parameters from data, which are directly connected to de novo methylation and fidelity probabilities. The model we construct is based on those proposed in [2,12], and it is a discrete-time Markov chain describing how the methylation progresses through generations at the population level (i.e. counting how many cells have a specific CpG site are unmethylated, hemi-methylated, or fully methylated).

The main challenge with this model is computational, as we need to perform the parameter estimation task genome-wide, for each CpG site and each biological sample. The model cannot be solved analytically, and it is too large for being solved with standard numerical techniques. Hence, we engineered two different strategies for computing the likelihoods required to train the model parameters, one based on an analytical approximation, and the other based on simulation. In this paper, we present and compare the two approaches, both theoretically and experimentally.

To validate the accuracy of the presented methods we simulated test data using our model, and we estimated the parameters for these test data sets. Furthermore, we used our the methods to estimate the parameters in real, experimentally derived, data sets.

The paper is organised as follows: in Sect. 2, we discuss the mathematical model, and in Sect. 3 we present the parameter estimation techniques. Preliminary results are shown and explained in detail in Sect. 4, and conclusions are drawn in Sect. 5.

2   Stochastic Model of DNA Methylation

We propose a stochastic model for the dynamics of DNA methylation of a cell population over a certain number of cell divisions. This model is an extension of previous models that have described the average state of a single CpG site within a certain cell population [2,12].

Single Cell Model. To describe the DNA methylation dynamics of a single CpG site, we consider three possible site states: unmethylated on both DNA-strands (*unmethylated, U*), methylated on both strands (*fully methylated, F*) or methylated on one out of the two strands (*hemimethylated, H*). This naturally leads to a (discrete-time) Markov model description where one time step corresponds to one cell cycle or the time between two cell divisions (in cultured cells, this time is often on the order of 24 h). Over the course of one cell cycle, the DNA methylation state of the CpG site changes in three phases. In phase one, the two strands of DNA are separated such that each daughter cell receives one strand, and a complementary stand is synthesised. This complementary strand is always unmethylated, such that this step dilutes the DNA methylation levels compared to the parent cell. Thus, the transitions for this phase are from U to U and from F to H with probability one, respectively, as well as from H to H or to U with probability 0.5, respectively (Fig. 1, left). In the second phase, which occurs during and after the synthesis of the new strand, a special class of enzymes, called DNA methyltransferases (DNMTs), try to maintain the pattern of the mother strand by methylating hemimethylated CpG sites. Maintenance methylation is a stochastic process [2], such that the state of a site changes from H to F with probability f_m and from U to U with probability f_u. Successful maintenance typically occurs with a relatively high probability. However, in both cases maintenance might fail with probability $1 - f_m$ (transition from H to H) and with probability $1 - f_u$ (transition from U to H). The third phase, which lasts from the end of a cell division to the beginning of the next, allows for de novo methylation, where methyl groups are transferred by DNMTs to sites that are in state U or H. Here, the assumption is that de novo methylation occurs at a given site and strand independently of the DNA methylation state of the CpG on the other strand [2]. Thus, with probability μ the state changes either from H to F or from U to H (Fig. 1, left). Note that we neglect the extremely rare transition from U to F through de novo methylation, in order to keep the model simple. Simulations of the model show no significant differences if the transition from U to F due to de novo methylation is added (results not shown).

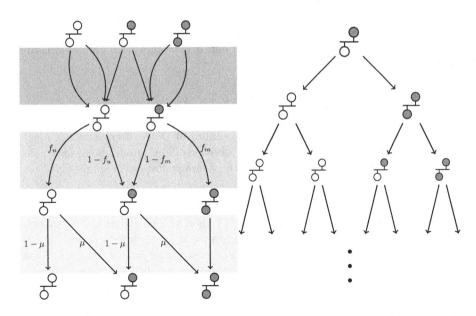

Fig. 1. The three phases of the single cell model (left) with cell-division (dark grey, the division of single cells into two new cells is represented by two arrows.), maintenance methylation (grey), and de novo methylation (light grey) as well as the population model (right) over 20 generations

Population Model. In the proposed population model we consider a fixed CpG site in a single allele, thus modelling independently CpG sites in different alleles, and start initially with a single cell. Then, after the next time step we consider all daughter cells (two at time $t = 1$, four at time $t = 2$, etc.). Thus, the resulting process is a branching process and after 20 generations we have to consider $2^{20} \approx 10^6$ cells. In order to describe the state of the CpG site in the cell population at time t, we compute the probabilities that, when a site in the parent cell is in state X, the two states of the daughter cells are Y and Z, where $X, Y, Z \in \{U, H, F\}$, according to the following matrix M

$$M = \begin{array}{c} \\ U \\ H \\ F \end{array} \begin{pmatrix} UU & UH & UF & HH & HF & FF \\ t_{UU}^2 & 2 \cdot t_{UU} t_{UH} & 2 \cdot t_{UU} t_{UF} & t_{UH}^2 & 2 \cdot t_{UH} t_{UF} & t_{UF}^2 \\ 0 & t_{UU} t_{HH} & t_{UU} t_{HF} & t_{UH} t_{HH} & t_{UF} t_{HH} + t_{UH} t_{HF} & t_{UF} t_{HF} \\ 0 & 0 & 0 & t_{HH}^2 & 2 \cdot t_{HH} t_{HF} & t_{HF}^2 \end{pmatrix},$$

with $(t_{UU}, t_{UH}, t_{UF}, t_{HH}, t_{HF}) = (f_u \cdot (1 - \mu), (1 - f_u)(1 - \mu) + f_u \mu, (1 - f_u)\mu,$ $(1 - f_m)(1 - \mu), f_m + (1 - f_m)\mu)$. Note that the entries for column UH considers the two symmetric cases that either the site in one daughter cell is in state U and the other one in state H or vice versa. The same holds for the columns UF and HF. The entries of M are computed by considering the corresponding paths in the diagram in Fig. 1, left. Consider for example the entry $M_{H,HF} = t_{UF} t_{HH} + t_{UH} t_{HF}$. During cell division a site in state H is divided into a site in state U and a site in state H (upper grey block in Fig. 1, left). Next we have to consider the paths from the nodes in the second line to those in the last line, i.e. from U to F ($t_{UF} = (1 - f_u) \cdot \mu$) and from H to H ($t_{HH} = (1 - f_m)(1 - \mu)$). Analogously, we consider the paths from U to H and from H to F which yields the second term $t_{UH} t_{HF}$.

In Fig. 1, right, we illustrate a trace of the population model over time. Assuming the state of the CpG site is H in the initial cell, then the two daughter cells of the next generation could be in state U and F, while in the following generation we could have states U,U,H and F, etc. After $n = 20$ generations, in the final population there are 2^{20} cells, in each of them, the state of the tracked CpG site will be in one of the three states. Note that we apply matrix M to all cells of the current generation to determine the possible daughter cells. In addition, we assume that DNA methylation in a cell and on a given allele occurs independently of other cells and alleles.

The model we are considering belongs to the well-known family of multi-type Galton-Watson branching process [8], giving us the possibility of exploiting the vast theory developed for them.

Our wet-lab data contain only information about how many methylated ($m = m(n)$) and unmethylated ($u = u(n)$) single strand sites are present in a subset of the final population (after 20 generations). However, this means that for the unknown full state we have the relationships

$$m(n) = Y_H(n) + 2 \cdot Y_F(n) \tag{1}$$
$$u(n) = Y_H(n) + 2 \cdot Y_U(n) \tag{2}$$

if $Y_U(n), Y_H(n)$, and $Y_F(n)$ are the numbers of unmethylated, hemimethylated, and fully methylated CpG sites in the final population. As $n = 20$ will be fixed, we will often omit this index it in the following.

3 Parameter Estimation

Since our ultimate goal is to do whole-genome studies and apply our model to different cell types, we are interested in parameter estimation procedures that are computationally efficient. Our model is parametric in $\Theta = (f_m, f_u, \mu)$ as well as in the state of the initial cell. In order to estimate these parameters we use a maximum likelihood approach and compute the likelihood of observing m and u (single strand observations). Computing this exactly is computationally very expensive due to the large size of the state space and the stiffness of the model (f_m and f_u being close to one). In the following we present two methods to approximate the maximum likelihood and estimate such parameters. The first one is based on stochastic simulations of the model and a statistical estimation of the likelihood of the observed data. The other approach uses a moment-based numerical method to approximate the likelihood.

Description of data. The wet-lab DNA methylation data comprise lists of integer pairs $\lambda = (u_e, m_e)$. Each pair describes the DNA methylation measurements for a given CpG site e, where one of the strands was observed u_e times unmethylated and m_e times methylated (the experimental setup does not allow to distinguish between upper and lower strand). Since it is known that certain groups of CpG sites behave similarly we will also use our model to describe the average behaviour of a CpG site within such a group and collect all observation pairs for these sites in a set Λ. If we only consider a single CpG site e, then $\Lambda = \{(u_e, m_e)\}$.

Likelihood for single data pairs. Consider a possible state of the model $\boldsymbol{Y} = \boldsymbol{Y}(n) = (Y_U(n), Y_H(n), Y_F(n))$ after $n = 20$ generations, whose entries sum up to $Y_U + Y_H + Y_F = 2^{20}$. From this vector it is possible to compute the numbers m and u of methylated and unmethylated strands (see Eqs. 1 and 2). In the following, we use m_X and u_X to denote the number of methylated and unmethylated strands conditional on the site of the initial cell being in state $X \in \{U, H, F\}$ and we define the two relative frequencies $p_{uX} = \frac{u_X}{u_X + m_X}$ and $p_{mX} = \frac{m_X}{u_X + m_X} = 1 - p_{uX}$.

We now want to compute the likelihood that a certain data pair $\lambda = (u_e, m_e)$ is observed given the parameter set Θ. We assume that the measured cells are randomly chosen and therefore we can reduce the computation of the emission probabilities to the urn problem (drawing with replacement). Hence, the emission probability for observation λ, if the final state is $\boldsymbol{y} = (y_U, y_H, y_F)$, is given by

$$P_{X,\Theta}(\lambda \mid \boldsymbol{y}) = \binom{m_e + u_e}{m_e} (p_{uX})^{u_e} \cdot (p_{mX})^{m_e}. \qquad (3)$$

Thus, the likelihood of the observation λ, conditional on a given initial state $X \in \{U, H, F\}$ and the parameters Θ, is

$$P(\lambda \mid X, \Theta) = \sum_{\boldsymbol{y}} P(\boldsymbol{Y}(n) = \boldsymbol{y}) \cdot P_{X,\Theta}(\lambda \mid \boldsymbol{y}) = \mathbb{E}_{\boldsymbol{Y}}[P(\lambda \mid X, \boldsymbol{Y})]. \qquad (4)$$

The exact computation of the expectation $\mathbb{E}_{\boldsymbol{Y}}[P(\lambda \mid X, \boldsymbol{Y})]$ requires the knowledge of the probability of each possible final stage $\boldsymbol{y}$, which is computationally expensive. Therefore, we approximate such an expectation in two possible ways, either statistically relying on simulations of the model, or by stochastic approximation.

Simulation-based approach. An estimator for the likelihood $\mathbb{E}_{\boldsymbol{Y}}[P(\lambda \mid X, \boldsymbol{Y})]$ can be obtained by taking the sample mean of the emission probabilities of all trajectories.[1] We compute the sample mean that approximates the likelihood by generating 10000 trajectories using the method explained below. Note that during the optimisation process, we vary this number for performance reasons.

To generate a trajectory of the process, we use a standard simulation algorithm for discrete time Markov chains. The simulation is initialized by computing the matrix M (see Sect. 2) as well as setting the initial state $\boldsymbol{Y}(0)$, i.e. the state of one site of the single initial cell of the zeroth generation. Then in each step we determine the state of the site of the next generation according to the distributions in M. Instead of repeatedly generating the two daughter cells for all parent cells, we draw samples from a multinomial distribution according to the number of sites in state $X \in \{U, M, F\}$ and the probabilities in the corresponding line of matrix M. For instance, if the state of the initial site is H (see Fig. 1, right) we set the initial counting vector $\boldsymbol{Y}(0) = (Y_U(0), Y_H(0), Y_F(0)) = (0, 1, 0)$. In the next step a new counting vector, say $\boldsymbol{Y}(1) = (1, 0, 1)$ as in Fig. 1, right, is determined according to the multinomial distribution as described above. We iterate this process until we reach generation $n = 20$, thus obtaining a sample of the final state $\boldsymbol{Y}(n)$.

Moment-based approach. An alternative to simulation is to try to approximate the likelihood $\mathbb{E}_{\boldsymbol{Y}}[P(\lambda \mid X, \boldsymbol{Y})]$ by resorting to ideas of stochastic approximation. Our approach is conceptually simple: first, we compute the first two moments of the distribution of $\boldsymbol{Y} = \boldsymbol{Y}(n)$, conditional on the initial site state being $X \in \{U, H, F\}$, namely its mean $\mathbf{e}^X = \mathrm{E}[\boldsymbol{Y}(n)]$, and the covariance matrix $\mathbf{C}^X = (C_{ij})$, $C_{ij} = \mathrm{Cov}[Y_i(n), Y_j(n)]$, $i, j \in \{U, H, F\}$. Then, we assume $\boldsymbol{Y}$ takes continuous values rather than integer ones, and invoke the maximum entropy principle [1] to approximate it by a 2-dimensional normal distribution with mean $\mathbf{e}^X$ and covariance matrix $\mathbf{C}^X$ (we can get rid of one dimension exploiting the fact that the population at generation n equals 2^n). By letting $f_{\mathbf{e}^X, \mathbf{C}^X}$ be the corresponding normal density, we then have

$$P(\lambda \mid X, \Theta) \approx \int_{u,h} \binom{m_e + u_e}{m_e} \left(\frac{y_U + 0.5 y_H}{2^{20}} \right)^{u_e} \left(\frac{1 - y_U - 0.5 y_H}{2^{20}} \right)^{m_e} f_{\mathbf{e}^X, \mathbf{C}^X}(\boldsymbol{y}) \, d\boldsymbol{y}.$$

---

[1] Note that the emission probabilities are dependent on the relative frequencies p_{uX} and p_{mX}, which are random variables as they depend on the random quantities u_X and m_X.

This integral is then numerically approximated by using the two-dimensional Simpson's rule [6].

In order to compute mean and covariance of $\boldsymbol{Y}(n) = (Y_U(n), Y_H(n), Y_F(n))$, $n = 0 \ldots 20$, we exploit the fact that $\boldsymbol{Y}$ is a multi-type Galton-Watson branching process [8]. Following [11], we define the expectation matrix $\mathbf{M}$ with elements

$$M_{ij} = \mathrm{E}\left[Y_j(1)|Y(0) = \mathbf{b}^i\right], \tag{5}$$

where $i, j \in \{U, H, F\}$ and $\mathbf{b}^U = (1, 0, 0)$, $\mathbf{b}^H = (0, 1, 0)$, $\mathbf{b}^F = (0, 0, 1)$. We also define the covariance matrices $\mathbf{V}^k, k \in \{U, H, F\}$ such that

$$V_{ij}^k = \mathrm{Cov}\left[V_i(1), V_j(1)\mid Y(0) = \mathbf{b}^k\right]. \tag{6}$$

Then, the following recurrence holds [11]:

$$[\mathbf{e}(n+1)\,\mathcal{C}(n+1)] = [\mathbf{e}(n)\,\mathcal{C}(n)]\begin{bmatrix} \mathbf{M} & \begin{matrix} \mathcal{V}^U \\ \mathcal{V}^H \\ \mathcal{V}^F \end{matrix} \\ \hline \mathbf{0} & \mathbf{M} \times \mathbf{M} \end{bmatrix} = [\mathbf{e}(n)\,\mathcal{C}(n)]\mathbf{T},$$

where $\mathbf{M} \times \mathbf{M}$ is the Kronecker product, $\mathcal{C}(n) = (C_{UU}(n), C_{UH}(n), C_{UF}(n), C_{HU}(n), C_{HH}(n), C_{HF}(n), C_{FU}(n), C_{FH}(n), C_{FF}(n))$ and $\mathcal{V}^i = (V_{UU}^i, V_{UH}^i, V_{UF}^i, V_{HU}^i, V_{HH}^i, V_{HF}^i, V_{FU}^i, V_{FH}^i, V_{FF}^i)$. For each initial state $k \in \{U, H, F\}$ we also compute

$$[\mathbf{e}^k\,\mathcal{C}^k] = [\mathbf{b}^k\,\mathbf{0}]\mathbf{T}^{20}.$$

Estimating the initial state. The previously discussed approach allows us to compute the likelihood for a single pair λ conditional on the initial state $X \in \{U, H, F\}$. In order to estimate such an initial configuration, we consider the estimated likelihoods $P(\lambda \mid X, \Theta)$ in a Bayesian context. We start by assuming a prior distribution $P(X \mid \Theta)$ over the initial states, and then compute the posterior distribution $P(X \mid \lambda, \Theta)$ according to Bayes theorem as

$$P(X \mid \lambda, \Theta) = \frac{P(\lambda \mid X, \Theta) \cdot P(X \mid \Theta)}{\sum_{X \in \{U, H, F\}} P(\lambda \mid X, \Theta) \cdot P(X \mid \Theta)}.$$

In order to fix the prior, we need to take into account that it is unlikely that the original cell has a hemimethylated site (which is very uncommon for living cells), so the prior should give it a small probability for $X = H$. Our solution is to consider as prior probability the state of the model after one generation, starting from the distribution $(U\ H\ F) = (0.5\ 0\ 0.5)$. For instance, we have $P(H \mid \Theta) = t_{UH}(t_{UU} + t_{UH} + t_{UF}) + t_{HH}(t_{HH} + t_{HF})$ (see also Sect. 2). Then, we can compute the model likelihood, for a given λ, independent of initial conditions, as $P(\lambda \mid \Theta) = \sum_{X \in \{U, H, F\}} P(\lambda \mid X, \Theta) \cdot P(X \mid \Theta)$.

Likelihood optimisation. The model likelihood for all data pairs $\Lambda = \{\lambda_1, \lambda_2, \ldots\}$ is finally obtained by taking the product of the likelihood of all individual pairs. By taking the logarithm, the model log-likelihood then is

$$\log(P(\Lambda \mid \Theta)) = \sum_{\lambda \in \Lambda} \log(P(\lambda \mid \Theta)).$$

To estimate the parameters we used a simple maximum likelihood approach. We computed the likelihoods $-\log(P(\Lambda \mid \Theta))$ for varying Θ and converged to a minimum using simple optimisation procedures. In the final version, we use the *Nelder-Mead* procedure which is a derivative-free optimisation that performed best in our tests [10].

4 Results

In order to validate the proposed estimation algorithms, we ran detailed tests with simulated data (Sect. 4.1). We also present preliminary results of the whole genome analysis based on real experimental data (Sect. 4.2).

4.1 Results for Simulated Data

Generation of Synthetic Data. In order to simulate realistic experimental data with our model, we need two additional parameters governing the behaviour of the experiment: *coverage*, which is the average number of measurements per CpG site, and *length*, which is the number of CpG sites in the simulated dataset Λ_{sim}. Given such information, synthetic experimental data is generated according to Algorithm 1. In order to vary the coverage and keep the variance of the coverage as realistic as possible, we determine for a fixed coverage the number of measurements per CpG site in such a way that it resembles this number in the truly measured data.[2]

1: prepare a list L_{real} of numbers of measurements per site as follows: choose randomly a sequence of measurement numbers from the real data with *length* entries and compute the average $\overline{C_{real}}$ over all entries of this list
2: set $\Lambda_{sim} = \emptyset$
3: **for** i := 1 **to** *length* **do**
4:    draw probabilistically the initial state X (as described in Sect. 3)
5:    run 20 generations from X
6:    compute $p(methylated) = (\#methylated\ sites)/(\#sites)$
7:    get C_{real} as the ith entry of L_{real}
8:    compute $C = Round((C_{real} \cdot (coverage - 0.5))/(\overline{C_{real}}) + 0.5)$
9:    draw a random number m from a binomial with $p = p(methylated)$ and C
10:   add $\lambda = (m, C - m)$ to Λ_{sim}
11: **end for**
Algorithm 1: Generation of synthetic data.

---

[2] We avoid the number of measurements C to be set to zero by subtracting 0.5 from the reduced coverage and add 0.5 to the quantity to round (Algorithm 1, line 8).

Scanning the Parameter Space of Simulated Data Sets. We first examine the likelihood landscape by deep sampling of the parameter space, fixing the coverage to 5 and the length to 1000, as these values are typical for some of the real data sets considered in the following section. We use the moment-based method described in Sect. 3 to approximate the likelihood.

We consider synthetic data obtained from the model with the arbitrarily chosen values of $\Theta_{sim} = (f_u, f_m, \mu)$ from table of Fig. 3. For each parameter set we generated a data set using Algorithm 1. To get an impression of the likelihood landscapes, we computed the likelihood for a fine grid of the parameter space Θ with the proposed approximative approach for parameter sets 1–3. In Fig. 2 we show the results. For better visualisation purposes, we report 2-dimensional plots. We represent for each pair of parameters the negative log-likelihood as a grey value and restrict to the maximum log-likelihood for each pair of parameters in the plot.

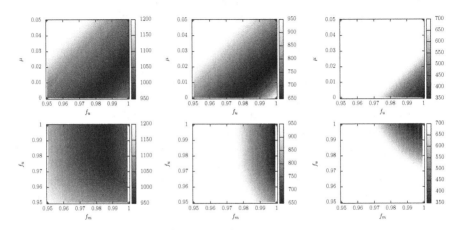

Fig. 2. Likelihood landscape for low coverage data. Parameter set 1 (left), 2 (middle), and 3 (right) from table of Fig. 3.

In all three cases the estimation resulted in a very flat likelihood landscape since a low coverage of 5 was used. For high coverage synthetic data (as we have it in the real data) the estimation was much more accurate (see below). In particular the sum $\alpha := \mu + (1 - f_m) + (1 - f_u)$, which reflects the probability of copy mistakes in the methylation pattern, is very close to the true value. Nevertheless, in the low coverage case for parameter set 1 we find high likelihood values if the sum $\mu + (1 - f_u) < 0.031$, in rough agreement with the value of the parameter set that was used to simulate the data. Also there is a tendency for f_m to be at a value > 0.97. In all three of the upper plots it can be seen that there is a dependency between f_u and μ. Increasing f_u seems to have the same effect on the likelihood as decreasing μ, which makes sense because in both cases the U state is copied more often and it is less often the case that the daughter cell has state H.

The parameters of the second and third set resulted in observations that are similar to real data as maintenance typically occurs with high probability. In the case of parameter set 3 the likelihood increases significantly when crossing the line $\mu = f_u - 0.97$ and becomes maximal at $f_u = 1$ and $\mu = 0.001$, while f_m is estimated to be smaller than 0.999.

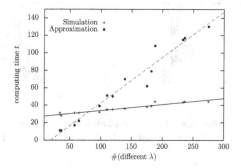

Θ_{sim}	f_u	f_m	μ	α
1	0.999	0.97	0.03	0.061
2	0.999	0.999	0.03	0.032
3	0.999	0.999	0.001	0.003
4	0.97	0.999	0.03	0.061
5	0.97	0.999	0.001	0.032
6	0.999	0.97	0.001	0.032

Fig. 3. Comparison of the performances of the simulation and the approximation approach (left). The parameter sets of the simulated data (right). Note that we also list $\alpha := \mu + (1 - f_m) + (1 - f_u)$ in this table, which is used as an indicator number later.

Comparison of Simulation and Approximation Approaches.

Next we compare the two approaches for approximating the likelihood, namely the simulation-based and the moment-based methods explained in Sect. 3 . We used the same three parameter sets 1–3 from the previous section (see table of Fig. 3) but instead of optimising the third parameter we fix $f_u = 0.999$ and plot the computed likelihood depending on μ and f_m. The results are shown in Fig. 4. It can be seen that the likelihoods obtained by the moment-based method (lower plots) are much smoother, being free from the random effects of the simulation approach (upper plots). Nevertheless, the results of both methods are very similar and the maximum likelihood points in the parameter spaces are very close. For example for parameter set 1, 87 % of the log-likelihoods differ by not more than 10 % from each other. Furthermore, for the plots in the middle of Fig. 4, both methods find as optimal f_m the true value of 0.999, while μ is optimal at 0.023 for the moment-based approach and 0.026 for the simulation approach (true value is 0.03). Note that the optimal parameters that are recovered with the different methods differ more for the plots in Fig. 4, left and less for the plots in Fig. 4, right, due to the different kinds of likelihood landscapes. For performance reasons, we restrict ourselves to the simulative approach for the remainder of the paper. In fact, as soon as there are more than approximatively 80 different data pairs λ in one data set, the numerical method becomes slower than the simulation. For 80 different data pairs both methods need approximately 30 s, while for a data set with 200 different data pairs, the simulation needs 40 s and the numerical method needs 100 s. (see Fig. 3 for a complete comparison). For huge

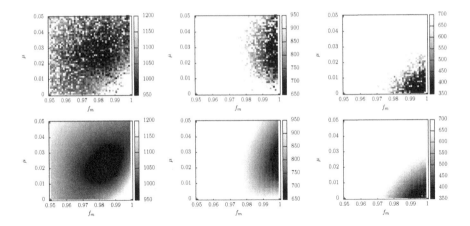

Fig. 4. Comparison between numerical (lower row of plots) and simulative (upper row of plots) approximation of the likelihood: Parameter set 1 (left), 2 (middle), and 3 (right) from table of Fig. 3.

data sets, which are common in real experimental data, the simulation would clearly outperform the numerical approach.

Parameter Estimation for Simulated Data. To explore the quality of our estimation procedure, data for all parameter sets Θ_{sim} listed in table of Fig. 3 with different coverages and lengths were generated. Then, we estimated ten times such parameters with the simulation approach (see Sect. 3). Since we have seen that the model cannot distinguish well between parameter sets with similar $\mu+(1-f_u)$, we concentrate in the following on the sum $\alpha = \mu+(1-f_m)+(1-f_u)$, which reflects the probability of copy mistakes in the methylation pattern. If α is zero, then each site in state U or F will be in state U or F in all daughter cells again. The higher α becomes, the more errors happen during one generation. The results for the (average of the ten) differences $\Delta\alpha$ between the estimated α and the true α with which the simulated data was generated is plotted as a function of the chosen lengths and coverages in Fig. 5.

It can be seen that the coverage of a certain data set plays a crucial role when estimating the parameters of the proposed model. For coverages between 16 and 64, a constant value for $\Delta\alpha$ is reached, which becomes greater again for coverages greater than 100. In contrast, raising the length of a data set is not leading to less accurate parameter estimations. From length 500 on the distances are converging. Since parameter estimation on real data is typically done for large genome regions with many CpG sites and each site produces one data pair of methylation data, it is common to have a data set of at least 500 entries.

The estimated α when only a single observation λ is given, is obviously rather inaccurate. In Fig. 5 we see that the estimation based on ten observations gives much better results. Since we only have one observation for each CpG site in the data of the real system and since CpG sites that belong to the same region typically show similar DNA methylation dynamics, we estimate in the next section the parameters of the average behaviour of a group of several CpG sites whose observations are collected by the set Λ.

4.2   Parameter Estimation for Real Data

The main motivation behind our work is the availability of huge datasets of DNA methylation data that we will use to investigate methylation fidelities, i.e. learn the parameters μ, f_m, and f_u. In the following, we use two data sets with human blood samples and solid tumour samples. We ran the simulation-based parameter estimation procedure for both cell types and examined the differences between the estimated fidelities in blood and in tumour cells. In order to estimate parameters for sets Λ of observations of CpG sites in close proximity, we grouped CpG sites in consecutive ranges of 5000 base pairs into genomic regions and used our model to describe the average behaviour of a site in each region. For the analysis, first a region was identified, the methylation data of this region were extracted from the data of the blood and tumour samples, and if there was information about at least 100 different CpG sites, the parameters were estimated for the extracted data. Figure 6 shows estimated parameters of different regions of chromosome 7. Both plots look broadly similar, but there is a certain number of regions where f_m is reduced in tumour samples. In these regions μ tends to be increased (the points are brighter). To visually investigate this observation, Fig. 7 shows for each region $f_x(cancer)$ as a function of $f_x(blood)$, with $x = m, u$. While f_u is distributed more or less equally over the whole pictured region, f_m is more clustered and in average reduced in cancer-cells. The corresponding plot for μ looks similar to the plot of f_u and is not reported. To validate this visual impression, we tested the null hypothesis $H_{0,1} : f_m(blood) \leq f_m(cancer)$ with a Mann-Whitney test. The resulting p-value was $p < 2^{-32}$, so we can safely reject this hypothesis. Although the plots for f_u and μ look very similar according to the Mann-Whitney test, we also have to reject $H_{0,2} : f_u(blood) \geq f_u(cancer)$ and $H_{0,3} : \mu(blood) \leq \mu(cancer)$. Hence, the Mann-Whitney statistically hints

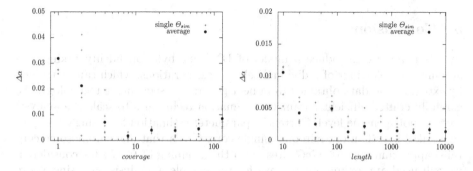

Fig. 5. Distances between estimated α and true α for a fixed length of 1000 (left). Distances for all six parameter sets for a fixed coverage of 6(right).

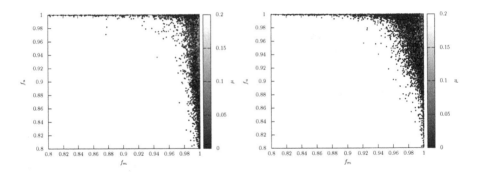

Fig. 6. Estimated parameter sets for groups of 5000 base pairs of chromosome 7. Left: data from 306 blood samples. Right: data from 101 solid tumor samples.

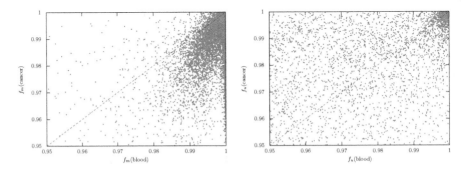

Fig. 7. Comparison of estimated single parameters. Left: f_m. Right: f_u

at the fact that blood and tumour samples behave differently in terms of DNA methylation fidelity, which is relevant for cancer biology, although the test fails to suggest a clear criterion which can be used as an indicator for detecting abnormal cells. Visual inspection of Fig. 7 points to f_m as a potential candidate. This issue will be further investigated when running the analysis on additional data sets.

5   Conclusions

In this paper we introduce a model of DNA methylation fidelity taking into account the behaviour of individual cells over generations, which can be trained by experimental data obtained from next generation sequencing technology. We carefully crafted efficient parameter estimation techniques to scale the analysis to the whole-genome level. Currently, parameter estimation for a single group of sites takes less than one min on a single core. Given that one chromosome contains approximately 10^9 CpG sites, with the grouping of CpG sites considered, we will need approximately 14 days for the complete analysis on a single core machine. However, as this code is fully and straightforwardly parallelizable, the whole-genome analysis is feasible on a high-performance cluster.

In this paper we also present preliminary tests on simulated data and on a real whole-genome dataset, trying to detect differences in methylation fidelities between human blood and tumour samples. Preliminary statistical analysis of data appears to support that there is indeed a systematic difference in the DNA methylation dynamics of the two samples. Deeper investigations are currently carried out on the whole-genome scale to better understand the statistical nature and biological significance of these differences. We will also investigate if and how differences in methylation fidelity reflect on differences in the shape of methylation profiles, comparing with state-of-the-art statistical tests [9].

References

1. Abramov, R.: A practical computational framework for the multidimensional moment-constrained maximum entropy principle. J. Comput. Phys. **211**(1), 198–209 (2006)
2. Arand, J., Spieler, D., Karius, T., Branco, M.R., Meilinger, D., Meissner, A., Jenuwein, T., Xu, G., Leonhardt, H., Wolf, V., et al.: In vivo control of CpG and non-CpG DNA methylation by DNA methyltransferases. PLOS Genet. **8**(6), e1002750 (2012)
3. Bird, A.: DNA methylation patterns and epigenetic memory. Genes Dev. **16**(1), 6–21 (2002)
4. Bock, C.: Analysing and interpreting DNA methylation data. Nat. Rev. Genet. **13**(10), 705–719 (2012)
5. Bock, C., Tomazou, E.M., Brinkman, A.B., Müller, F., Simmer, F., Gu, H., Jäger, N., Gnirke, A., Stunnenberg, H.G., Meissner, A.: Quantitative comparison of genome-wide DNA methylation mapping technologies. Nat. Biotechnol. **28**(10), 1106–1114 (2010)
6. Burden, R.L., Faires, J.D.: Numerical Analysis. Brooks/Cole, Cengage Learning, Boston (2011)
7. Farlik, M., Sheffield, N.C., Nuzzo, A., Datlinger, P., Schönegger, A., Klughammer, J., Bock, C.: Single-cell DNA methylome sequencing and bioinformatic inference of epigenomic cell-state dynamics. Cell Rep. **10**(8), 1386–1397 (2015)
8. Harris, T.E.: The Theory of Branching Processes. Courier Corporation, North Chelmsford, Massachusetts, USA (2002)
9. Mayo, T.R., Schweikert, G., Sanguinetti, G.: M^3D: a kernel-based test for spatially correlated changes in methylation profiles. Bioinformatics **31**(6), 809–816 (2015)
10. Nelder, J.A., Mead, R.: A simplex method for function minimization. Comput. J. **7**(4), 308–313 (1965)
11. Quine, M.: A note on the moment structure of the multitype Galton-Watson process. Biometrika **57**(1), 219–222 (1970)
12. Sontag, L.B., Lorincz, M.C., Luebeck, E.G.: Dynamics, stability and inheritance of somatic DNA methylation imprints. J. Theor. Biol. **242**(4), 890–899 (2006)

Studying Emergent Behaviours in Morphogenesis Using Signal Spatio-Temporal Logic

Ezio Bartocci[1], Luca Bortolussi[2,3,4], Dimitrios Milios[5(✉)],
Laura Nenzi[6], and Guido Sanguinetti[5,7]

[1] Faculty of Informatics, Vienna University of Technology, Vienna, Austria
[2] Department of Maths and Geosciences, University of Trieste, Trieste, Italy
[3] CNR/ISTI, Pisa, Italy
[4] Modelling and Simulation Group, Saarland University, Saarbrücken, Germany
[5] School of Informatics, University of Edinburgh, Edinburgh, UK
dmilios@inf.ed.ac.uk
[6] IMT Lucca, Lucca, Italy
[7] SynthSys, Centre for Synthetic and Systems Biology, University of Edinburgh,
Edinburgh, UK

Abstract. Pattern formation is an important spatio-temporal emergent behaviour in biology. Mathematical models of pattern formation in the stochastic setting are extremely challenging to execute and analyse. Here we propose a formal analysis of the emergent behaviour of stochastic reaction diffusion systems in terms of Signal Spatio-Temporal Logic, a recently proposed logic for reasoning on spatio-temporal systems. We present a formal analysis of the spatio-temporal dynamics of the Bicoid morphogen in *Drosophila melanogaster*, one of the most important proteins in the formation of the horizontal segmentation in the development of the fly embryo. We use a recently proposed framework for statistical model checking of stochastic systems with uncertainty on parameters to characterise the parametric dependence and robustness of the French Flag pattern, highlighting non-trivial correlations between the parameter values and the emergence of the patterning.

1 Introduction

One of the most fascinating questions in biology is how regular patterns can emerge from biochemical processes acting at the cellular level, a process known as *morphogenesis* in developmental biology. Some evident examples of these patterns can be observed in the stripes of a zebra, the spots on a leopard, the filament structure of the cyanobacteria Anabaena or the square pattern of the sulfur bacteria T. rosea. Mathematical and computational methods hold enormous promise in the quest to unveil the underlying mechanisms of morphogenesis and reproducing, using computer-based simulations, the patterns observed in nature. Alan Turing, mostly known as the father of computer science, was also a pioneer in developing a first mathematical model [28] that provides *the chemical basis of*

© Springer International Publishing Switzerland 2015
A. Abate and D. Šafránek (Eds.): HSB 2015, LNBI 9271, pp. 156–172, 2015.
DOI: 10.1007/978-3-319-26916-0_9

morphogenesis. This model, also referred as the *Turing's reaction-diffusion system*, is able to reproduce the formation of some complex patterns in nature such as the stripes seen in the animal skin.

Formal analysis of how patterns arise from mathematical models is however challenging due to the high computational burden of spatio-temporal modelling, as well as the intrinsic difficulty of defining spatio-temporal patterns in a suitable language. Pattern recognition is generally considered as a branch of machine learning [6], where patterns are classified according statistical descriptors (or features) [21] or the structural relationship among them [25]. This approach, despite its success and popularity, lacks of a rigourous foundation to specify such patterns and to reason about them in a systematic way. On the other end, formal methods provide logic-based languages [2,10,17,18] with a well-defined syntax and semantics to specify in a precise and concise way emergent behaviours and the necessary techniques to automatically detect them.

Related Work. In the last year, two novel spatio-temporal logics, SpaTeL [18] and SSTL [10,24] have made their appearance almost at the same time in the realm of formal methods to specify the emergence of spatio-temporal patterns.

The *Spatial-Temporal Logic* (SpaTeL) in [18] is the unification of Signal Temporal Logic [23] (STL) and Tree-Spatial-Superposition-Logic (TSSL) introduced in [2] to classify and detect spatial patterns. TSSL reasons over quad trees, spatial data structures that are constructed by recursively partitioning the space into uniform quadrants. TSSL is derived from Linear Spatial-Superposition-Logic (LSSL) [17], where the notion of superposition provides a way to describe statistically the distribution of discrete states in a particular partition of the space and the spatial operators correspond to *zooming in and out* of particular areas. In [17] the authors show also that by nesting these operators they are able to specify self-similar and fractal-like structures that generally characterize the patterns emerging in nature. SpaTeL is equipped with a qualitative (yes/no answer) and a quantitative semantics that provide a measure or robustness of how much the property is satisfied or violated. In [18] this measure of robustness is used as a fitness function to guide the parameter synthesis process for a deterministic reaction diffusion system using particle swarm optimisation (PSO) algorithms. However, the authors do not consider stochastic reaction-diffusion systems and PSO techniques generally do not provide any guarantee for reaching the global optimum. Hence, in this paper we will adopt a method with proved convergence guarantees, introduced previously in [3,4] for the system design of stochastic processes using the robustness of temporal properties.

The *Signal Spatio-Temporal Logic* (SSTL) [10,24] is the extension of STL [23] with three spatial modalities, *somewhere*, *everywhere* and *surround*, which can be nested arbitrarily with the original STL temporal operator. In [10,24], the authors provide a qualitative and quantitative semantics of SSTL and efficient monitoring algorithms for both semantics. A more detailed description of SSTL is provided in Sect. 3. While in this paper we adopt SSTL to specify spatio-temporal patterns, the overall method for robust parameter synthesis for stochastic reaction diffusion systems presented here can be performed also using SpaTeL.

Contribution. In this work, we combine formal methods with statistical machine learning by presenting a novel analysis of a stochastic model of the *spatio-temporal behaviour* of the Bicoid protein in the Drosophila's Embryo. The spatial gradient of this molecule has been shown to be at the basis of the subdivision of the embryo along its main axis, as specific concentration thresholds in its gradient are detected by cells and lead to the expression of distinct set of target genes.

The main technical contribution of the paper is the combination of SSTL within the statistical machine learning framework of [4,7–9], in order to efficiently perform parameter space exploration and system design of spatio-temporal properties.

From a system biology perspective, instead, we present a detailed spatio-temporal analysis of the French Flag pattern on the gradient of the Bicoid protein. This analysis permits novel insights as to how the various model parameters interact to give rise to the patterning behaviour.

Paper Structure. The rest of the paper is organised as follows. In Sect. 2 we discuss the spatial pattern formation in the Drosophila embryos. In Sect. 3 we first recall the syntax and semantics of SSTL and then use it to specify the French Flag Property. The smoothed model checking and the parameter estimation is presented in Sect. 4. In Sect. 5, we present the results and we conclude with final remarks and directions for future work in Sect. 6.

2   Spatial Pattern Formation and the French Flag Model

In this section, we describe a model of segmentation in *Drosophila melanogaster* and the spatio temporal pattern characterising it, known as the French Flag model.

2.1   Pattern Formation and Reaction-Diffusion Systems

Patterning is a ubiquitous feature of biological organisms, and the presence of regular geometric motifs on many organisms has long fascinated scientists. Pattern formation is also the subject of one of the earliest, and most influential, computational systems biology works, Alan Turing's pioneering work on morphogenesis [28]. Turing's insight was that biological patterns can be viewed as emergent behaviour (in modern terminology) arising from local interactions of microscopic agents. More precisely, Turing considered spatially distributed systems whose local concentration vector $\mathbf{u}$ obeys a *reaction-diffusion* partial differential equation (PDE)

$$\frac{\partial \mathbf{u}}{\partial t} = D\nabla^2 \mathbf{u} + f(\mathbf{u}). \tag{2.1}$$

Equation (2.1) defines the time evolution of the local concentration $\mathbf{u}$ as the sum of two terms: a dispersal or diffusion term $D\nabla^2 \mathbf{u}$, which globally drives the system towards a uniform equilibrium, and a reaction term $f(\mathbf{u})$, which accounts for local interactions of the chemicals. Turing then proved that, under certain conditions on the reaction/diffusion parameters, these two counteracting processes could give rise to regular patterns of concentration, providing a plausible mechanistic model of biological pattern formation.

Turing's ideas have been empirically demonstrated in many areas of biochemistry (see [22] for a recent review), and are still influential in particular in the field of developmental biology (see e.g. [16] for a recent paper building on these ideas). The crucial idea in the application of reaction-diffusion systems to development is that these mechanisms would underpin the local concentration patterns of regulatory proteins, which would instruct different genetic programs to be executed at different spatial locations. These special regulatory proteins are called *morphogens* in developmental biology, as they are believed to be responsible for the establishment of the shape of an organism in higher organisms. One of the most widely studied models of morphogenesis is the establishment of spatial patterning (stripes) along the body of the fruit fly *Drosophila melanogaster*. Several morphogens are known in *Drosophila*; mostly, these are maternal proteins that are produced in a localised area of the embryo (in correspondence to a maternal deposit of messenger RNA), and then establish a concentration gradient during development, effectively providing cells within an embryo with a spatial reference. An important morphogen is the protein *Bicoid*, which is the central object of study in this paper and is described in detail in the next subsection.

Before closing this whirlwind review of developmental biology, it is worth remarking on a fundamental shift of perspective that has happened since Turing's pioneering work, the realisation of the importance of stochasticity in biology. Numerous lines of evidence indicate that biology at the single cell level is intrinsically stochastic. Stochasticity cannot be ignored when modelling early embryogenesis, when only a handful of cells are present. Morphogenetic reaction-diffusion models can therefore be modified to account for the intrinsic discreteness of biology at the microscopic level. The natural analogue, systems of agents moving in continuous space, is however prohibitively expensive computationally; an approach that is more amenable to analysis is to discretise space into a number of cells (voxels) which are assumed to be spatially homogenous, and to replace spatial diffusion with transitions between different cells. Morphogenetic systems, and in particular the *Bicoid* system, have already been analysed from a simulation perspective in [31] and from a statistical perspective in [12]. In this paper, we present a first analysis of this system from the point of view of (spatio) temporal logic, to analyse directly the system's behaviour at the level of the emergent properties of the trajectories.

2.2 The Bicoid Gradient

The *Bicoid* (Bcd) molecule was the first protein to be identified among the morphogens. In the *Drosophila* embryos, the Bcd protein is distributed along the Anterior-Posterior axis (A-P axis). The Bcd mRNA is translated at the anterior pole of the embryo, and the synthesised protein spreads through the A-P axis by *diffusion* accompanied by *decay*.

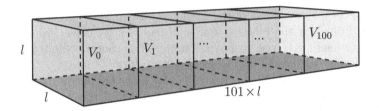

Fig. 1. A schematisation of the *Drosophila* embryo volume. The volume is divided in 101 cubic subvolumes, $V_0, ..., V_{100}$, with side $l = 5\,\mu m$.

We will describe the dynamics of the Bcd protein by a stochastic *reaction-diffusion* system, as reported in [31]. Given a certain volume where the Bcd protein is distributed, we can divide it into a series of subvolumes or voxels that are small enough to be regarded as well mixed. Then, we can consider the *decay* reaction as a transition that happens inside the subvolumes and the *diffusion* as exchange of molecules between neighbouring voxels. In particular, we consider 101 homogeneous cubic subvolumes with side $l = 5\,\mu m$ that comprise the entire volume as in Fig. 1. The length of the side l and the number of subvolumes were chosen in light of those of actual Drosophila embryos, which are $500\,\mu m$ long. The first subvolume ($j = 0$), corresponds to the anterior pole of the embryo and it is the only subvolume where the Bcd protein is synthesised.

We can describe the set $\mathcal{R}$ of reactions governing the stochastic dynamics of Bcd as:

$$\nu_p: \quad \varnothing \rightarrow B_0 \qquad \text{at rate } J, \qquad \qquad \text{(production)}$$

$$\nu_{deg_j}: \quad B_j \rightarrow \varnothing \qquad \text{at rate } w, \quad \text{for } j = 0, ...100, \qquad \text{(degradation)}$$

$$\nu_{dif_j^+}: \quad B_j \rightarrow B_{j+1} \quad \text{at rate } \frac{D}{l^2}, \quad \text{for } j = 0, ...99, \quad \text{(diffusion to the right)}.$$

$$\nu_{dif_j^-}: \quad B_j \rightarrow B_{j-1} \quad \text{at rate } \frac{D}{l^2}, \quad \text{for } j = 1, ...100, \quad \text{(diffusion to the left)}.$$

where B_j is a Bcd protein in the jth subvolume.

The state vector of the system is then $\mathbf{x}_B = (x_{B_0}, ..., x_{B_{100}})$ where x_{B_j} is the number of Bcd molecules in the jth subvolume. From the set $\mathcal{R}$ we can derive the infinitesimal generator matrix of the CTMC that formally represents the dynamics of the system. The CTMC can then be simulated with a standard algorithm, like SSA or tau-leaping.

Note that, from the set of reactions $\mathcal{R}$, we can easily revert the discretisation process and obtain a semantics in terms of Reaction-Diffusion Rate Equation (RDRE). This is obtained by converting variables into concentrations, taking the length of voxels to zero, and interpreting each rate as a flow, both in the degradation and in the diffusion reactions. In this way, we can define the system

$$\frac{\partial u}{\partial t} = D\frac{\partial^2 u}{\partial y} - wu, \tag{2.2}$$

where $u(y,t)$ is the concentration of Bcd at time t in position y, measured in μm, $y \in [0,500]$, giving the boundary conditions $\frac{\partial u}{\partial y}\big|_{y=0} = -\frac{J}{\Delta}$ and $\frac{\partial u}{\partial y}\big|_{y=500} = 0$, where $\Delta = l^3$.

2.3 Segmentation and the French Flag Model

The spatial distribution of the Bicoid protein has a crucial role in the formation of the horizontal segmentation in the development of the Drosophila's embryo. One of the most important interpretations of this distribution is given by the *French Flag model* [29], and more generally by the theory of *gap genes* [19,30]. The body of the fruit fly *Drosophila melanogaster*, as in most arthropods, exhibits a particular type of spatial patterning called *segmentation*, whereby the main body is composed of several segments. Gap genes were discovered and named following mutagenetic experiments, whereby biologists observed that deletion of certain genes resulted in the omission of a segment in the fly's body, as if the mutant organism had a gap. This observation implies that gap genes must be expressed in a precisely spatially co-ordinated manner, i.e., the biochemistry of the fruit fly must possess a way of measuring distances.

The French Flag model is a simplified model of gap gene regulation in early embryogenesis involving only four genes, the Bicoid morphogen protein and three target genes. The underlying assumption is that the spatial distribution of Bicoid protein, which as we have seen tends to decrease along the A-P axis (see Fig. 2), provides the ruler with which the *Drosophila* embryo measures distances. Gap genes are activated in a concentration dependent manner by Bicoid, so that a set of genes are activated at the high concentrations near the anterior part of the embryo (the blue in the French Flag), a different set of genes is activated in the central part (the white) and a third set is activated a low concentrations near the posterior end (red). This model has survived with some modifications [20] until this day, its beauty providing a paradigm for pattern development in many areas of biology. From our point of view, this model is particularly interesting because it refocuses attention from local intensive quantities (local concentrations) towards the importance of a global emergent property of the system (the establishment of a gradient), which is ideally suited for reasoning upon in terms of spatio-temporal logics. We will see in the next section this how to describe the French Flag pattern using a spatio-temporal logic.

3 Formula Specification of Spatio-Temporal Behaviour

In this section, we describe *Signal Spatio-Temporal Logic* (SSTL) which will then be used to specify the spatio-temporal behaviour of the French Flag pattern.

3.1   Signal Spatio-Temporal Logic

The *Signal Spatio-Temporal Logic* (SSTL) [10,24] is a linear time logic suitable to specify spatio-temporal behaviours of traces generated from simulations. It is an extension of *Signal Temporal Logic* (STL) [23] with two spatial modalities.

The space is described as a weighted graph $G = (L, E, w)$ where L is a set of locations, E is a set of edges and $w : E \to \mathbb{R}_{\geq 0}$ is the function that returns the cost/weight of each edge, typically encoding the distance between two nearby locations.

The syntax of SSTL is given by

$$\varphi := \mathtt{true} \mid \mu \mid \neg\varphi \mid \varphi_1 \wedge \varphi_2 \mid \varphi_1 \, \mathcal{U}_I \, \varphi_2 \mid \diamondsuit_{[w_1,w_2]} \varphi \mid \varphi_1 \, \mathcal{S}_{[w_1,w_2]} \varphi_2,$$

where the STL operators are the atomic proposition μ, the standard boolean connectives conjunction and negation and the *bounded until* operator $\mathcal{U}_I$, with I a dense-real interval. The new spatial operators are the *somewhere* operator, $\diamondsuit_{[w_1,w_2]}$, and the *bounded surround* operator $\mathcal{S}_{[w_1,w_2]}$, where $[w_1, w_2]$ is a closed real interval with $w_1 < w_2$. The spatial somewhere operator $\diamondsuit_{[w_1,w_2]}\varphi$ requires φ to hold in a location reachable from the current one with a total cost greater than or equal to w_1 and less than or equal to w_2. The surround formula $\varphi_1 \mathcal{S}_{[w_1,w_2]}\varphi_2$, instead, is true in a location ℓ when ℓ belongs to a subset of locations A, a region, satisfying φ_1, such that its external boundary $B^+(A)$ (i.e., all the nearest neighbours of locations in A) contains only locations satisfying φ_2. Furthermore, locations in $B^+(A)$ must be reached from ℓ by a shortest path of cost between w_1 and w_2, i.e. they have to be at distance between w_1 and w_2 from ℓ. There are also three derivable operators: the *eventually* operator $\mathcal{F}_I \varphi := \mathtt{true} \, \mathcal{U}_I \, \varphi$, the *always* operator $\mathcal{G}_I \varphi := \neg \mathcal{F}_I \neg \varphi$ and the *everywhere* operator $\boxdot_{[w_1,w_2]} \varphi := \neg \diamondsuit_{[w_1,w_2]} \neg \varphi$ that requires φ to hold in all the locations reachable from the current one with a total cost between w_1 and w_2.

SSTL is interpreted on spatio-temporal traces $\mathbf{x} : \mathbb{T} \times L \to \mathbb{R}^n$, where $\mathbb{T}$ is the time domain, usually a real interval $[0, T]$, with $T > 0$; we can write the trace as $\mathbf{x}(t, \ell) = (x_1(t, \ell), \cdots, x_n(t, \ell))$, where each $x_i : \mathbb{T} \times L \to \mathbb{R}$, for $i = 1, ..., n$, is the projection on the ith coordinate/variable.

Similarly to STL, SSTL has two semantics, the classical *boolean* semantics and a *quantitative* semantics.

The boolean semantics returns true or false depending on whether the trace satisfies the SSTL property, i.e. $(\mathbf{x}, t, \ell) \vDash \varphi$ is true if and only if the trace $\mathbf{x}(t, \ell)$ satisfies φ. By convention, the whole trace satisfies a property in location ℓ iff it satisfies the property at time zero, i.e. $(\mathbf{x}, \ell) \vDash \varphi \Leftrightarrow (\mathbf{x}, 0, \ell) \vDash \varphi$.

The quantitative semantics, instead, returns a real value $\rho(\varphi, \mathbf{x}, t, \ell)$ that quantifies the level of satisfaction of the formula by the trajectory $\mathbf{x}$ at time t in location ℓ. The absolute value $|\rho(\varphi, \mathbf{x}, t, \ell)|$ can be interpreted as measure of the robustness of the satisfaction or dissatisfaction. Furthermore, the sign of $\rho(\varphi, \mathbf{x}, t, \ell)$ is related to the truth of the formula: if $\rho(\varphi, \mathbf{x}, t, \ell) > 0$, then $(\mathbf{x}, t, \ell) \vDash \varphi$, and similarly if $\rho(\varphi, \mathbf{x}, t, \ell) < 0$, then $(\mathbf{x}, t, \ell) \nvDash \varphi$. The definition of this quantitative measure ρ is based on [13,14], and it is a reformulation of the robustness degree of [15]. In accordance with the boolean semantics, the quantitative value of the whole trace in location ℓ is given by its value at time zero, i.e. $\rho(\mathbf{x}, \ell) = \rho(\mathbf{x}, 0, \ell)$.

SSTL is equipped with efficient monitoring algorithms for both the boolean and the quantitative semantics, whose description, together with a formalisation of the semantics, can be found in [10, 24].

3.2   The French Flag Property

To describe the French Flag pattern we have first to define the trajectories that we want to characterise and its related graph.

Let consider a trace (a simulation) $(\mathbf{x}_B(t))_{t \in [0,T]} = (x_{B_0}(t), ..., x_{B_{100}}(t))_{t \in [0,T]}$ of the Bicoid model described in the previous section, where $[0, T]$ is the time domain, with $T > 0$. We can transform the temporal trace in a spatio-temporal trajectory defining $x_B : L \times [0, T] \rightarrow \mathbb{R}$ s.t. $x_B(V_i, t) := x_{B_i}(t)$, where $L = \{V_0, ..., V_{100}\}$ is the set of locations. The graph $G = (L, E, w)$ of the system is a one-dimensional graph where each V_i is connected only to V_{i-1} and V_{i+1}, with $w(V_i, V_{i+1}) = 1$, i.e. all the edges have weight equal to 1. The weight between two arbitrary locations is given by the weight of the shortest path connecting them.

We can now use the logic to specify the French Flag model. As we described in Sect. 2, this pattern is used to represent the effect of a morphogen in the expression of different genes, i.e. to represent the correlation between the concentration of the morphogen and the activation or repression of other genes. In particular, the spatial distribution of the morphogen, at the steady state, is divided in three regions: a blue, a white and a red region, as shown in Fig. 2 (left), that activate different target genes.

We can describe this behaviour with the property

$$\psi_{flag} := \varphi_{blue} \wedge \varphi_{white} \wedge \varphi_{red} \tag{3.3}$$

$$
\begin{aligned}
\varphi_{blue} &:= \boxdot_{[0, w_{blue}]}(x_B > K_{blue} - h_{bw}) \\
\varphi_{white} &:= \boxdot_{[w_{blue}, w_{white}]}((x_B < K_{blue} + h_{bw}) \wedge (x_B > K_{white} - h_{wr})) \\
\varphi_{red} &:= \boxdot_{[w_{white}, w_{max}]}(x_B < K_{white} + h_{wr})
\end{aligned} \tag{3.4}
$$

The verification of the formula is done in the location V_0. $(x, V_0) \vDash \psi_{flag}$ iff it satisfies each subformulae $\varphi_{blue}, \varphi_{white}, \varphi_{red}$; $(x, V_0) \vDash \varphi_{blue}$ iff, in all the locations V_i s.t. $w(V_0, V_i) \leq w_{blue}$, the number of Bicoid molecules is higher than $K_{blue} - h_{bw}$, i.e. $x_B > K_{blue} - h_{bw}$. In a similar way we can describe φ_{white} and φ_{red}. The meaning of the property is that the spatial distribution of the Bicoid protein is divided in three regions, the blue, where the $x_B > K_{blue} - h_{bw}$, the white, where $K_{blue} + h_{bw} > x_B > K_{white} - h_{wr}$, and the red, where $x_B < K_{white} + h_{wr}$. Note that h_{bw} and h_{wr} parameters have the role to relax the thresholds that define different regions, to properly deal with noise in Bcd expression, we will discuss this point more in detail in the Sect. 5.1.

At steady state, the concentration of the Bicoid protein is exponentially distributed along the anterior-posterior (A-P) axis, with higher concentrations towards the anterior. We can identify the insurgence time of this pattern, and if it remains stable, combining the spatial property with temporal operators as follows:

$$\psi_{stableflag} := \mathcal{F}_{[T_{flag}, T_{flag}+\delta]}(\mathcal{G}_{[0, T_{end}]}\psi_{flag}) \qquad (3.5)$$

$\psi_{stableflag}$ means that eventually, in a time between T_{flag} and $T_{flag} + \delta$, the property ψ_{flag} remains true for at least T_{end} time units.

4   Methodologies

The main objective of this work is to study the effects of the Bicoid parameters on the satisfaction of the French Flag property. Exhaustive parameter exploration is particularly expensive for the model in question, due to the high cost of stochastic simulation. In this section, we briefly introduce the methodologies that we use to perform parameter synthesis and model checking in presence of parametric uncertainty.

4.1   Smoothed Model Checking

The Smoothed Model Checking algorithm [7] relies on the characterisation of the satisfaction probability of a formula φ as a function of the parameters. Given a CTMC $\mathcal{M}_\theta$, whose transition rates depend on a set of parameters θ, the satisfaction function of φ is defined as follows:

$$f(\theta) \equiv p(\varphi = \text{true}|\mathcal{M}_\theta)$$

It has been proven in [7] that, if the transition rates of $\mathcal{M}_\theta$ depends smoothly on the parameters θ and polynomially on the state of the system, then the satisfaction function of φ is a smooth function of the parameters.

The smoothed model checking approach leverages of the smoothness of the satisfaction function and transfers information across nearby parameter values. More specifically, we place a Gaussian Process (GP) prior over the space of possible functions, and we evaluate the satisfaction function for a set of parameter values. We then calculate the GP posterior under the light of these observations, which constitutes analytical approximation to the satisfaction function. This implies that we can estimate the satisfaction probability at any point in the parameter space with no additional cost.

The premise is that fewer samples are required to achieve a given level of accuracy. In the experiments of [7], it has been possible to accurately approximate the satisfaction function over a wide range of parameters using less than 10 % of the simulation runs required to obtain the same result with exhaustive parameter exploration. This resulted in a decrease of the total analysis time nearly by 90 %.

4.2   Robust Parameter Synthesis

The problem of robust parameter synthesis constitutes of identifying the model parameters that maximise the robustness of some desired property. According to the quantitative semantics of SSTL, the robustness value $\rho(\varphi, \mathbf{x}, t, \ell)$ expresses the level of satisfaction of φ by a trajectory $\mathbf{x}$ at time t in location ℓ. Since trajectories are random for a stochastic system, we designate the robustness of φ for a CTMC as a random variable R_φ. We are therefore interested in maximising the expected quantitative score:

$$E[R_\varphi] = \int \rho(\varphi, \mathbf{x}, t, \ell) p(\mathbf{x}) d\mathbf{x} \qquad (4.6)$$

where $p(\mathbf{x})$ is the probability density of trajectory $\mathbf{x}$. For a specified time t and location ℓ, the expectation $E[R_\varphi]$ constitutes an objective function, for which we can obtain noisy estimates by generating samples from the trajectory space via stochastic simulation.

Since evaluating the expected robustness is computationally expensive, we employ the Gaussian process optimisation algorithm described in [9]. In short, the objective function is approximated by a *Gaussian Process* (GP). The algorithm is initialised with a random grid of points, for each of which $E[R_\varphi]$ is approximated via statistical means. Using these points as a training set, a GP is used to make predictions regarding the $E[R_\varphi]$ value at different parts of the search space, without exhaustive exploration of the parameter space. We calculate the GP posterior for a set of test points; that involves calculating an estimate of the expected robustness and its associated variance. The GP optimisation algorithm dictates that the point that maximises the an upper quantile of the GP posterior is added to the training set, after being evaluated for its associated robustness via SMC. A high value for the upper quantile at any point in the parameter space indicates the possibility of an undiscovered maximum nearby. This feature allows us to direct the search towards areas of the parameter space that appear to be more promising. This process is repeated for a number of iterations, and the training set is progressively updated with new potential maxima. For a smooth objective function, the algorithm is proved to converge to the global optimum in [27].

5   Results

In this section, we perform a series of experiments to explore the sensitivity and robustness of the French Flag property w.r.t. changes in the rates of production J and degradation w, and the diffusion rate parameter D. The size of the cubic subvolumes is known, that is $l = 5\mu m$, as it is one of the main modelling assumptions.

5.1   Experimental Data

Following [26,31], we chose as parameters of the $\psi_{stableflag}$ property (3.5), specified in Sect. 3.2, $T_{flag} = 3950$, $\delta = 10$, $T_{end} = 1000$, $w_{blue} = 35.5$, $w_{white} = 67.5$ and

$w_{max} = 101$. The w_{blue} and w_{while} parameters mean that the blue area involves the subvolumes between V_0 and V_{35}, the white area extends from volume V_{36} to V_{67}, and finally the red one from V_{68} to V_{100}; the time is in terms of seconds.

In order to fix the thresholds parameters K_{blue}, K_{white} and h_{bw}, h_{wr} we use the Bicoid fluorescence concentration at cycle 13 (where the gradient is considered to be in the steady state) downloaded from the FlyEx database [1]. The choice of the data follows the analysis doing in [31]. To the best of our knowledge, all the quantifications of the Bicoid protein in the Drosophila embryo refers to the measurements of fluorescence concentrations, rather than direct observations of the Bicoid molecular population. From [31], we define the fluorescence concentration $I = m \times x_B$, where m is a scaling factor that denotes the fluorescence-to-molecule ratio. Our approach is to rescale the thresholds reported in terms of fluorescence concentrations with the m factor.

The data has been given originally in the form of two-dimensional coordinates paired, the A-P and D-V coordinate, from the central 10 % strip. As in [31], we choose the embryos where the variation inside each spatial subregions is low, in particular in these embryos the inverse of the spatial exponential coefficient varied by less that 1 %. We have transformed the data so that we have a single concentration value for each of the 101 discretised locations. Figure 2 depicts the result. On the left-side figure, we see how the different locations lie within the areas prescribed by the French Flag property. Although the shape of the data is apparently negative exponential, there is a considerable amount of noise, which has to be taken into consideration in terms of the French Flag property. We therefore define the thresholds in the form regions, rather than strict values. On the right-side of Fig. 2, we see a magnified version of the figure, where only the white area is depicted. The majority of the concentrations recorded for volumes from V_{36} to V_{67} are between 60 and 2. In the same way, we can empirically derive zones of desired concentration levels for the blue and read areas. Therefore we have $K_{blue} = 45/m$, $h_{bw} = 15/m$, $K_{white} = 6/m$, and $h_{wr} = 4/m$.

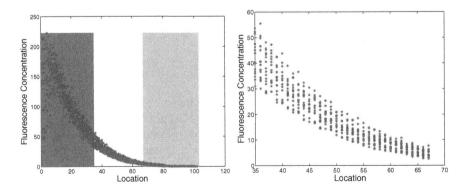

Fig. 2. Left: Fluorescence concentrations of the Bicoid protein for 17 embryos during the cycle 13. Right: The same concentrations in the area between locations 35 and 67, which define the white area in the French flag property.

5.2   Optimisation of Expected Robustness

We now explore how the model parameters (including the scaling factor m) can be tuned to increase the robustness of the French Flag pattern.

We applying the GP optimisation algorithm discussed in Sect. 4.2, for a four-dimensional space that involves the parameters: $w \in [0.001, 0.01]$, $J \in [10, 400]$, $D \in [1, 40]$, and $m \in [0.01, 1]$. The parameter ranges have been selected so that the resulting space is a superset of the explored space in [31]. Regarding the fluorescence-to-molecule ratio in particular, we note that the extremes considered in [31] have been 0.07 and 0.7.

For each evaluation of the expected robustness, the system has been simulated up to time $t = 4000\,$s, which is when the steady-state is approached according to [31]. The robustness expectation has been approximated statistically using 12 simulation runs for each parameter set. The algorithm has been initialised by 80 evaluations of the objective function at random points; a number of 282 evaluations were performed at points selected by the optimisation process, until convergence was detected. Convergence has been determined when no significant improvement of the expected robustness has been observed for 200 iterations. An improvement is considered significant, if it is more than 1 % increase over the previously recorded maximum robustness.

In the end, a total of 362 function evaluations have been performed, which is arguably a small number of samples to explore a four-dimensional space. The execution times have been 85 min for the initial 80 evaluations, and 263 min for the actual optimisation process. Stochastic simulations have been performed in parallel using 12 threads. The experiments have been performed on an Intel® Xeon® CPU E5-2680 v3 2.50 GHz. The majority of the computational effort was spent in simulation, despite the fact that only 12 trajectories have been generated for each parameter set considered. Therefore the idea of reducing the number of samples by exploiting the smoothness of the objective function has been a sensible practice.

The values returned by the optimisation process have been: $w^* = 0.0038$, $J^* = 390$, $D^* = 32.5$, and $m^* = 0.048$. The robustness of the optimum returned has been 2.99, implying that the property is robustly satisfied for the given solution. In Fig. 3, we present a sample trajectory for the given parameter configuration, and the average of 40 random trajectories, along with the associated 99.8 % confidence bounds. The sample trajectory is plotted against the experimental data that were used to adjust the threshold parameters of the French Flag property. We see that the optimised model has a behaviour very similar to the one observed in real-world experiments. However, it appears that the simulation results are much less noisy, when compared to the actual observations. This finding is in agreement with the result of [31], where it was argued that the intrinsic noise as modelled by the stochastic dynamics of the master equation is not sufficient to explain the variability in the data, i.e. the noise in the fluorescence measurement as a crucial role that has to be taken into account.

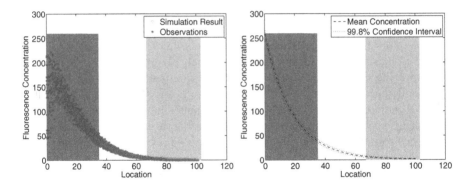

Fig. 3. Left: Sample trajectory for the parameter configuration that maximises the robustness of the French Flag property. Right: Average of 40 random trajectories; the dotted lines indicate the 99.8 % confidence interval.

5.3   Parameter Exploration with Smoothed Model Checking

In this section, we perform a more thorough exploration of the parameter space. Our objective is to discover dependencies among the parameters, considering the satisfaction probability of the French Flag property. On that respect, the fluorescence-to-molecule ratio m is not significant, as this will have an obvious effect on the thresholds for the property. We fix the fluorescence-to-molecule ratio m to 0.048, which is the optimal value reported by the optimisation algorithm in the previous section. The rest of the model parameters, $w \in [0.001, 0.01]$, $J \in [10, 400]$, and $D \in [1, 40]$, are explored via the smoothed model checking approach.

During the initialisation step of the algorithm, we have performed 216 evaluations of the satisfaction function of (3.3), for a regularly distributed set of values. As in the previous section, the satisfaction probability is approximated by statistical model checking using 12 simulation runs for each parameter configuration, where the system is simulated up to time $t = 4000\,\mathrm{s}$.

The duration of this initial statistical model checking process has been nearly 170 min, on an Intel® Xeon® CPU E5-2680 v3 2.50 GHz, using 12 threads in parallel. The hyperparameter optimisation that is required to tune the GP probit regression model subsequently required only 20 s, which is a trivial price to pay compared to the massive simulation cost. The final GP probit regression for a grid of 4096 points required only 1.2 s. Most importantly, it is only this last cost that we are required to pay to produce any further estimations of the satisfaction function.

Figure 4 depicts the satisfaction function for the French Flag property for parameters $\theta = \{w, J, D\}$, as this has been approximated by smoothed model checking. Each of the depicted subfigures shows the satisfaction probability as function of the production rate J and the diffusion parameter D, for a different value of the degradation rate w. Regarding the confidence of the estimated probabilities, we report that the 73.6 % of the values are associated with 95 % confidence intervals of width less than 0.2.

As a general remark, it appears that the manifestation of the gradient pattern, as this is captured by the French Flag property, is associated with a fine balance among the model parameters. There is a small area in the parameter space for which the property is satisfied with high probability. As we increase the decay parameter w however, we observe two behaviour regarding this area: its size is being increased, and its location is being shifted to the right. This implies that w is positively correlated with the production rate J. In other words, a particular ratio between protein production and decay is required for the formation of the particular pattern. At the same time, increasing the decay rate means that the formula may be satisfied for a wider range of the diffusion parameter.

It also appears that there is a negative correlation between the production rate J and the diffusion parameter D. This behaviour is present for the entire range of w examined, but it tends to become more obvious as w is increased. It is reasonable to conclude that a simultaneous increase of J and D would destroy the exponential shape of the Bicoid distribution across space.

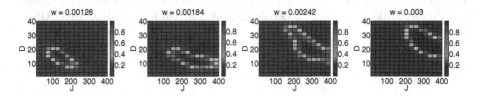

Fig. 4. Emulated satisfaction probability of the French Flag property as function of $\theta = \{w, J, D\}$. Each subfigure has the w parameter fixed.

6   Conclusions

We present a framework for the formal analysis under parametric uncertainty and the robust parameter synthesis of spatio-temporal properties emerging in a stochastic reaction-diffusion system. These properties are specified using the spatio-temporal logic SSTL. The framework combines statistical machine learning techniques based on Gaussian processes with the algorithm for monitoring SSTL properties.

As a case study, we analyse the occurrence of the French Flag pattern in the Bicoid gradient, during the development of Drosophila embryo. Analysing how this property depends on the parameters of the model is challenging due to the very high computational cost of simulating a spatio-temporal model, and has only been possible by adopting recent efficient verification techniques that employ machine learning methodologies [8]. Furthermore, the combination of these new techniques with SSTL permits exploring behaviours that are extremely difficult to express (and monitor) with standard temporal logics, where each individual location would need to be accounted.

The natural extension of this work is the analysis of more complicated models and properties, for example adding to this model the proteins of the target genes related with the spatial distribution of the Bicoid protein, enabling the study of the spatial dependency between proteins. To be independent from the spatial approximation, we plan also to consider different discretisation of the Drosophila's volume. Another future work could be the consideration of a model rescale with a random factor that mimics the extrinsic noise due to the fluoresce measurements. We plan also to extend our previous result in mining temporal logic properties [5,11] for the spatio-temporal case. Finally, we are considering an extension of the logic to continuous spaces and we would like to compare the expressiveness of SSTL with SpaTeL.

Acknowledgements. L.B. acknowledges partial support from the EU-FET project QUANTICOL (nr. 600708) and by FRA-UniTS. G.S. and D.M. acknowledge the support from the ERC under grant MLCS306999. E.B. acknowledges the partial support of the Austrian National Research Network S 11405-N23 (RiSE/SHiNE) of the Austrian Science Fund (FWF), the ICT COST Action IC1402 Runtime Verification beyond Monitoring (ARVI) and the IKT der Zukunft of Austrian FFG project HARMONIA (nr. 845631).

References

1. Flyex database. http://urchin.spbcas.ru/flyex/
2. Aydin Gol, E., Bartocci, E., Belta, E.: A formal methods approach to pattern synthesis in reaction diffusion systems. In: Proceedings of CDC 2014: the 53rd IEEE Conference on Decision and Control, pp. 108–113. IEEE (2014)
3. Bartocci, E., Bortolussi, L., Nenzi, L., Sanguinetti, G.: On the robustness of temporal properties for stochastic models. In: Proceedings of HSB 2013: The Second International Workshop on Hybrid Systems and Biology, pp. 3–19 (2013)
4. Bartocci, E., Bortolussi, L., Nenzi, L., Sanguinetti, G.: System design of stochastic models using robustness of temporal properties. Theor. Comput. Sci. **587**, 3–25 (2015)
5. Bartocci, E., Bortolussi, L., Sanguinetti, G.: Data-driven statistical learning of temporal logic properties. In: Legay, A., Bozga, M. (eds.) FORMATS 2014. LNCS, vol. 8711, pp. 23–37. Springer, Heidelberg (2014)
6. Bishop, C.M.: Pattern Recognition and Machine Learning. Springer, Heidelberg (2006)
7. Bortolussi, L., Milios, D., Sanguinetti, G.: Smoothed model checking for uncertain continuous time Markov chains. CoRR ArXiv 1402.1450 (2014)
8. Bortolussi, L., Milios, D., Sanguinetti, G.: U-Check: model checking and parameter synthesis under uncertainty. In: Campos, J., Haverkort, B.R. (eds.) QEST 2015. LNCS, vol. 9259, pp. 89–104. Springer, Heidelberg (2015)
9. Bortolussi, L., Sanguinetti, G.: Learning and designing stochastic processes from logical constraints. In: Joshi, K., Siegle, M., Stoelinga, M., D'Argenio, P.R. (eds.) QEST 2013. LNCS, vol. 8054, pp. 89–105. Springer, Heidelberg (2013)

10. Bortolussi, L., Nenzi, L.: Specifying and monitoring properties of stochastic spatio-temporal systems in signal temporal logic. In: Proceedings of VALUETOOLS 2014: The 8th International Conference on Performance Evaluation Methodologies and Tools, pp. 66–73. ICST (2014)
11. Bufo, S., Bartocci, E., Sanguinetti, G., Borelli, M., Lucangelo, U., Bortolussi, L.: Temporal logic based monitoring of assisted ventilation in intensive care patients. In: Margaria, T., Steffen, B. (eds.) ISoLA 2014, Part II. LNCS, vol. 8803, pp. 391–403. Springer, Heidelberg (2014)
12. Dewar, M.A., Kadirkamanathan, V., Opper, M., Sanguinetti, G.: Parameter estimation and inference for stochastic reaction-diffusion systems: application to morphogenesis in D. melanogaster. BMC Syst. Biol. **4**, 21 (2010)
13. Donzé, A., Maler, O.: Robust satisfaction of temporal logic over real-valued signals. In: Chatterjee, K., Henzinger, T.A. (eds.) FORMATS 2010. LNCS, vol. 6246, pp. 92–106. Springer, Heidelberg (2010)
14. Donzé, A., Ferrère, T., Maler, O.: Efficient robust monitoring for STL. In: Sharygina, N., Veith, H. (eds.) CAV 2013. LNCS, vol. 8044, pp. 264–279. Springer, Heidelberg (2013)
15. Fainekos, G.E., Pappas, G.J.: Robustness of temporal logic specifications for continuous-time signals. Theor. Comput. Sci. **410**(42), 4262–4291 (2009)
16. Fried, P., Iber, D.: Dynamic scaling of morphogen gradients on growing domains. Nat. Commun. **5**, 5077 (2014)
17. Grosu, R., Smolka, S.A., Corradini, F., Wasilewska, A., Entcheva, E., Bartocci, E.: Learning and detecting emergent behavior in networks of cardiac myocytes. Commun. ACM **52**(3), 97–105 (2009)
18. Haghighi, I., Jones, A., Kong, Z., Bartocci, E., Grosu, R., Belta, C.: SpaTeL: a novel spatial-temporal logic and its applications to networked systems. In: Proceedings of HSCC 2015: The 18th International Conference on Hybrid Systems: Computation and Control, pp. 189–198. ACM (2015)
19. Jaeger, J.: The gap gene network. Cell. Mol. Life Sci. **68**(2), 243–274 (2010)
20. Jaeger, J., Martinez-Arias, A.: Getting the Measure of Positional Information. PLoS Biol. **7**(3), e1000081 (2009)
21. Jain, A., Duin, R., Mao, J.: Statistical pattern recognition: a review. IEEE Trans. Pattern Anal. Mach. Learn. **22**, 4–37 (2000)
22. Maini, P.K., Woolley, T.E., Baker, R.E., Gaffney, E.A., Lee, S.S.: Turing's model for biological pattern formation and the robustness problem. Interface Focus **2**(4), 487–496 (2012)
23. Maler, O., Nickovic, D.: Monitoring temporal properties of continuous signals. In: Lakhnech, Y., Yovine, S. (eds.) FORMATS/FTRTFT 2004. LNCS, vol. 3253, pp. 152–166. Springer, Heidelberg (2004)
24. Nenzi, L., Bortolussi, L., Ciancia, V., Loreti, M., Massink, M.: Qualitative and quantitative monitoring of spatio-temporal properties. In: Bartocci, E., et al. (eds.) RV 2015. LNCS, vol. 9333, pp. 21–37. Springer, Heidelberg (2015)
25. Pavlidis, T.: Structural Pattern Recognition. Springer, New York (1980)
26. Phillips, R., Kondev, J., Theriot, J.: Physical biology of the cell. Garland Science, New York (2009)
27. Srinivas, N., Krause, A., Kakade, S., Seeger, M.: Information-theoretic regret bounds for Gaussian process optimisation in the bandit setting. IEEE Trans. Inf. Theory **58**(5), 3250–3265 (2012)
28. Turing, A.: The chemical basis of morphogenesis. Phil. Trans. R. Soc. Lond. **237**, 37–72 (1952)

29. Wolpert, L.: The french flag problem: a contribution to the discussion on pattern development and regulation. Towards Theor. Biol. **1**, 125–133 (1968)
30. Wolpert, L., Tickle, C., Arias, A.M.: Principles of Development. Oxford University Press, London (2015)
31. Wu, Y.F., Myasnikova, E., Reinitz, J.: Master equation simulation analysis of immunostained bicoid morphogen gradient. BMC Syst. Biol. **1**, 52 (2007)

Efficient Reduction of Kappa Models by Static Inspection of the Rule-Set

Andreea Beica[1], Calin C. Guet[2], and Tatjana Petrov[2]([✉])

[1] ENS Paris, Rue d'Ulm 45, 75005 Paris, France
[2] IST Austria, Am Campus 1, Klosteneuburg, Austria
tatjana.petrov@gmail.com

Abstract. When designing genetic circuits, the typical primitives used in major existing modelling formalisms are gene interaction graphs, where edges between genes denote either an activation or inhibition relation. However, when designing experiments, it is important to be precise about the low-level mechanistic details as to how each such relation is implemented. The rule-based modelling language Kappa allows to unambiguously specify mechanistic details such as DNA binding sites, dimerisation of transcription factors, or co-operative interactions. Such a detailed description comes with complexity and computationally costly executions. We propose a general method for automatically transforming a rule-based program, by eliminating intermediate species and adjusting the rate constants accordingly. To the best of our knowledge, we show the first automated reduction of rule-based models based on equilibrium approximations.

Our algorithm is an adaptation of an existing algorithm, which was designed for reducing reaction-based programs; our version of the algorithm scans the rule-based Kappa model in search for those interaction patterns known to be amenable to equilibrium approximations (e.g. Michaelis-Menten scheme). Additional checks are then performed in order to verify if the reduction is meaningful in the context of the full model. The reduced model is efficiently obtained by static inspection over the rule-set. The tool is tested on a detailed rule-based model of a λ-phage switch, which lists 92 rules and 13 agents. The reduced model has 11 rules and 5 agents, and provides a dramatic reduction in simulation time of several orders of magnitude.

1 Introduction

One of the main goals of synthetic biology is to design and control genetic circuits in an analogous way to how electronic circuits are manipulated in human made computer systems. The field has demonstrated success in engineering simple genetic circuits that are encoded in DNA and perform their function in the

This research was supported by the People Programme (Marie Curie Actions) of the European Union's Seventh Framework Programme (FP7/2007-2013) under REA grant agreement no. 291734, and the SNSF Early Postdoc.Mobility Fellowship, the grant number P2EZP2_148797.

© Springer International Publishing Switzerland 2015
A. Abate and D. Šafránek (Eds.): HSB 2015, LNBI 9271, pp. 173–191, 2015.
DOI: 10.1007/978-3-319-26916-0_10

cellular environment [1,2]. However, there remains a need for rigorous quantitative characterisation of such small circuits and their mutual compatibility [3]. The important ingredient towards such characterisation is having an appropriate language for capturing model requirements, for prototyping the circuits, and for predicting their quantitative behaviour before committing to the time-intensive experimental implementation.

Quantitative modelling of biomolecular systems is particularly challenging, because one deals with stochastic, highly dimensional, non-linear dynamical systems. For these reasons, modellers often immediately apply ad-hoc simplifications which neglect the mechanistic details, but allow to predict (simulate) the system's behaviour as a function of time. For example, the fact that *protein A activates protein P* is often modelled immediately in terms of a reaction $A \to A + P$ with the Hill kinetic coefficient (e.g. $\frac{k[A]^n}{1+k[A]^n}$), while the mechanism in fact includes the formation of a macromolecular complex and its binding to a molecular target. While such models are easier to execute, the simplification makes models hard to edit or refine. For example - a new experimental insight about an interaction mechanism cannot be easily integrated properly into the model, since several mechanistic steps are merged into a single kinetic rate. Moreover, an abstract model does not provide precise enough design guide for circuit synthesis, and sometimes, only the more detailed models explain certain behaviours (e.g., in [4], it is shown that only when incorporating the mRNA, the model explains certain experimentally observed facts).

Rule-based languages, such as Kappa [5] or BioNetGen [6], are designed to naturally capture the protein-centric and concurrent nature of biochemical signalling: the internal protein structure is maintained in form of a *site-graph*, and interactions can take place upon testing only *patterns*, local contexts of molecular species. A site-graph is a graph where each node contains different types of sites, and edges can emerge from these sites. Nodes typically encode proteins and their sites are the protein binding-domains or modifiable residues; the edges indicate bonds between proteins. Then, every species is a connected site-graph, and a reaction mixture is a multi-set of connected site-graphs. The executions of rule-based models are traces of a continuous-time Markov chain (CTMC), defined according to the principles of chemical kinetics. In general, rule-based models are advantageous to the classical reaction models (Petri nets) for two major reasons. First, the explicit graphical representation of molecular complexes makes models easy to read, write, edit or compose (by simply merging two collections of rules). For example, the reaction of dimerization between two λ CI molecules is classically written $2CI \to CI2$, where the convention is that CI represents a free monomer, and $CI2$ represents a free dimer. On the other hand, the same reaction written in Kappa amounts to:

$$\text{`}CI2\text{:'} \; \texttt{CI (ci,or)}, \texttt{CI (ci,or)} \leftrightarrow \texttt{CI (ci!1,or)}, \texttt{CI (ci!1,or)} \; @\texttt{k}_{2+}, \texttt{k}_{2-},$$

where the binding sites ci and or are binding sites of the protein CI, and CI (ci!1,or) denotes that the identifier of the rule-based bond accounting for the physical interaction between the two CI monomers, is 1. Secondly, a rule set can

$$\boxed{T \text{@} \text{@} Op} \underset{k_2}{\overset{k_1}{\rightleftharpoons}} \boxed{T \text{@} \text{@} Op}$$

$$\boxed{T \text{@} \text{@} Op} \xrightarrow{k_3} \boxed{T \text{@} \text{@} Op} + \boxed{P}$$

```
T(a),Op(x) ↔ T(a!1), Op(x!1)
T(a!1),Op(x!1) ↔ T(a!1), Op(x!1), P()
```

Fig. 1. An example of a rule-based model. The transcription factor T binds to the operator's site x via site a and, when bound, it initiates the production of protein P.

be executed, or subjected to formal static analysis: for example, it provides efficient simulations [7,8], automated answers about the reachability of a particular molecular complex [9], or about causal relations between rule executions [10].

The downside of incorporating too many mechanistic details in the model, is that they lead to computationally costly execution. For this reason, we define and implement an efficient method for automatically detecting and applying equilibrium approximations. As a result, one obtains a smaller model, where some species are eliminated, and the kinetic rates are appropriately adjusted. In this way, the experimentalist can choose to obtain the predictions more efficiently but less accurately, however without losing track of the underlying low-level mechanisms.

In related works [11,12], the authors propose an algorithm for reducing a reaction-based model, by searching for interaction schemes amenable to equilibrium approximations. In this paper, we adapt this algorithm to rule-based models.

Implementation and Testing. The tool is implemented in OCaml, and it is tested on a detailed rule-based model of a λ-phage switch [13,14]. Simulations were carried out on the complete chemical reaction genetic circuit model which contains 92 rules, 13 agents and 61 species. The model is reduced to only 11 rules and 5 agents.

Related Work. The principle of obtaining conclusions about system's dynamics by analysing their model description, originates from, and is exhaustively studied in the field of formal program verification and model checking [15,16], while it is recently gaining recognition in the context of programs used for modeling biochemical networks. An example is the related work of detecting fragments for reducing the deterministic or stochastic rule-based models [17–19], detecting the information flow for ODE models of biochemical signaling [20,21], or the reaction network theory [22].

2 Stochastic Chemical Reaction Networks

For a well-mixed reaction system with molecular species $S = \{S_1, \ldots, S_n\}$, the state of a system can be represented as a multi set of those species, denoted by $\mathbf{x} = (x_1, ..., x_n) \in \mathbb{N}^n$. The dynamics of such a system is determined by a set of reactions $\mathsf{R} = \{\mathsf{r}_1, ..., \mathsf{r}_r\}$. Each reaction is a triple $\mathsf{r}_j \equiv (\mathbf{a}_j, \boldsymbol{\nu}_j, c_j) \in \mathbb{N}^n \times \mathbb{N}^n \times \mathbb{R}_{\geq 0}$, written down in the following form:

$$a_{1j}S_1, \ldots, a_{nj}S_n \xrightarrow{k_j} a'_{1j}S_1, \ldots, a'_{nj}S_n,$$

$$\text{such that } a'_{ij} = a_{ij} + \nu_{ij}.$$

The vectors $\mathbf{a}_j$ and $\mathbf{a}'_j$ are often called respectively the *consumption* and *production* vectors due to reaction r_j, and k_j is the *kinetic rate* of reaction r_j. If the reaction r_j occurs, after being in state $\mathbf{x}$, the next state will be $\mathbf{x}' = \mathbf{x} + \boldsymbol{\nu}_j$. This will be possible only if $x_i \geq a_{ji}$ for all $i = 1, \ldots, n$. Under certain physical assumptions [23], the species multiplicities follow a continuous-time Markov chain (CTMC) $\{X(t)\}$, defined over the state space $S = \{\mathbf{x} \mid \mathbf{x} \text{ is reachable from } \mathbf{x}_0 \text{ in } \mathsf{R}\}$. Hence, the probability of moving to the state $\mathbf{x} + \boldsymbol{\nu}_j$ from $\mathbf{x}$ after time Δ is

$$P(X(t + \Delta) = \mathbf{x} + \nu_k \mid X(t) = \mathbf{x}) = \lambda_j(\mathbf{x})\Delta + o(\Delta),$$

with λ_j the propensity of jth reaction, assumed to follow the principle of mass-action: $\lambda_j(\mathbf{x}) = k_j \prod_{i=1}^{n} \binom{x_i}{a_{ij}}$. The binomial coefficient $\binom{x_i}{a_{ij}}$ reflects the probability of choosing a_{ij} molecules of species S_i out of x_i available ones.

In the continuous, deterministic model of a chemical reaction network, the state $\mathbf{z}(t) = (z_1, \ldots, z_n)(t) \in \mathbb{R}^n$ is represented by listing the concentrations of each species. The dynamics is given by a set of differential equations in form

$$\frac{\mathrm{d}}{\mathrm{d}t} z_i = \nu_{ij} \sum_{j=1}^{r} c_j \prod_{i=1}^{n} z_i(t)^{a_{ij}}, \tag{1}$$

where c_j is a deterministic rate constant, computed from the stochastic one and the volume N from $c_j = k_j N^{|\mathbf{a}_j|-1}$ ($|\mathbf{x}|$ denotes the 1-norm of the vector $\mathbf{x}$). The deterministic model is a limit of the stochastic model when all species in a reaction network are highly abundant [24].

2.1   Deterministic Limit

Denote by $R_j(t)$ the number of times that the j-th reaction had happened until the time t. Then, the state of the stochastic model at time t is

$$\mathbf{X}(t) = \mathbf{X}(0) + \sum_{j=1}^{r} R_j(t)\boldsymbol{\nu}_j. \tag{2}$$

The value of $R_j(t)$ is a random variable, that can be described by a non-homogenous Poisson process, with parameter $\int_0^t \lambda_j(X(s))\mathrm{d}s$, that is, $R_j(t) = \xi_j(\int_0^t \lambda_j(\mathbf{X}(s))\mathrm{d}s)$. Then, the evolution of the state $\mathbf{X}(t)$ is given by the expression

$$\mathbf{X}(t) = \mathbf{X}(0) + \sum_{j=1}^{r} \xi_j \left(\int_0^t \lambda_j(\mathbf{X}(s))\mathrm{d}s \right) \boldsymbol{\nu}_j. \tag{3}$$

By scaling the species multiplicities with the volume: $Z_i(t) = X_i(t)/N$, adjusting the propensities accordingly, in the limit of infinite volume $N \to \infty$, the scaled process $\mathbf{Z}(t)$ follows an ordinary differential equation (1) [24].

It is worth mentioning here that the above scaling from stochastic to the deterministic model is a special case of a more general framework presented in [25], referred to as the *multiscale stochastic reaction networks*. Intuitively, the deterministic model is a special case where all species are scaled to concentrations and reaction rates are scaled always in the same way, depending on their arity. The reductions shown in this paper can be seen as a variant of multiscale framework, where some species are scaled to concentrations and others are kept in copy numbers, and where reaction rates have varying scales as well.

3   Rule-Based Models

We introduce the rule-based modeling language Kappa, which is used to specify chemical reaction networks, by explicitly describing chemical species in form of site-graphs. A simple example of a Kappa model is presented in Fig. 1.

For the stochastic semantics of Kappa, that is a continuous-time Markov chain (CTMC) assigned to a rule-based model, we refer to [19] or [26]. Intuitively, any rule-based system can be expanded to an equivalent reaction system (with potentially infinitely many species and reactions). The stochastic semantics of a Kappa system is then the CTMC $\{X(t)\}$ assigned to that equivalent reaction system. Even though the semantics of a Kappa system is defined as the semantics of the equivalent reaction system, in practice, using Kappa models can be advantageous for several reasons - they are easy to read, write, edit or compose, they can compactly represent potentially infinite set of reactions or species, and, perhaps most importantly, they can be symbolically executed.

We present Kappa in a process-like notation. We start with an operational semantics.

Given a set X, $\wp(X)$ denotes the power set of X (i.e. the set of all subsets of X). We assume a finite set of agent names $\mathcal{A}$, representing different kinds of proteins; a finite set of sites $\mathcal{S}$, corresponding to protein domains; a finite set of internal states $\mathbb{I}$, and Σ_ι, Σ_β two signature maps from $\mathcal{A}$ to $\wp(\mathcal{S})$, listing the domains of a protein which can bear respectively an internal state and a binding state. We denote by Σ the signature map that associates to each agent name $A \in \mathcal{A}$ the combined interface $\Sigma_\iota(A) \cup \Sigma_\beta(A)$.

Definition 1 *(Kappa agent).* A *Kappa agent* $A(\sigma)$ is defined by its type $A \in \mathcal{A}$ and its *interface* σ. In $A(\sigma)$, the interface σ is a sequence of sites s in $\Sigma(A)$, with internal states (as subscript) and binding states (as superscript). The internal state of the site s may be written as s_ϵ, which means that either it does not have internal states (when $s \in \Sigma(A) \setminus \Sigma_\iota(A)$), or it is not specified. A site that bears an internal state $m \in \mathbb{I}$ is written s_m (in such a case $s \in \Sigma_\iota(A)$). The binding state of a site s can be specified as s^ϵ, if it is *free*, otherwise it is bound (which is possible only when $s \in \Sigma_\beta(A)$). There are several levels of information about the binding partner: we use a binding label $i \in \mathbb{N}$ when we know the binding partner, or a wildcard bond $-$ when we only know that the site is bound. The detailed description of the syntax of a Kappa agent is given by the following grammar:

$$
\begin{aligned}
a &::= \mathbb{N}(\sigma) & \text{(agent)} \\
N &::= A \in \mathcal{A} & \text{(agent name)} \\
\sigma &::= \varepsilon \mid s,\sigma & \text{(interface)} \\
s &::= n_\iota^\lambda & \text{(site)} \\
n &::= x \in \mathcal{S} & \text{(site name)} \\
\iota &::= \epsilon \mid m \in \mathbb{I} & \text{(internal state)} \\
\lambda &::= \epsilon \mid - \mid i \in \mathbb{N} & \text{(binding state)}
\end{aligned}
$$

We generally omit the symbol ϵ.

Definition 2 *(Kappa expression).* *Kappa expression* E is a set of agents $\mathbb{A}(\sigma)$ and fictitious agents $\emptyset$. Thus the syntax of a Kappa expression is defined as follows:

$$ E ::= \varepsilon \mid a \, , \, E \mid \emptyset \, , \, E. $$

The structural equivalence $\equiv$, defined as the smallest binary equivalence relation between expressions that satisfies the rules given as follows

$$ E \, , \; A(\sigma,s,s',\sigma') \, , \; E' \equiv E \, , \; A(\sigma,s',s,\sigma') \, , \; E' $$
$$ E \, , \, a \, , \, a' \, , \, E' \equiv E \, , \, a' \, , \, a \, , \, E' $$
$$ E \equiv E \, , \, \emptyset $$
$$ \frac{i,j \in \mathbb{N} \text{ and } i \text{ does not occur in } E}{E[i/j] \equiv E} $$
$$ \frac{i \in \mathbb{N} \text{ and } i \text{ occurs only once in } E}{E[\epsilon/i] \equiv E} $$

stipulates that neither the order of sites in interfaces nor the order of agents in expressions matters, that a fictitious agent might as well not be there, that binding labels can be injectively renamed and that *dangling bonds* can be removed.

Definition 3 *(Kappa pattern, mixture and species).* A *Kappa pattern* is a Kappa expression which satisfies the following five conditions: (i) no site name occurs more than once in a given interface; (ii) each site name s in the interface of the agent A occurs in $\Sigma(A)$; (iii) each site s which occurs in the interface of the agent A with a non empty internal state occurs in $\Sigma_\iota(A)$; (iv) each site s which occurs in the interface of the agent A with a non empty binding state occurs in $\Sigma_\lambda(A)$; and (v) each binding label $i \in \mathbb{N}$ occurs exactly twice if it does at all $-$ there are no dangling bonds. A *mixture* is a pattern that is fully specified, i.e. each agent A documents its full interface $\Sigma(A)$, a site can only be free or tagged with a binding label $i \in \mathbb{N}$, a site in $\Sigma_\iota(A)$ bears an internal state in $\mathbb{I}$, and no fictitious agent occurs. A *species* is a connected mixture, i.e. for each two agents A_0 and A there is a finite sequence of agents $A_1, \ldots, A_k$ s.t. there is a bond between a site of A_k and of A and for $i = 0, 1, \ldots, k-1$, there is a site of agent A_i and a site of agent A_{i+1}.

Definition 4 *(species occurring in a pattern).* Given Kappa patterns E_s and E_p, if E_s defines a Kappa species, and E_s is a substring of E_p, we say that a species E_s *occurs* in a pattern E_p.

Definition 5 *(Kappa rule).* A Kappa rule r is defined by two Kappa patterns E_ℓ and E_r, and a rate $k \in \mathbb{R}_{\geq 0}$, and is written: $r = E_\ell \rightarrow E_r @ k$.

A rule r is well-defined, if the expression E_r is obtained from E_ℓ by finite application of the following operations: (i) creation (some fictitious agents $\emptyset$ are replaced with some fully defined agents of the form $\mathbf{A}(\sigma)$, moreover σ documents all the sites occurring in $\Sigma(A)$ and all site in $\Sigma_\iota(A)$ bears an internal state in $\mathbb{I}$), (ii) unbinding (some occurrences of the wild card and binding labels are removed), (iii) deletion (some agents with only free sites are replaced with fictitious agent $\emptyset$), (iv) modification (some non-empty internal states are replaced with some non-empty internal states), (v) binding (some free sites are bound pair-wise by using binding labels in $\mathbb{N}$).

In this work, we assume that a rule-based model is such that all left-hand-side and right-hand-side represent mixtures, that is, each rule is equivalent to one reaction. The extension to the case where this assumption does not hold is subject to future work. Hence, in our static inspection of rules, we test species (fully defined connected mixtures). To this end, we adopt the terminology of *reactant, modifier* and *product* from [12].

Definition 6 *(reactant, modifier, product).* Given a rule (E_l, E_r), a Kappa species s is called

- a *reactant*, if it occurs in pattern E_l and does not occur in pattern E_r,
- a *modifier*, if the number of occurrences in pattern E_l equals the number of occurrences in pattern E_r,
- a *product*, if it does not occur in pattern E_l, and it occurs in pattern E_r.

Definition 7 *(Kappa system).* A *Kappa system* $\mathcal{R}(\mathbf{x}_0, \mathcal{O}, \{r_1, \ldots, r_n\})$ is given by an initial mixture $\mathbf{x}_0$, a set of Kappa patterns $\mathcal{O}$ called *observables*, and a finite set of rules $\{r_1, \ldots, r_n\}$.

4   Model Approximation

In this section, we provide some mathematical background for the approximation algorithms[1]. The reductions are based on three reduction schemes: enzymatic catalysis reduction, generalized enzymatic catalysis reduction and fast dimerization reduction.

Enzymatic Reduction. Assume the elementary enzymatic transformation from a substrate S to a product P, through the intermediate complex $E : S$:

---

[1] for explicit algorithms, please consult the Appendix of the online paper version [27].

$$E + S \underset{k_2}{\overset{k_1}{\rightleftharpoons}} E : S \overset{k_3}{\rightarrow} E + P, \tag{4}$$

which our algorithm will convert to the well-known Michaelis-Menten form

$$S \overset{\frac{k_3 E_T K}{1+K x_S}}{\longrightarrow} P, \tag{5}$$

where $E_T = x_E(t) + x_{E:S}(t)$ denotes the total concentration of the enzyme, and $K = \frac{k_1}{k_2+k_3}$.

The above approximation is generally considered to be sufficiently good under different assumptions, such as, for example, that the rate of dissociation of the complex to the substrate is much faster than its dissociation to the product (i.e. $k_2 \gg k_3$), also known as the *equilibrium approximation*. Even if the equilibrium condition is not satisfied, it can be compensated in a situation where the total number of substrates significantly outnumbers the enzyme concentration - $x_S(0) + K \gg E_T$, known as the *quasi-steady-state assumption*.

Whenever one of the above assumptions holds, the quantity of the intermediate complex can be assumed to be rapidly reaching equilibrium, that is, $\frac{d}{dt} x_{E:S}(t) = 0$. Then, it is straightforward to derive the rate of direct conversion from substrate to product:

$$\frac{d}{dt} x_P = \frac{k_3 E_T K}{1 + K x_S} x_S,$$

which exactly corresponds to the equation for the rule (5).

In our reduction algorithm, we will apply the reduction whenever the pattern (4) is detected and the additional requirements with respect to the context of other rules are met.

The informal terminology of being 'significantly faster', motivated the rigorous study of the limitations of the approximations based on separating time scales. While the enzymatic (Michaelis-Menten) approximation has been first introduced and subsequently studied in the context of deterministic models (e.g. [28], Chap. 6), it was more recently that the time-scale separation was investigated in the stochastic context [29–34]. Notably, the following result from [35] (also shown as a special case of the multi scale stochastic analysis from [25]), shows that, under an appropriate scaling of species' abundance and reaction rates, the original model and the approximate model converge to the same process.

Theorem 1 *(Darden* [35], *Kang* [25]*).* Consider the reaction network (4) (equivalently the rule-based system depicted in Fig. 2), and denote by $X_S(t)$, $X_E(t)$, $X_{E:S}(t)$ and $X_P(t)$ the copy numbers of the respective species due to the random-time change model (2). Assume that $N = \mathcal{O}(X_S)$ and denote by $E_T = X_{E:S}(t) + X_E(t)$ and $V_E(t) = \int_0^t N^{-1} X_E(s) ds$. Assume that $k_1 \rightarrow \gamma_1$,

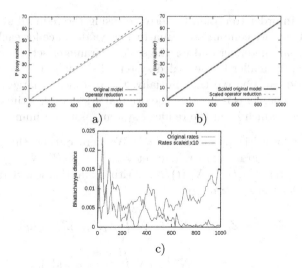

Fig. 2. Example shown in Fig. 1. The mean protein expression for one hundred sampled traces, before and after the enzymatic catalysis reduction. (a) Parameters $k_1 = 0.2156$, $k_2 = 1$, $k_3 = 0.014$ and there are initially 50 transcription factors. The mean and standard deviation (not shown) are computed for each time point, for the original (full line) and reduced model (dotted line). (b) Parameters k_2, k_3, and the initial number of transcription factors T are scaled up by factor $N = 10$. Same notation as for (a) (c) The Bhattacharyya distance between the distributions of the protein level with a model before and after the reduction. Red plot refers to the parameter values shown in (a), and the green plot to the scaled parameter values shown in (b) (Color figure online).

$k_2/N \to \gamma_2$, $k_3/N \to \gamma_3$, $N \to \infty$, and $\frac{X_S(0)}{N} \to x_S(0)$. Then $(\frac{X_S(t)}{N}, v_E(t))$ converges to $(x_S(t), v_E(t))$ and

$$\frac{d}{dt}v_E(s) = \frac{E_T}{1 + \hat{K}x_E(s)} \quad \text{and} \quad \frac{d}{dt}x_S = -\frac{E_T\gamma_3\hat{K}x_S(t)}{1 + \hat{K}x_S(t)},$$

where $\hat{K} = \frac{\gamma_1}{\gamma_2 + \gamma_3}$.

The assumptions listed in the theorem capture the that: (i) X_S and X_P are scaled to concentrations, while X_E and $X_{E:S}$ remain in copy numbers; (ii) the stochastic reaction rate k_1 is an order of magnitude smaller than the rates k_2 and k_3 (as a consequence of being related to the bimolecular, and not unimolecular reaction).

Example 1. To illustrate the meaning of the Theorem 1, we apply our reduction method on a small example shown in Fig. 1. We plot the mean and we compute the standard deviation of the protein level for the original and for the reduced model. Then, we scale up the parameters k_2 and k_3, as well as the initial concentration of transcription factor T, in order to mimic the effect of choosing a larger N in Theorem 1. The deviation between the curves is decreased, as can be seen in Fig. 2. In order to obtain the error of using the reduced system instead of the original one, we compute the Bhattacharyya distance for each time point, for the actual parameter set and for the scaled parameter set. As expected, the distance is overall smaller in the scaled system. Especially in the scaled system (green line), we can observe that initially, the distance is larger, and then it decreases with time. This is because the original system takes time to reach the equilibrium state which is, in the reduced system, assumed immediately.

A complete proof is provided in [25]. We here outline the general idea. Let $N > 0$ be a natural number, and let $Z_S(t) = X_S(t)/N$, $Z_E(t) = X_E(t)$, $Z_{S:E}(t) = X_{S:E}(t)$, $Z_P(t) = X_P(t)/N$. Writing out the scaled random time-change model for the substrate gives:

$$Z_S(t) = Z_S(0) - N^{-1}\xi_1(N \int_0^t \gamma_1 Z_S(s)Z_E(s)ds)$$
$$+ N^{-1}\xi_2(N \int_0^t \gamma_2 Z_{S:E}(s)ds),$$

and writing out the scaled random time-change model for the complex gives:

$$Z_{E:S}(t) = Z_{E:S}(0) + \xi_1(N \int_0^t \gamma_1 Z_S(s)Z_E(s)ds)$$
$$- \xi_2(N \int_0^t \gamma_2 Z_{S:E}(s)ds)$$
$$- \xi_3(N \int_0^t \gamma_3 Z_{S:E}(s)ds).$$

After dividing the latter with N, and applying the law of large numbers, we obtain the balance equations analogous to assuming that the complex is at equilibrium. This equation implies the expression for $\frac{d}{dt}v_E(s)$. The equation for $\frac{d}{dt}x_S$ follows from the model of $Z_S(t)$: we first use the conservation law $Z_{S:E}(s) = N^{-1}E_T - Z_E(t)$ and then substitute the obtained value of $\frac{d}{dt}v_E(s)$.

In order to confirm that the reduction is appropriate, our goal is now to show that the scaled versions of the original model (4) and the reduced model (5) are equivalent in the limit when $N \to \infty$. Let $Z_P(t) := N^{-1}X_P(t)$ be the scaled random time change for the product in the original model, and $\hat{Z}_P(t) := N^{-1}\hat{X}_P(t)$ in the reduced model. Notice that, from the balance equations, $\frac{d}{dt}x_P = -\frac{d}{dt}x_S$. According to the reduced system (5), the random time change for the product is given by

$$\hat{Z}_P(t) = \hat{Z}_P(0) + N^{-1}\xi(\int_0^t \frac{k_3 E_T K}{1 + K N \hat{Z}_S(s)} N \hat{Z}_S(s)ds)$$
$$= \hat{Z}_P(0) + N^{-1}\xi(\int_0^t N \frac{\gamma_3 E_T \hat{K}}{1 + \hat{K}\hat{Z}_S(s)} \hat{Z}_S(s)ds).$$

Passing to the limit, we obtain the desired relation $\frac{d}{dt}\hat{z}_P(t) = \frac{d}{dt}z_P(t)$.

The above Theorem does not provide the means of computing the approximation error, or an algorithm which suggests which difference in time-scales is good enough for an approximation to perform well. Rather, this result shows that the enzymatic approximation is justified in the limit when the assumptions about the reaction rates and species' abundance are met. In other words, when $N \to \infty$, the scaled versions of the original and reduced models – e.g. $Z_P(t) = N^{-1}X_P(t)$ and $\hat{Z}_P = N^{-1}\hat{X}_P$ – both converge to at the same, well-behaved process. This provides confidence that the actual process $\hat{X}_P$ is a good approximation of the process X_P.

Generalised Enzymatic Reduction. The enzymatic approximation can be generalized to a situation where many sets of substrates compete for binding to the same enzyme. Consider a sub-network of n reactions where the i-th such reaction reads:

$$E + S_{i,1} + \ldots + S_{i,m_i} \underset{k_i^-}{\overset{k_i}{\rightleftharpoons}} E : S_{i,1} : \ldots : S_{i,m_i} \overset{\hat{k}_i}{\to} E + P_i.$$

The resulting approximation is

$$S_{i,1} + \ldots + S_{i,m_i} \xrightarrow{\frac{\hat{k}_i E_T K_1 x_{S_i}}{Z}} P_i,$$

where $x_{S_i} = \prod_{j \in \{1,\ldots,m_i\}} x_{S_{i,j}}$, $Z = 1 + \sum_{j \in \{1,\ldots,n\}} x_{S_j}$ and $E_T = x_E(t) + \sum_{i=1}^{n} x_{E:S_{i,1}:\ldots:S_{i,m_i}}(t)$. The latter expression follows from $\frac{\mathrm{d}}{\mathrm{d}t} x_{E:S_{i,1}:\ldots:S_{i,m_i}}(t) = 0$ for all $i = 0, \ldots, n$.

Fast Dimerization Reduction. Consider now the dimerisation reaction $M + M \underset{k^-}{\overset{k}{\rightleftharpoons}} M_2$. Assuming that both rates k and k^- are fast comparing to other reactions involving M or M_2, it is common to assume that the reaction is equilibrated, that is, $kx_M(t)^2 - k^- x_{M2}(t) = 0$, where $x_M(t)$ and $x_{M2}(t)$ denote the copy number at time t, of monomers and dimers respectively. Such assumption allows us to eliminate the dimerization reactions, and only the total amount of molecules M needs to be tracked in the system. The respective monomer and dimer concentrations can be expressed as fractions of the total concentration:

$$x_M(t) = \frac{1}{4K}\left(\sqrt{8KM_T(t) + 1} - 1\right), \quad \text{and}$$

$$x_{M_2}(t) = \frac{M_T(t)}{2} - \frac{1}{2}x_M(t),$$

where $K = \frac{k}{k^-}$ and $M_T(t) = x_M(t) + 2x_{M_2}(t)$.

5   Reduction Algorithm

We now present the reduction algorithm, which is equivalent to that shown in [11,12], except for the adaptations which come as the data structure used to represent species in rule-based models are site-graphs, different to vectors of species' multiplicities used in reaction-based models. Also, unlike in the original algorithm, there is no need to check the form of the reaction rate function, as in Kappa rule-based models one implicitly assumes chemical kinetics to follow the mass-action rule.

5.1 Top-Level Algorithm

Our reduction of a Kappa system $\mathcal{R} = (\mathbf{x}_0, \mathcal{O}, \{r_1, \ldots, r_n\})$ is performed by static analysis over $\mathcal{R}$, in search for one of the interaction patterns which are consequences of the theory shown in Sect. 4: (i) the *modifier elimination*, applied first, in order to reduce complexity without losing accuracy, followed by (ii) the *competitive enzymatic*, (iii) *operator site* and (iv) *fast dimerization* reductions, and ending with the (v) *similar reaction composition* to combine the structurally similar reactions that are often generated after the operator abstraction of the model. The abstraction methods are applied until they generate no more changes in the model, as presented in the top-level algorithm shown in Algorithm 1.

Input : A Kappa system $\mathcal{R} = (\mathbf{x}_0, \mathcal{O}, \{r_1, \ldots, r_n\})$ over a set of species S and
           observables $\mathcal{O}$.
Output: A Kappa model $\mathcal{R}'$ over a set of species S′ and observables $\mathcal{O}$.

1 **repeat**
2      $M' \longleftarrow M$
3      $M \longleftarrow$ Modifier elimination(M)
4      $M \longleftarrow$ Competitive enzymatic reduction(M)
5      $M \longleftarrow$ Operator-site reduction(M)
6      $M \longleftarrow$ Fast dimerization reduction(M)
7      $M \longleftarrow$ Similar reaction composition(M)
8 **until** $M' = M$;

Algorithm 1. Top-level approximation algorithm.

5.2 Similar Reaction Composition and Modifier Elimination

In *similar reaction composition*, reactions that have the same reactants, modifiers and products are combined into a single reaction, by summing their rate laws.

The *modifier elimination* abstraction can be applied when a species only appears as a modifier throughout a model; such a species will never change its copy number throughout the dynamics, and therefore, its quantity will be constant. In this case, the species can be eliminated from the reactions and each rate law will be multiplied by the initial copy number of this species.

Notice that this reductions are exact, that is, applying them does not change the semantics of the rule-based system.

5.3   Fast dimerization reduction

The algorithm searches for dimerisation rules. Suppose that a pair of reversible reactions $M + M \leftrightarrow M_2$ is detected. Before proceeding to the reduction, we check whether a dimer is produced elsewhere, or if the monomer is a modifier elsewhere. These checks are necessary because they prevent from deviating from the assumed equilibrium. Finally, if all checks passed, the dimerization reaction can be eliminated. A new species M_T is introduced, and, wherever the monomer M or dimer M_2 were involved, they are replaced by the species M_T, and the rate is adapted accordingly, by the expressions shown in Sect. 4.

5.4   Generalised Enzymatic Reduction

The algorithm searches for the scheme described in Sect. 4, by searching for candidate enzymes. Each pattern is tested as to whether it is catalyzing some enzymatic reduction. If a pattern s indeed is an enzyme (operator) in an enzymatic reaction scheme, a set of all patterns c which compete to bind to s is formed, as well as the set of their complexes sc. Then, before proceeding with the reduction, additional tests must be performed: (i) pattern s must be a species, and it is not an observable, (ii) s must be small in copy number, that is, its initial copy number is smaller than a threshold, (iii) s can neither be produced, nor degraded, (iv) complex sc is not an observable and is never appearing in another rule of $\mathcal{R}$ and has initially zero abundance.

These tests are equivalent to those shown in [11,12]. Then, the patterns s and sc can be eliminated from the rule-set and the reaction rates are adjusted according to the description in Sect. 4.

Often times, enzymatic catalysis reduction is appropriate to eliminate the binding of the transcription factor to the operator site. In this context, the operator site takes the role of the enzyme, and transcription factor(s) the role of the substrate. Whenever a candidate enzyme is detected, and the other algorithm checks pass, the rates are appropriately scaled. The competitive enzymatic reduction is suitable in a situation when more transcription factors compete for binding the enzyme, each in a different reaction. In other words, the algorithm finds k rules where k different substrates compete for the same enzyme.

Example 2. We illustrate the competitive enzymatic transformation on a small subnetwork of the λ-phage model, which will be introduced in Sect. 6. The four rules presented below model the binding of the agent RNAP to the operator site of the agent PRE and subsequent production of protein CI. Agent PRE binds either only RNAP (at rate k_{1+} and k_{1-}), or simultaneously with CII (at rate k_{2+} and k_{2-}). The protein can be produced whenever PRE and PRE are bound, but the rates will be different depending on whether only RNAP is bound to the operator (rate k_b), or, in addition, CII is bound to the operator (rate k_a):

PRE (cii,rnap) , RNAP (p1,p2)
$\leftrightarrow$ PRE (cii,rnap!1) , RNAP (p1!1,p2) @k_{1+}, k_{1-}

PRE (cii,rnap) , CII (pre) , RNAP (p1,p2)
$\leftrightarrow$ PRE (cii!1,rnap!2) , CII (pre!1) , RNAP (p1!2,p2) @k_{a+}, k_{a-}

PRE (cii,rnap!1) , RNAP (p1!1,p2)
$\rightarrow$ PRE (cii,rnap!1) , RNAP (p1!1,p2) , 10CI (ci,or) @k_b

PRE (cii!1,rnap!2) , CII (pre!1) , RNAP (p1!2,p2)
$\rightarrow$ PRE (cii!1,rnap!2) , CII (pre!1) , RNAP (p1!2,p2) , 10CI (ci,or) @k_a

After the competitive enzymatic reduction, the operator PRE is eliminated from each of the two competing enzymatic catalysis patterns. Finally, the production of CI is modelled only as a function of RNAP and CII, and the rate is appropriately modified:

RNAP (p1,p2) , CII (pre)
$\rightarrow$ RNAP (p1,p2) , CII (pre) , 10CI (pr,ci) @ k_{new}.

6 λ-phage Decision Circuit

The phage λ is a virus that infects *E.coli* cells, and replicates using one of the two strategies: *lysis* or *lysogeny*. In the *lysis* strategy, phage λ uses the machinery of the *E.coli* cell to replicate itself and then lyses the cell wall, killing the cell and allowing the newly formed viruses to escape and infect other cells, while in the *lysogeny* scenario, it inserts its DNA into the host cell's DNA and replicates through normal cell division, remaining in a latent state in the host cell (it can always revert to the lysis strategy). The decision between *lysis* and *lysogeny* is known to be influenced by environmental parameters, as well as the multiplicity of infection and variations in the average phage input [36]. The key element controlling the decision process is the O_R operator (shown in Fig. 3), which is composed of three operator sites (O_{R1}, O_{R2}, O_{R3}) to which transcription factors can bind, in order to activate or repress the two promoters (P_{RM} and P_R) overlapping the operator sites. When RNAP (RNA polymerase, an enzyme that produces primary transcript RNA) binds to P_{RM}, it initiates transcription to the left, to produce mRNA transcripts from the *cI* gene; RNAP bound to the P_R promoter, on the other hand, initiates transcription to the right, producing transcripts from the *cro* gene. The two promoters form a genetic switch, since transcripts can typically only be produced in one direction at a time.

The *cI* gene codes for the CI protein, also known as the λ *repressor*: in its dimer form (two CI monomers react to form a dimer, CI_2), it is attracted to the O_R operator sites in the phage's DNA, repressing the P_R promoter from which

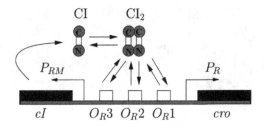

Fig. 3. CI monomers are produced from the cI gene; two monomers can form a dimer, that can bind to one of the O_R operator sites (the Figure is taken from [11]).

Cro production is initiated and further activating CI production. Similarly, the *cro* gene codes for the Cro protein, which also dimerizes in order to bind to O_R operator sites and prevent production from P_{RM}, or even its own production.

While CI_2 and Cro_2 can bind to any of the three operator sites at any time, they have a different affinity to each site. The CI_2 has its strongest affinity to the O_{R1} operator site, next to the O_{R2} site, and finally to the O_{R3} site (in other words, CI_2 first turns off P_R, then activates P_{RM}, and finally, represses its own production), while Cro_2 has the reverse affinity (it first turns off CI production, then turns off its own production).

The feedback through the binding of the products as transcription factors coupled with the affinities described makes the O_R operator behave as a genetic bistable switch. In one state, Cro is produced locking out production of CI. In this state, the cell follows the *lysis* pathway since genes downstream of Cro produce the proteins necessary to construct new viruses and lyse the cell. In the other state, CI is produced locking out production of Cro. In this state, the cell follows the *lysogeny* pathway since proteins necessary to produce new viruses are not produced. Instead, proteins to insert the DNA of the phage into the host cell are produced.

What's more, in the lysogeny state, the cell develops an immunity to further infection: the cro genes found on the DNA inserted by further infections of the virus are also shut off by CI_2 molecules that are produced by the first virus to commit to lysogeny. Once a cell commits to lysogeny, it becomes very stable and does not easily change over to the lysis pathway. An induction event is necessary to cause the transition from lysogeny to lysis. For example, lysogens (i.e., cells with phage DNA integrated within their own DNA) that are exposed to UV light end up following the lysis pathway.

7   Results and Discussion

We applied our reduction algorithm to a Kappa model of the phage λ decision circuit that we built using the reaction-based model presented in [11,12]. Simulations were carried out on the complete chemical reaction genetic circuit model which contains 92 species, 13 rules and 61 species (the contact map is shown in

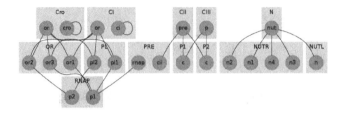

Fig. 4. The contact map of the full λ-phage model. The model consists of 92 species, 13 rules and 61 species. The reduced model has 11 rules and 5 proteins.

Fig. 4). After applying the reduction, the Kappa model is reduced to 11 rules and 5 proteins.

In Fig. 5(left), we plot the mean for the CI copy number obtained from 100 runs of the original and of the reduced model, and the graphs show agreement.

In Fig. 5(right), we compared the probability of lysogeny before and after the reduction of the model (lysogeny profile is detected if there are 328 molecules of CI before there are 133 molecules of Cro). The graphs show overall agreement in predicting the lysogeny profile. More precisely, for two and less MOI's (multiplicities of infection), the probability of lysogeny is almost negligible; For three MOIs, both graphs show that lysogeny and lysis are equally probable (the reduced model reports slightly larger probability), and for five or more MOI's, both graphs show that lysogeny is highly probable. While one simulation of

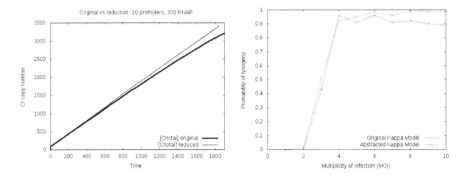

Fig. 5. (left) Average trace of 100 simulations of the original model (solid) and the reduced model (thin) after the reduction, for initially 10 λ phage cells (multiplicities of infection – MOI's). The simulation time for one simulation trace of the original model is ≈ 40 minutes of CPU time, and of the reduced model is 5 seconds of CPU time. The initial number of proteins CI, Cro, CII and CIII and N is set to 100. (right) Comparison of the probability of lysogeny before and after the reduction of the model (lysogeny profile is detected if there are 328 molecules of CI before there are 133 molecules of Cro). The profile was obtained by running 1000 simulations of the model for one cell cycle (2100 time units), for MOIs ranging from 1 to 10.

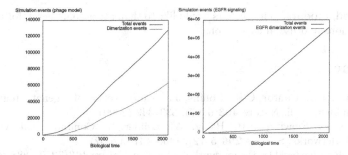

Fig. 6. (a) The ratio of dimerisation events vs. total events in lambda phage model. The number of dimerisation events takes roughly half of the total events over the whole cell cycle. (b) The ratio of dimerisation events vs. total events in EGFR/insulin model. The number of dimerisation events takes only a small fraction of the total events over the whole cell cycle.

the original model takes about 40 mins, one simulation of the abstracted model takes about 5 seconds. Once again, the results are similar, with a significant improvement in simulation speed.

The tool is available for download [37].

8   Conclusion and Future Work

The presented method can be seen as the first step towards a systematic time-scale separation of stochastic rule-based models. We plan to extend this work in several directions. First, we plan to develop the theory for formally comparing the original and reduced model, and subsequently to develop efficient algorithms for assessing the error between the systems; our major interest is in bounding the approximation error *without* executing the original system. To this end, the error can be measured with respect to a given observable, or, more generally, with respect to a given qualitative property specified in, for example, linear temporal logic (LTL). Second, we plan to investigate how the algorithm can exploit the specificities of rule-based models and potentially result in more efficient pattern recognition. Finally, we plan to test the applicability of the reduction algorithm on other case studies.

In particular, we plan to extend the set of approximation patterns so to obtain good reductions for complex models of signaling pathways. More precisely, while our tool is applicable to any rule-based model, the chosen set of approximation patterns are tailored for GRNs and may not provide significant reductions when applied to the signaling pathways. To illustrate this, we applied the reduction to the EGF/insuling crosstalk model, and we observe that the number of dimerisation events does not take the significant portion of all events (see Fig. 6), at least not as radically as it was the case with the λ-phage example. To this end, we plan to include more patterns for reducing signaling pathways, by, for example, approximating multiple phosphorylation events.

Acknowledgements. The authors would like to thank to Jérôme Feret, for the inspiring discussions and useful suggestions.

References

1. Gardner, T.S., Cantor, C.R., Collins, J.J.: Construction of a genetic toggle switch in escherichia coli. Nature **403**(6767), 339–342 (2000)
2. Guet, C.C., Elowitz, M.B., Hsing, W., Leibler, S.: Combinatorial synthesis of genetic networks. Science **296**(5572), 1466–1470 (2002)
3. Kwok, R.: Five hard truths for synthetic biology. Nature **463**(7279), 288–290 (2010)
4. Del Vecchio, D.: Design and analysis of an activator-repressor clock in e. coli. In: American Control Conference. ACC 2007, pp. 1589–1594. IEEE (2007)
5. Feret, J., Krivine, J.: Kasim: a simulator for kappa (2008–2013). http://www.kappalanguage.org
6. Blinov, M.L., Faeder, J.R., Goldstein, B., Hlavacek, W.S.: Bionetgen: software for rule-based modeling of signal transduction based on the interactions of molecular domains. Bioinformatics **20**(17), 3289–3291 (2004)
7. Danos, V., Feret, J., Fontana, W., Krivine, J.: Scalable simulation of cellular signaling networks. In: Shao, Z. (ed.) APLAS 2007. LNCS, vol. 4807, pp. 139–157. Springer, Heidelberg (2007)
8. Hogg, J.S., Harris, L.A., Stover, L.J., Nair, N.S., Faeder, J.R.: Exact hybrid particle/population simulation of rule-based models of biochemical systems. PLoS Comput. Biol. **10**(4), e1003544 (2014)
9. Danos, V., Feret, J., Fontana, W., Krivine, J.: Abstract interpretation of reachable complexes in biological signalling networks. In: Proceedings of the 9th International Conference on Verification, Model Checking and Abstract Interpretation (VMCAI 2008), vol. 4905, pp. 42–58 (2008)
10. Danos, V., Feret, J., Fontana, W., Harmer, R., Krivine, J.: Rule-based modelling of cellular signalling. In: Caires, L., Vasconcelos, V.T. (eds.) CONCUR 2007. LNCS, vol. 4703, pp. 17–41. Springer, Heidelberg (2007)
11. Myers, C.J.: Engineering Genetic Circuits. CRC Press, Boca Raton (2011)
12. Kuwahara, H., Myers, C.J., Samoilov, M.S., Barker, N.A., Arkin, A.P.: Automated abstraction methodology for genetic regulatory networks. In: Priami, C., Plotkin, G. (eds.) Transactions on Computational Systems Biology VI. LNCS (LNBI), vol. 4220, pp. 150–175. Springer, Heidelberg (2006)
13. Ptashne, M.: A Genetic Switch: Gene Control and Phage Lambda. Blackwell Scientific Publications, Palo Alto (1986)
14. Ptashne, M.A.: Genes and Signals. Cold Spring Harbor Laboratory Press, New York (2001)
15. Cousot, P.: Abstract interpretation based formal methods and future challenges. In: Wilhelm, R. (ed.) Informatics: 10 Years Back, 10 Years Ahead. LNCS, vol. 2000, pp. 138–156. Springer, Heidelberg (2001)
16. Burch, J.R., Clarke, E.M., McMillan, K.L., Dill, D.L., Hwang, L.-J.: Symbolic model checking: 10^{20} states and beyond. Inf. Comput. **98**(2), 142–170 (1992)
17. Feret, J., Danos, V., Krivine, J., Harmer, R., Fontana, W.: Internal coarse-graining of molecular systems. Proc. Nat. Acad. Sci. **106**(16), 6453–6458 (2009)
18. Ganguly, A., Petrov, T., Koeppl, H.: Markov chain aggregation and its applications to combinatorial reaction networks. J. Math. Biol. **69**, 767–797 (2013)
19. Feret, J., Henzinger, T., Koeppl, H., Petrov, T.: Lumpability abstractions of rule-based systems. Theor. Comput. Sci. **431**, 137–164 (2012)

20. Conzelmann, H., Saez-Rodriguez, J., Sauter, T., Kholodenko, B.N., Gilles, E.D.: A domain-oriented approach to the reduction of combinatorial complexity in signal transduction networks. BMC Bioinf., 7:34 (2006)

21. Borisov, N.M., Markevich, N.I., Hoek, J.B., Kholodenko, B.N.: Signaling through receptors and scaffolds: independent interactions reduce combinatorial complexity. Biophys. J. **89**(2), 951–966 (2005)

22. Craciun, G., Feinberg, M.: Multiple equilibria in complex chemical reaction networks: Ii. the species-reaction graph. SIAM J. Appl. Math. **66**(4), 1321–1338 (2006)

23. Gillespie, D.T.: Exact stochastic simulation of coupled chemical reactions. J. Phys. Chem. **81**(25), 2340–2361 (1977)

24. Kurtz, T.G.: Limit theorems for sequences of jump Markov processes approximating ordinary differential processes. J. Appl. Probab. **8**(2), 344–356 (1971)

25. Kang, H.-W., Kurtz, T.G.: Separation of time-scales and model reduction for stochastic reaction networks. Ann. Appl. Prob. **23**(2), 529–583 (2013)

26. Danos, V., Feret, J., Fontana, W., Krivine, J.: Abstract interpretation of cellular signalling networks. In: Logozzo, F., Peled, D.A., Zuck, L.D. (eds.) VMCAI 2008. LNCS, vol. 4905, pp. 83–97. Springer, Heidelberg (2008)

27. Beica, A., Guet, C., Petrov, T.: Efficient reduction of kappa models by static inspection of the rule-set (2015). arXiv preprint arXiv:1501.00440

28. Murray, J.D.: Mathematical Biology I: an Introduction, vol. 17 of Interdisciplinary Applied Mathematics. Springer, New York (2002)

29. Rao, C.V., Arkin, A.P.: Stochastic chemical kinetics and the quasi-steady-state assumption: application to the gillespie algorithm. J. Chem. Phys. **118**(11), 4999–5010 (2003)

30. Haseltine, E.L., Rawlings, J.B.: Approximate simulation of coupled fast and slow reactions for stochastic chemical kinetics. J. Chem. Phys. **117**(15), 6959–6969 (2002)

31. Crudu, A., Debussche, A., Radulescu, O.: Hybrid stochastic simplifications for multiscale gene networks. BMC Syst. Biol. **3**(1), 89 (2009)

32. Hepp, B., Gupta, A., Khammash, M.: Adaptive hybrid simulations for multiscale stochastic reaction networks (2014). arXiv preprint arXiv:1402.3523

33. Sanft, K.R., Gillespie, D.T., Petzold, L.R.: Legitimacy of the stochastic michaelis-menten approximation. IET Syst. Biol. **5**(1), 58–69 (2011)

34. Gillespie, D.T., Cao, Y., Sanft, K.R., Petzold, L.R.: The subtle business of model reduction for stochastic chemical kinetics. J. Chem. Phys. **130**(6), 064103 (2009)

35. Darden, T.A.: A pseudo-steady-state approximation for stochastic chemical kinetics. PhD thesis, University of California, Berkeley (1979)

36. Arkin, A., Ross, J., McAdams, H.H.: Stochastic kinetic analysis of developmental pathway bifurcation in phage λ-infected escherichia coli cells. Genetics **149**(4), 1633–1648 (1998)

37. Beica, A., Petrov, T.: Kared (2014). http://pub.ist.ac.at/~tpetrov/KappaRed.tar.gz

Application of Advanced Models on Case Studies

Model Checking Tap Withdrawal in C. Elegans

Md. Ariful Islam[1]([✉]), Richard De Francisco[1], Chuchu Fan[3], Radu Grosu[1,2],
Sayan Mitra[3], and Scott A. Smolka[1]

[1] Department of Computer Science, Stony Brook University, New York, USA
mdaislam@cs.stonybrook.edu
[2] Department of Computer Engineering, Vienna University of Technology,
Vienna, Austria
[3] Department of Electrical and Computer Engineering,
University of Illinois Urbana Champaign, Champaign, USA

Abstract. We present what we believe to be the first formal verification
of a biologically realistic (nonlinear ODE) model of a neural circuit in a
multicellular organism: Tap Withdrawal (TW) in C. Elegans, the common roundworm. TW is a reflexive behavior exhibited by C. Elegans in
response to vibrating the surface on which it is moving; the neural circuit
underlying this response is the subject of this investigation. Specially, we
perform reach-tube-based reachability analysis on the TW circuit model
of Wicks et al. (1996) to estimate key model parameters. Underlying our
approach is the use of Fan and Mitra's recently developed technique for
automatically computing local discrepancy (convergence and divergence
rates) of general nonlinear systems.

The results we obtain are a significant extension of those of Wicks
et al. (1996), who equip their model with fixed parameter values that
reproduce the predominant TW response they observed experimentally
in a population of 590 worms. In contrast, our techniques allow us to
much more fully explore the model's parameter space, identifying in the
process the parameter ranges responsible for the predominant behavior
as well as the non-dominant ones. The verification framework we developed to conduct this analysis is model-agnostic, and can thus be re-used
on other complex nonlinear systems.

1 Introduction

Although neurology and brain modeling/simulation is a popular field of biological study, formal verification has yet to take root. There has been cursory study
into neurological model checking (see Sect. 2), but not with the nonlinear ODE
models used by biologists. The application of verification technology to hardware circuits has played a key role in the *Electronic Design Automation* (EDA)
industry; perhaps it will play a similar role with neural circuits.

For our initial neurological study, we have selected the round worm,
Caenorhabditis Elegans, due to the simplicity of its nervous system (302 neurons,
$\sim$5,000 synapses) and the breadth of research on the animal. The complete con-
nectome of the worm is documented, and there have been a number of interesting
experiments on its response to stimuli.

© Springer International Publishing Switzerland 2015
A. Abate and D. Šafránek (Eds.): HSB 2015, LNBI 9271, pp. 195–210, 2015.
DOI: 10.1007/978-3-319-26916-0_11

For model-checking purposes, we were particularly interested in the *tap with-drawal* (TW) neural circuit. The TW circuit governs the reactionary motion of the animal when the petri dish in which it swims is perturbed. (A related circuit, *touch sensitivity*, controls the reaction of the worm when a stimulus is applied to a single point on the body.) Studies of the TW circuit have traditionally involved using lasers to ablate the different neurons in the circuit of multiple animals and measuring the results when stimuli are applied.

A model of the TW circuit was presented by Wicks, Roehrig, and Rankin in [16]. Their model is in the form of a system of nonlinear ODEs with an indication of polarity (inhibitory or excitatory) of each neuron in the TW circuit. Additionally, Wicks and Rankin had a previous paper in which they measure the three possible reactions of the animals to TW with various neurons ablated [15]; see also Fig. 3. The three behaviors—acceleration, reversal of movement, and no response—are logged with the percentage of the experimental population to display that behavior.

The [16] model has a number of circuit parameters, such as gap-junction conductance, capacitance, and leakage current, that crucially affect the behavior of the organism. A single value for each parameter is given in [16]. With this single set of parameter values, the model produces predominant behavior in most ablation groups with a few exceptions.

While the experimental work in [15,16] and the model presented in [16] were by no means insubstantial, the exploration of the model is vastly incomplete. The fixed parameter values fit through experimentation cause the model to replicate the predominant behavior seen in said experiments, but little can be said about the model beyond that. The ranges that can produce the predominant behavior, as well as the two other behaviors, are completely missing. This is not to fault the authors of [16], however, as the technology needed to uncover these ranges simply did not exist at the time.

The missing technology was the ability to automatically generate local discrepancy functions [2], and has only recently been developed [5]. With this technique, we can theoretically compute reach tubes used in verification. In reality, this is not a simple plug-and-play situation. To make use of [5], we needed to create the verification framework in Fig. 1. Through careful model engineering Fig. 1 (1–3) and verification engineering Fig. 1 (4–6) we were able to explore and verify the full parameter ranges in the Wicks et al. model to produce all three behaviors in the TW circuit. Such an understanding of the model is critical to morphospace exploration [14] of the animal. A detailed description of our framework and its application to the [16] model Fig. 1(b) is given in Sect. 4.

This verification framework has the additional benefit of being model agnostic. It can be reused to verify other complex nonlinear ODE models.

The rest of the paper develops as follows. Section 2 reviews related work. Section 3 provides requisite background material on the TW neural circuit, its reactionary behavior, and the ODE model of [16]. Section 4 describes our reach-tube reachability analysis and associated property checking. Section 5 presents our extensive collection of model-checking/parameter-estimation results. Section 6 offers our concluding remarks and directions for future work.

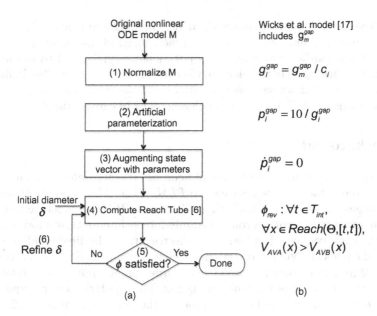

Fig. 1. Verification framework of nonlinear ODE model based on automatic computation of discrepancy function. (a) The general framework, (b) Application to [16] model.

2 Related Work

Iyengar et al. [10] present a Pathway Logic (PL) model of neural circuits in the marine mollusk *Aplysia*. Specifically, the circuits they focus on are those involved in neural plasticity and memory formation. PL systems do not use differential equations, favoring qualitative symbolic models. They do not argue that they can replace traditional ODE systems, but rather that their qualitative insights can support the quantitative analysis of such systems. Neurons are expressed in terms of rewrite rules and data types. Their simulations, unlike our reachability analysis, do not provide exhaustive exploration of the state space. Additionally, PL models are abstractions usually made in collaboration between computer scientists and biologists. Our work meets the biologists on their own terms, using the pre-existing ODE systems developed from physiological experiments.

Tiwari and Talcott [13] build a discrete symbolic model of the neural circuit Central Pattern Generator (CPG) in *Aplysia*. The CPG governs rhythmic foregut motion as the mollusk feeds. Working from a physiological (non-linear ODE) model, they abstract to a discrete system and use the Symbolic Analysis Laboratory (SAL) model checker to verify various properties of this system. They cite the complexity of the original model and the difficulty of parameter estimation as motivation for their abstraction. Neuronal inputs can be positive, negative, or zero and outputs are boolean: a pulse is generated or not. Our approach uses the original biological model of the TW circuit of *C. Elegans* [16], and through reachability analysis, we obtain the parameter ranges of interest.

We have extensive experience with model checking and reachability analysis in the cardiac domain, e.g. [6,8,9,12]. In fact, much of our previous work has focused on the cardiac myocyte, a computationally similar cell to the neuron. This is not surprising as both belong to the class of *excitable cells*. The similarities are so numerous that we have used a variation of the Hodgkin-Huxley model of the squid giant axon [7] to model ion channel flow in cardiac tissue.

3   Background

In *C. Elegans*, there are three classes of neurons: *sensory*, *inter*, and *motor*. For the TW circuit, the sensory neurons are *PLM, PVD, ALM,* and *AVM*, and the inter-neurons are *AVD, DVA, PVC, AVA,* and *AVB*. The model we are using abstracts away the motor neurons as simply forward and reverse movement.

Neurons are connected in two ways: electrically via bi-directional *gap junctions*, and chemically via uni-directional chemical *synapses*. Each connection has varying degrees of throughput, and each neuron can be *excitatory* or *inhibitory*, governing the polarity of transmitted signals. These polarities were experimentally determined in [16], and used to produce the circuit shown in Fig. 2.

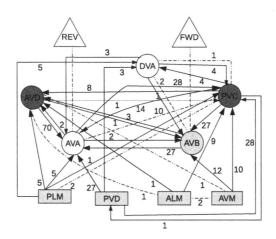

Fig. 2. Tap Withdrawal Circuit of C. Elegans. Rectangle: Sensory Neurons; Circle: Inter-neurons; Dashed Undirected Edge: Gap Junction; Solid Directed Edge: Chemical Synapse; Edge Label: Number of Connections; Dark Gray: Excitatory Neuron; Light Gray: Inhibitory Neuron; White: Unknown Polarity. FWD: Forward Motor system; REV: Reverse Motor System.

The TW circuit produces three distinct locomotive behaviors: *acceleration*, *reversal* of movement, and a *lack of response*. In [15], Wicks et al. performed a series of laser ablation experiments in which they knocked out a neuron in a group of animals (worms), subjected them to a tapped surface, and recorded the magnitude and direction of the resulting behavior. Figure 3 shows the response types for each of their experiments.

The dynamics of a neuron's membrane potential, V, is determined by the sum of all input currents, written as:

$$C\dot{V} = \frac{1}{R}(V^{leak} - V) + \sum I^{gap} + \sum I^{syn} + I^{stim}$$

where C is the *membrane capacitance*, R is the *membrane resistance*, V^{leak} is the *leakage potential*, I^{gap} and I^{syn} are gap-junction and the chemical synapse currents, respectively, and I^{stim} is the applied external *stimulus* current. The summations are over all neurons with which this neuron has a (gap-junction or synaptic) connection.

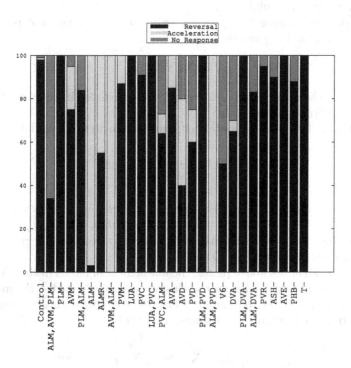

Fig. 3. Effect of ablation on Tap Withdrawal reflex (experimental results). The length of the bars indicate the fraction of the population demonstrating the particular behavior [15].

The current flow between neuron i and j via a gap-junction is given by:

$$I^{gap}_{ij} = n^{gap}_{ij} g^{gap}_m (V_j - V_i)$$

where the constant g^{gap}_m is the *maximum conductance* of the gap junction, and n^{gap}_{ij} is the number of gap-junction connections between neurons i and j. The conductance g^{gap}_m is one of the key circuit parameters of this model that dramatically affects the behavior of the animal.

The synaptic current flowing from pre-synaptic neuron j to post-synaptic neuron i is described as follows:

$$I_{ij}^{syn} = n_{ij}^{syn} g_{ij}^{syn}(t)(E_j - V_i)$$

where $g_{ij}^{syn}(t)$ is the time-varying synaptic conductance of neuron i, n_{ij}^{syn} is the number of synaptic connections from neuron j to neuron i, and E_j is the *reversal potential* of neuron j for the synaptic conductance.

The chemical synapse is characterized by a synaptic sign, or polarity, specifying if said synapse is excitatory or inhibitory. The value of E_j is assumed to be constant for the same synaptic sign; its value is higher if the synapse is excitatory rather than inhibitory.

Synaptic conductance is dependent only upon the membrane potential of presynaptic neuron V_j, given by:

$$g_{ij}^{syn}(t) = g_{\infty}^{syn}(V_j)$$

where g_{∞}^{syn} is the steady-state *post-synaptic conductance* in response to a pre-synaptic membrane potential.

The steady-state post-synaptic membrane conductance is modeled as:

$$g_{\infty}^{syn}(V_j) = \frac{g_m^{syn}}{1 + \exp(k\frac{V_j - V_j^{eq}}{V_{Range}})}$$

where g_m^{syn} is the *maximum post-synaptic membrane conductance* for the synapse, V_j^{eq} is the *pre-synaptic equilibrium potential*, and V_{Range} is the *pre-synaptic voltage range* over which the synapse is activated. k is an experimentally derived constant, valued at -4.3944.

Combining all of the above pieces, the mathematical model of the TW circuit is a system of nonlinear ODEs, with each state variable defined as the membrane potential of a neuron in the circuit. Consider a circuit with N neurons. The dynamics of the ith neuron of the circuit is given by:

$$C_i \dot{V_i} = \frac{V_{l_i} - V_i}{R_i} + \sum_{j=1}^{N} I_{ij}^{gap} + \sum_{j=1}^{N} I_{ij}^{syn} + I_i^{stim} \tag{1}$$

$$I_{ij}^{gap} = n_{ij}^{gap} g_m^{gap}(V_j - V_i) \tag{2}$$

$$I_{ij}^{syn} = n_{ij}^{syn} g_{ij}^{syn}(E_j - V_i) \tag{3}$$

$$g_{ij}^{syn} = \frac{g_m^{syn}}{1 + \exp(k\frac{V_j - V_j^{eq}}{V_{Range}})}. \tag{4}$$

The equilibrium potentials (V^{eq}) of the neurons are computed by setting the left-hand side of Eq. (1) to zero. This leads to a system of linear equations, that can be solved as follows:

$$V^{eq} = A^{-1}b \tag{5}$$

where matrix A is given by:

$$A_{ij} = \begin{cases} -R_i n_{ij}^{gap} g_m^{gap} & \text{if } i \neq j \\ 1 + R_i \sum_{j=1}^{N} n_{ij}^{gap} g_{ij}^{gap} g_m^{syn}/2 & \text{if } i = j \end{cases}$$

and vector b is written as:

$$b_i = V_{l_i} + R_{m_i} \sum_{j=1}^{N} E_j n_{ij}^{syn} g_m^{syn}/2.$$

The potential of the motor neurons AVB and AVA determine the observable behavior of the animal. If the integral of the difference between V_{AVA} - V_{AVB} is large, the animal will reverse movement. By extension, if the difference is a large negative value, the animal will accelerate, and if the difference is close to zero there will be no response. The equation that converts the membrane potential of AVB and AVA to a behavioral property, (e.g. *reversal*), is given by:

$$\text{Propensity to Reverse} \propto \int (V_{AVA} - V_{AVB})dt \qquad (6)$$

where the integration is computed from the beginning of tap stimulation until either the simulation ends or the integrand changes sign. To allow initial transients after the tap, the test for a change of integrand sign occurs only after a grace period of 100 ms.

For the purpose of reachability analysis (Sect. 4), we normalize the system of equations with respect to the capacitance. This correlates to step (1) in Fig. 1. Combining Eqs. (1) and (4) and taking C_{m_i} to the right-hand side, we have:

$$\dot{V}_i = \frac{V_{l_i} - V_i}{R_i C_i} + \frac{g_m^{gap}}{C_i} \sum_{j=1}^{N} n_{ij}^{gap}(V_j - V_i) + \frac{g_m^{syn}}{C_i} \sum_{j=1}^{N} \frac{n_{ij}^{syn}(E_j - V_i)}{1 + \exp{(k \frac{V_j - V^{EQ}}{V_{Range}})}} + \frac{1}{C_i} I_i^{stim}$$

Now letting $g_i^{leak} = \frac{1}{R_i C_i}$, $g_i^{gap} = \frac{g_m^{gap}}{C_{m_i}}$, $g_i^{syn} = \frac{g_m^{syn}}{C_{m_i}}$ and $I_i^{ext} = \frac{1}{C_{m_i}}$ the system dynamics can be written as:

$$\dot{V}_i = g_i^{leak}(V_{l_i} - V_i) + g_i^{gap} \sum_{j=1}^{N} n_{ij}^{gap}(V_j - V_i) + g_i^{syn} \sum_{j=1}^{N} \frac{n_{ij}^{syn}(E_j - V_i)}{1 + \exp{(k \frac{V_j - V_j^{eq}}{V_{Range}})}} + I_i^{ext} \qquad (7)$$

This is the 9 dimensional ODE model of the TW circuit. The key circuit parameters are the gap conductances, g_i^{gap}, and we aim to characterize the ranges of these conductances that produce acceleration, reversal, and no response.

4 Reachability Analysis of Nonlinear TW Circuit

Reachability analysis for verifying properties for general nonlinear dynamical systems is a well-known hard problem. The verification framework introduced in Fig. 1 combines model and verification engineering to perform reachability analysis on the Wicks et al. [16] model, discovering crucial parameter ranges to produce all three behaviors of the TW circuit. Our framework can be applied to any nonlinear ODE model.

4.1   Background on Reachability Using Discrepancy

Consider an n-dimensional *autonomous dynamical system*:

$$\dot{x} = f(x), \tag{8}$$

where $f : \mathbb{R}^n \to \mathbb{R}^n$ is a Lipschitz continuous function. A *solution* or a *trajectory* of the system is a function $\xi : \mathbb{R}^n \times \mathbb{R}_{\geq 0} \to \mathbb{R}^n$ such that for any initial point $x_0 \in \mathbb{R}^n$ and at any time $t > 0$, $\xi(x_0, t)$ satisfies the differential Eq. (8). A state x in $\mathbb{R}^n$ is *reachable from the initial set* $\Theta \subseteq \mathbb{R}^n$ *within a time interval* $[t_1, t_2]$ if there exists an initial state $x_0 \in \Theta$ and a time $t \in [t_1, t_2]$ such that $x = \xi(x_0, t)$. The set of all reachable states in the interval $[t_1, t_2]$ is denoted by $\mathsf{Reach}(\Theta, [t_1, t_2])$. If $t_1 = 0$, we write $\mathsf{Reach}(t_2)$ when set Θ is clear from the context. If we can compute or approximate the reach set of such a model, then we can check for invariant or temporal properties of the model. Specifically, *C. Elegans* TW properties such as accelerated forward movement or reversal of movement fall into these categories. Our core reachability algorithm [2,3,8] uses a simulation engine that gives sampled numerical simulations of (8).

Definition 1. *A* (x_0, τ, ϵ, T)-*simulation of (8) is a sequence of time-stamped sets* (R_0, t_0), $(R_1, t_1) \ldots, (R_n, t_n)$ *satisfying:*

1. *Each* R_i *is a compact set in* $\mathbb{R}^n$ *with* $dia(R_i) \leq \epsilon$.
2. *The last time* $t_n = T$ *and for each* i, $0 < t_i - t_{i-1} \leq \tau$, *where the parameter* τ *is called the* sampling period.
3. *For each* t_i, *the trajectory from* x_0 *at* t_i *is in* R_i, *i.e.,* $\xi(x_0, t_i) \in R_i$, *and for any* $t \in [t_{i-1}, t_i]$, *the solution* $\xi(x_0, t) \in hull(R_{i-1}, R_i)$.

The algorithm for reachability analysis uses a key property of the model called a *discrepancy function*.

Definition 2. *A uniformly continuous function* $\beta : \mathbb{R}^n \times \mathbb{R}^n \times \mathbb{R}_{\geq 0} \to \mathbb{R}_{\geq 0}$ *is a* discrepancy function *of (8) if*

1. *for any pair of states* $x, x' \in \mathbb{R}^n$, *and any time* $t > 0$,

$$\|\xi(x, t) - \xi(x', t)\| \leq \beta(x, x', t), \text{ and} \tag{9}$$

2. *for any* t, *as* $x \to x'$, $\beta(., ., t) \to 0$.

If a function β meets the two conditions for any pair of states x, x' in a compact set K then it is called a K-*local discrepancy function*. Uniform continuity means that $\forall \epsilon > 0, \forall x, x' \in K, \exists \delta$ such that for any time t, $\|x - x'\| < \delta \Rightarrow \beta(x, x', t) < \epsilon$. The verification results in [2–4,8] required the user to provide the discrepancy function β as an additional input for the model. A Lipschitz constant of the dynamic function f gives an exponentially growing β, contraction metrics [11] can give tighter bounds for incrementally stable models, and sensitivity analysis gives tight bounds for linear systems [1], but none of these give an algorithm for computing β for general nonlinear models. Therefore, finding the discrepancy can be a barrier in the verification of large models like the TW circuit.

Here, we use Fan and Mitra's recently developed approach that automatically computes local discrepancy along individual trajectories [5]. Using the simulations and discrepancy, the reachability algorithm for checking properties proceeds as follows: Let the U be the set of states that violate the invariant in question. First, a δ-cover C of the initial set Θ is computed; that is, the union of all the δ-balls around the points in C contain Θ. This δ is chosen to be large enough so that the cardinality of C is small. Then the algorithm iteratively and selectively refines C and computes more and more precise over-approximations of $\mathsf{Reach}(\Theta, T)$ as a union $\cup_{x_0 \in C}\mathsf{Reach}(B_\delta(x_0), T)$. Here, $\mathsf{Reach}(B_\delta(x_0), T)$ is computed by first generating a (x_0, τ, ϵ, T)-simulation and then bloating it by a factor that maximizes $\beta(x, x', t)$ over $x, x' \in B_\delta(s_0)$ and $t \in [t_{i-1}, t_i]$. If $\mathsf{Reach}(B_\delta(x_0), T)$ is disjoint from U or is (partly) contained in U, then the algorithm decides that $B_\delta(x_0)$ satisfies and violates U, respectively. Otherwise, a finer cover of $B_\delta(x_0)$ is added to C and the iterative selective refinement continues. We refer to this in this paper as δ-refinement. In [2], it is shown that this algorithm is sound and relatively complete for proving bounded time invariants.

4.2   Applying Local Discrepancy to TW Circuit

Fan and Mitra's algorithm (see details in [5]) for automatically computing local discrepancy relies on the Lipschitz constant and the Jacobian of the dynamic function, along with simulations. The Lipschitz constant is used to construct a coarse, one-step over-approximation S of the reach set of the system along a simulation. Then the algorithm computes an upper bound on the maximum eigenvalue of the symmetric part of the Jacobian over S, using a theorem from matrix perturbation theory. This gives a piecewise exponential β, but the exponents are tight as they are obtained from the maximum eigenvalue of the linear approximation of the system in S. This means that for models with convergent trajectories, the exponent of β over S will be negative, and the $\mathsf{Reach}(T)$ approximation will quickly become very accurate. In the rest of this section, we describe key steps involved in making this approach work with the TW circuit.

The model of the TW circuit from Sect. 3 can be written as $\dot{V} = f(V)$, where $V \in \mathbb{R}^9$ has components V_i giving the membrane potential of neuron i. The Jacobian of the system is the matrix of partial derivatives with the ij^{th} term given by:

$$\frac{\partial f_i}{\partial V_j} = -g_i^{leak} - g_i^{gap} \sum_{j=1, j \neq i}^{N} n_{ij}^{gap} - g_i^{syn} \sum_{j=1, j \neq i}^{N} \frac{n_{ij}^{syn}}{1 + \exp(k\frac{V_j - V_j^{eq}}{V_{Range}})}$$

$$= g_i^{gap} n_{ij}^{gap} - g_i^{syn} n_{ij}^{syn} \frac{\frac{k}{V_{Range}} \exp(k\frac{V_j - V_j^{eq}}{V_{Range}})(E_j - V_i)}{(1 + \exp(k\frac{V_j - V_j^{eq}}{V_{Range}}))^2} \tag{10}$$

For parameter-range estimation of the TW circuit, each parameter p of interest is added as a new variable with constant dynamics ($\dot{p} = 0$). Computing the reach-set from initial values of p is then used to verify or falsify invariant properties for a continuous range of parameter values, and therefore a whole family

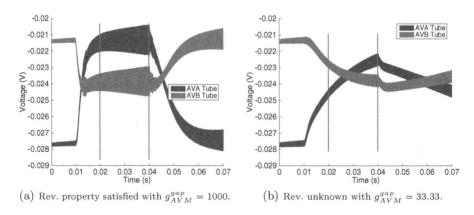

(a) Rev. property satisfied with $g_{AVM}^{gap} = 1000$.   (b) Rev. unknown with $g_{AVM}^{gap} = 33.33$.

Fig. 4. Model Checking Reversal Property of Control Group, with $\delta = 5 \times 10^{-5}$, varying g_{AVM}^{gap}.

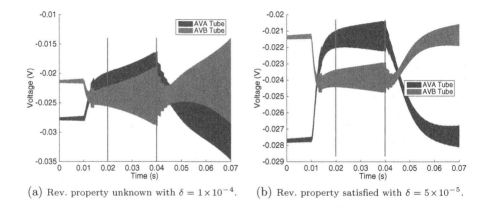

(a) Rev. property unknown with $\delta = 1 \times 10^{-4}$.   (b) Rev. property satisfied with $\delta = 5 \times 10^{-5}$.

Fig. 5. Model Checking Reversal Property of Control Group by refining δ.

of models, instead of analyzing just a single member of that family. Here the parameters of interest are the quantities $p_i^{leak} = 1/g_i^{leak}, p_i^{gap} = 10/g_i^{gap}, p_i^{syn} = 1/g_i^{syn}$.

Consider, for example, $1/g_i^{leak}$ as a parameter:

$$\begin{bmatrix} \dot{V} \\ 1/g_i^{leak} \end{bmatrix} = \begin{bmatrix} f(V) \\ 0 \end{bmatrix}.$$

In this case the Jacobian matrices for the system with parameters will be singular because of the all-zero rows that come from the parameter dynamics. The zero eigenvalues of these singular matrices are taken into account automatically by the algorithm for computing local discrepancy. In this paper we focus on p_i^{gap}, leaving the others for future work.

4.3   Checking Properties

Once the reach sets are computed, checking the *acceleration, reversal*, and *no-response* properties are conceptually straightforward. For instance, Eq. (6) gives a method to check reversal movement. Instead of computing the integral of $(V_{AVA} - V_{AVB})$, we use the following sufficient condition to check it:

$$\phi_{rev} : \forall\, t \in T_{int}, \forall\, x \in \mathsf{Reach}(\Theta, [t, t]), V_{AVA}(x) > V_{AVB}(x).$$

Here, T_{int} is a specific time interval after the stimulation time, Θ is the initial set with parameter ranges, and recall that $\mathsf{Reach}(\Theta, [t, t])$ is the set of states reached at time t from Θ. We implement this check by scanning the entire reach-tube and checking that its projection on $V_{AVB}(x)$ is above that of $V_{AVA}(x)$ over all intervals. If this check succeeds (as in Fig. 4(a)), we conclude that the range of parameter values produce the reversal movement. If the check fails, then the reversal movement is not provably satisfied (Fig. 5(a)) and in that case we δ-refine the initial partition (Fig. 5(b)). In some cases, such as Fig. 4(b), δ-refinement can not prove the property satisfied or unsatisfied. This often occurs when two tubes intersect within the interval of interest. In this case, the property is considered to be unknown.

Figure 6 helps paint a picture of how the δ-refinement process works with two parameters. We consider 4 refinement steps: $\delta = 7 \times 10^{-5}$, $\delta = 6 \times 10^{-5}$, $\delta = 5.5 \times 10^{-5}$, and $\delta = 5 \times 10^{-5}$. For $\delta = 7 \times 10^{-5}$, the property of interest is unknown at all points. With $\delta = 6 \times 10^{-5}$ the property is considered unknown for all red areas in the figure, including red and blue areas. Blue areas show where $\delta = 5.5 \times 10^{-5}$ are satisfied, and in the blue and yellow area both $\delta = 6 \times 10^{-5}$ and $\delta = 5.5 \times 10^{-5}$ have a satisfied property. The property is satisfied for the

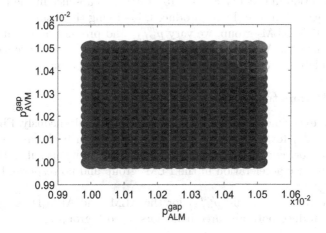

Fig. 6. Example of 2-D Parameter Refinement. Red Regions are Unknown for both $\delta = 6 \times 10^{-5}$ and $\delta = 5.5 \times 10^{-5}$, Red/Blue Regions are Unknown for $\delta = 6 \times 10^{-5}$, but Satisfied for $\delta = 5.5 \times 10^{-5}$, and Yellow/Blue Regions are Satisfied for both (Color figure online).

entire range of the graph when $\delta = 5 \times 10^{-5}$. Thus, the refinement process stops at $\delta = 5 \times 10^{-5}$, and the entire range of the parameter space is characterized.

5   Experimental Results

In this section, we apply our verification framework to the [16] model to estimate parameter ranges that produce three different behaviors (*reversal, acceleration, no response*) in the control and four ablation groups. We vary the gap-junction conductance of the sensory neurons (g_i^{gap}, $i \in \{AVM, ALM, PLM\}$) and keep all other parameters constant, as per [16]. Additionally, in the case of the no response behavior, we must lower the gap-junction conductance of the other neurons by a factor of 10^3.

In Sect. 4, we explain that we use p_i^{gap} as our parameter in the state vector instead of g_i^{gap}, where $p_i^{gap} = 10/g_i^{gap}$. The parameter space we explore can be considered a bounding box, where each p_i^{gap} ranges over $[0.01, 1]$. As exploring the entire parameter space is computationally intensive, we intelligently select a subspace to cover that lets us estimate contiguous ranges of parameters for each behavior. In Table 1, we present these ranges in terms of g_i^{gap}.

In the following subsections, we will present our results for parameter range estimation for all three behaviors of the control and ablation groups. This process requires three experiments per group.

5.1   1-D Parameter Space

Here we vary p_{AVM}^{gap} in all groups, except the AVM,ALM- group. By varying this parameter, we are able to produce reversal behavior in all four groups. We are also able to produce acceleration in all groups but PLM-. The PLM neuron drives acceleration in the TW circuit [15]. Hence, its absence in the PLM- group prevents acceleration from being produced, justifying the result.

For the AVM,ALM- group, we vary p_{PLM}^{gap} and produce acceleration and no response behaviors. As both AVM and ALM, responsible for reversal of movement, are ablated, reversal cannot be produced by this group.

5.2   2-D Parameter Space

In this set of experiments, we vary two parameters simultaneously. First we vary p_{AVM}^{gap} and p_{ALM}^{gap} for the control and PLM- groups. In both cases we produce reversal behavior. For the same reasons given in the previous subsection, we are unable to produce acceleration in the PLM- group and no response behavior in both these groups.

Next, we vary p_{AVM}^{gap} and p_{PLM}^{gap} for the ALM- and ALM,DVA- groups. We are able to produce both all three behaviors in both groups.

5.3   3-D Parameter Space

Since the ablation groups we have used in this paper all feature at least one of the primary sensory neurons (*ALM, AVM,* and *PLM*) ablated, we can only show the 3-D case for the original animal.

Table 1. Parameter ranges for all experiments, including δ and runtime information. REV=Reversal, ACC=Acceleration, NR=No Response.

Group Name	Property	Parameters	Ranges	δ	Runtime (sec)
Control	REV	g_{AVM}^{gap}	$[46.2, 1000]$	1×10^{-6}	6324.4
	REV	$g_{AVM}^{gap}, g_{ALM}^{gap}$	$[952.38, 1000]^2$	2×10^{-5}	776.5
	REV	$g_{AVM}^{gap}, g_{ALM}^{gap}, g_{ALM}^{gap}$	$[990.01, 1000]^3$	2×10^{-5}	314.23
	ACC	g_{AVM}^{gap}	$[15.87, 10]$	1×10^{-5}	1110.01
	ACC	$g_{AVM}^{gap}, g_{ALM}^{gap}$	$[15.86, 15.87]^2$	2×10^{-5}	1619.8
	ACC	$g_{AVM}^{gap}, g_{ALM}^{gap}, g_{ALM}^{gap}$	$[15.85, 15.87]^3$	2×10^{-5}	320.12
	NR	g_{AVM}^{gap}	-	-	-
	NR	$g_{AVM}^{gap}, g_{ALM}^{gap}$	-	-	-
	NR	$g_{AVM}^{gap}, g_{ALM}^{gap}, g_{ALM}^{gap}$	$[10.005, 10]^3$	5×10^{-5}	124.23
PLM-	REV	g_{AVM}^{gap}	$[467.3, 1000]$	1×10^{-5}	718.08
	REV	$g_{AVM}^{gap}, g_{ALM}^{gap}$	$[952.38, 1000]^2$	2×10^{-5}	775.12
	ACC	g_{AVM}^{gap}	-	-	-
	ACC	$g_{AVM}^{gap}, g_{ALM}^{gap}$	-	-	-
	NR	g_{AVM}^{gap}	-	-	-
	NR	$g_{AVM}^{gap}, g_{ALM}^{gap}$	$[15.84, 15.87]^2$	5×10^{-5}	124.23
ALM-	REV	g_{AVM}^{gap}	$[467.3, 1000]$	1×10^{-5}	718.08
	REV	$g_{AVM}^{gap}, g_{PLM}^{gap}$	$[952.38, 1000]^2$	2×10^{-5}	785.01
	ACC	g_{AVM}^{gap}	$[15.38, 15.87]$	$2e - 5$	660.87
	ACC	$g_{AVM}^{gap}, g_{PLM}^{gap}$	$[14.91, 14.93]^2$	2×10^{-5}	782.3
	NR	g_{AVM}^{gap}	-	-	-
	NR	$g_{AVM}^{gap}, g_{PLM}^{gap}$	$[10, 10.05]^2$	5×10^{-5}	125.01
ALM,DVA-	REV	g_{AVM}^{gap}	$[250, 500]$	1×10^{-5}	1085.74
	REV	$g_{AVM}^{gap}, g_{PLM}^{gap}$	$[487.80, 500]^2$	2×10^{-5}	779.75
	ACC	g_{AVM}^{gap}	$[13.88, 14.28]$	1×10^{-5}	1084.23
	ACC	$g_{AVM}^{gap}, g_{PLM}^{gap}$	$[15.84, 15.87]^2$	2×10^{-5}	782.3
	NR	g_{AVM}^{gap}	-	-	-
	NR	$g_{AVM}^{gap}, g_{PLM}^{gap}$	$[15.86, 15.87]^2$	2×10^{-5}	779.01
ALM,AVM-	REV	g_{PLM}^{gap}	-	-	-
	ACC	g_{PLM}^{gap}	$[33.33, 1000]$	5×10^{-5}	3619.19
	NR	$g_{PLM}^{gap}, g_{ALM}^{gap}$	$[10, 13.33]$	5×10^{-5}	3118.45

For the 3-D case, in addition to p_{AVM}^{gap} and p_{ALM}^{gap}, we have the p_{PLM}^{gap} conductance. Finally, we get a non-zero value for *no response* in the control, but Table 1 shows that this value is an order of magnitude smaller than *acceleration* and several orders smaller than *reversal*.

5.4 Runtime and Memory Complexity Analysis

The time and memory needed for the procedure depends upon the value of δ used and the size of the parameter space. Assume L_d to be the interval length in

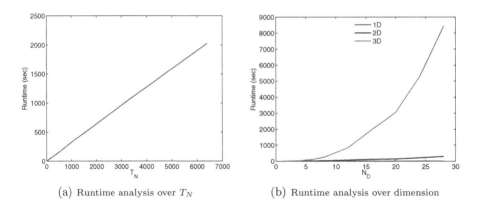

(a) Runtime analysis over T_N (b) Runtime analysis over dimension

Fig. 7. Experiment on runtime analysis.

the dth dimension. The total number of δ-balls required to cover the parameter space completely is:

$$T_N = \Pi_{d=1}^{D} N_d$$

where D is the number of parameters added to the state vector and $N_d = 2L_d/\delta$. If L_d is the same in all dimensions, $T_N = N_d^D$. We can analyze both runtime and memory complexity based on T_N. If we consider the time and memory required for verifying each δ-ball to be $O(1)$, then the time and memory complexity will both be $O(T_N) = O(N_d^D)$. Note that the complexity also depends on the value of the δ-refinement loop counter. Since we can safely assume that the loop will iterate only a constant number of times, this is not an issue.

Figure 7 illustrates how runtime relates to T_N in one (a) and multiple (b) dimensions. The graph from (a) is the same as the 1D line in (b), but for a larger range of T_N. This increased range more clearly illustrates the linear relationship of runtime to T_N when $D = 1$. Part (b) shows the rates for $D = 1$, $D = 2$ and $D = 3$ over a much smaller range of T_N but helps to demonstrate the effect of dimensionality on time complexity. Since runtime grows at a trinomial rate when $d = 3$, we use the largest δ values (smallest T_N) that correctly cover the parameter space. This is what makes the δ-refinement process imperative; it allows us to correctly verify a property while avoiding runtime blow-up.

6   Conclusions

In this paper, we performed reachability analysis with discrepancy to automatically determine parameter ranges for three fundamental reactions by *C. Elegans* to tap-withdrawal stimulation: reversal of movement, acceleration, and no response. We followed the lead of the *in vivo* experimental results of [15] to obtain parameter-estimation results for gap-junction conductances for a number of neural-ablation groups. The ranges we present are a significant expansion of the results in [16], where all of the parameters are constant and only the predominant behavior is produced. To the best of our knowledge, these results represent the first formal verification of a biologically realistic (nonlinear ODE) model of a neural circuit in a multicellular organism.

The verification framework we develop is model-agnostic, and allows the techniques of [5] to be applied to general nonlinear ODE models. This is only possible through the careful model and verification engineering developed in this paper.

As alluded in Sect. 5, our results cannot necessarily cover the entire parameter space due to the T_N required, but still enough to verify the properties in question. A potential solution to the incomplete coverage is parallelizing our approach. Luckily, calculating reach-tubes is a data-parallel computation and considered "trivially parallel" for the GPGPU (General-Purpose computing on a Graphics Processing Unit) architecture. This should allow us to run verification experiments in a fraction of the current required time, giving us a potential expansion of coverage.

Acknowledgments. We would like to thank Junxing Yang, Heraldo Memelli, Farhan Ali, and Elizabeth Cherry for their numerous contributions to this project. Our research is supported in part by the following grants: NSF IIS 1447549, NSF CAR 1054247, AFOSR FA9550-14-1-0261, AFOSR YIP FA9550-12-1-0336, CCF-0926190, and NASA NNX12AN15H.

References

1. Donzé, A., Maler, O.: Systematic simulation using sensitivity analysis. In: Bemporad, A., Bicchi, A., Buttazzo, G. (eds.) HSCC 2007. LNCS, vol. 4416, pp. 174–189. Springer, Heidelberg (2007)
2. Duggirala, P.S., Mitra, S., Viswanathan, M.: Verification of annotated models from executions. In: Proceedings of the International Conference on Embedded Software, EMSOFT 2013, Montreal, Canada. IEEE, September–October 2013
3. Duggirala, P.S., Mitra, S., Viswanathan, M., Potok, M.: C2E2: a verification tool for stateflow models. In: Baier, C., Tinelli, C. (eds.) TACAS 2015. LNCS, vol. 9035, pp. 68–82. Springer, Heidelberg (2015)
4. Duggirala, P.S., Wang, L., Mitra, S., Viswanathan, M., Muñoz, C.: Temporal precedence checking for switched models and its application to a parallel landing protocol. In: Jones, C., Pihlajasaari, P., Sun, J. (eds.) FM 2014. LNCS, vol. 8442, pp. 215–229. Springer, Heidelberg (2014)
5. Fan, C., Mitra, S.: Bounded verification using on-the-fly discrepancy computation. Technical report UILU-ENG-15-2201, Coordinated Science Laboratory, University of Illinois at Urbana-Champaign, February 2015
6. Grosu, R., Batt, G., Fenton, F.H., Glimm, J., Le Guernic, C., Smolka, S.A., Bartocci, E.: From cardiac cells to genetic regulatory networks. In: Gopalakrishnan, G., Qadeer, S. (eds.) CAV 2011. LNCS, vol. 6806, pp. 396–411. Springer, Heidelberg (2011)
7. Hodgkin, A.L., Huxley, A.F.: A quantitative description of membrane current and its application to conduction and excitation in nerve. J. Physio. **117**, 500–544 (1952)
8. Huang, Z., Fan, C., Mereacre, A., Mitra, S., Kwiatkowska, M.: Invariant verification of nonlinear hybrid automata networks of cardiac cells. In: Biere, A., Bloem, R. (eds.) CAV 2014. LNCS, vol. 8559, pp. 373–390. Springer, Heidelberg (2014)

9. Islam, M.A., Murthy, A., Girard, A., Smolka, S.A., Grosu, R.: Compositionality results for cardiac cell dynamics. In: Proceedings of the 17th International Conference on Hybrid Systems: Computation and Control. ACM (2014)

10. Iyengar, S.M., Talcott, C., Mozzachiodi, R., Cataldo, E., Baxter, D.A.:Executable symbolic models of neural processes. Netw. tools appl. biol. NETTAB07 (2007)

11. Lohmiller, W., Slotine, J.J.E.: On contraction analysis for non-linear systems. Automatica **34**, 683–696 (1998)

12. Murthy, A., Islam, M.A., Grosu, R. Smolka, S.A.: Computing bisimulation functions using SOS optimization and delta-decidability over the reals. In: Proceedings of the 18th International Conference on Hybrid Systems: Computation and Control. ACM (2015)

13. Tiwari, A., Talcott, C.: Analyzing a discrete model of *aplysia* central pattern. In: Heiner, M., Uhrmacher, A.M. (eds.) CMSB 2008. LNCS (LNBI), vol. 5307, pp. 347–366. Springer, Heidelberg (2008)

14. Varshney, L.R.: Individual differences (2015). http://blog.openworm.org/post/107263481195/individual-differences

15. Wicks, S.R., Rankin, C.H.: Integration of mechanosensory stimuli in caenorhabditis elegans. J. Neurosci. **15**(3), 2434–2444 (1995)

16. Wicks, S.R., Roehrig, C.J., Rankin, C.H.: A dynamic network simulation of the nematode tap withdrawal circuit: predictions concerning synaptic function using behavioral criteria. J. Neurosci. **16**(12), 4017–4031 (1996)

Solving General Auxin Transport Models with a Numerical Continuation Toolbox in Python: PyNCT

Delphine Draelants[(✉)], Przemysław Kłosiewicz, Jan Broeckhove,
and Wim Vanroose

Department of Mathematics and Computer Science, Universiteit Antwerpen,
Middelheimlaan 1, 2020 Antwerpen, Belgium
{Delphine.Draelants,Przemyslaw.Klosiewicz,Jan.Broeckhove,
Wim.Vanroose}@uantwerpen.be
http://www.uantwerpen.be/applied-mathematics

Abstract. Many biological processes are described with coupled non-linear systems of ordinary differential equations that contain a plethora of parameters. The goal is to understand these systems and to predict the effect of different influences. This asks for a dynamical systems approach where numerical continuation methods and bifurcation analysis are used to detect the solutions and their stability as a function of the parameters. We developed PyNCT – Python Numerical Continuation Toolbox – an open source Python package that implements numerical continuation methods and can perform bifurcation analysis based on sparse linear algebra. The software gives the user the choice of different solvers (direct and iterative) and allows the use of preconditioners to reduce the number of iterations and guarantee the convergence when working with complex non-linear models.

In this paper we demonstrate the usefulness of the toolbox with a class of models pertaining to auxin transport between cells in plant organs. We show how easy it is to compute the steady state solutions for different parameter values, to calculate how they depend on each other and to map parts of the solution landscape.

An interactive model development and discovery cycle is key in bio-systems research. It allows one to investigate and compare different model parameter settings and even different models and gauge the model's usefulness. Our toolbox allows for such quick experimentation and has a low entry barrier for non-technical users.

Although PyNCT was developed particularly for the study of transport models in biology, its implementation is generic and extensible, and can be used in many other dynamical system applications.

Keywords: Continuation methods · Bifurcation analysis · Transport models · PyNCT

1 Introduction

In recent years, many molecular-genetic experiments have been performed to increase our insights in how molecular processes in plants work. Due to the

© Springer International Publishing Switzerland 2015
A. Abate and D. Šafránek (Eds.): HSB 2015, LNBI 9271, pp. 211–225, 2015.
DOI: 10.1007/978-3-319-26916-0_12

restricted set of experiments that can be performed today, the amount of available experimental data is still limited. For this reason researchers try to expand their knowledge of biological processes by means of an interaction between experimental research and mathematical modelling.

Formerly, biologists used the available data to mathematically describe the biological processes (of plants). The incredible amount of uncertainties about how these systems work gave birth to different hypotheses and, as a result, different kinds of models with lots of parameters. These models were initially solved with simple numerical methods for a limited set of parameters.

Today the biological models become very large and complex. Moreover biologists are interested in understanding the whole solution landscape instead of one single particular solution. They want to understand the effect of different influences (parameters) and compare current models based on different hypotheses. In order to solve, analyse and compare these models, state-of-the-art numerical methods are necessary. Unfortunately, these methods, solvers and algorithms are often not easily accessible outside of their application niche. This makes it difficult for systems biologists to directly apply state-of-the-art numerical mathematics to their specific problems. The end result is a growing demand for software packages that combine biological models with numerical tools.

We developed a toolbox, PyNCT, that solves and analyses the non-linear coupled systems of equations that appear in a wide range of models for the transport of chemicals through networks of cells. PyNCT contains numerical continuation methods and can perform bifurcation analysis in order to find parts of the solution landscape as a function of the different parameters. The toolbox is based on sparse linear algebra which enables its users to solve very large systems. With the resulting simulation tools biologists can now explore, analyse and compare various models, test new hypotheses, ...without the need to understand the inner details of mathematics behind the numerical methods. Although we developed PyNCT specifically to study transport models, it works well for all dynamical systems.

The paper is organized as follows. First, in Sect. 2 we give more information about the application domain. We describe a tissue of cells mathematically, we present a general class of transport models already implemented in PyNCT, and we discuss the type of solutions biologists are interested in. We also discuss the state-of-the-art tools that are currently available. Then, in Sect. 3 we present the numerical algorithms in PyNCT in detail. Readers who are not interested in the mathematical details can skip this section. A motivation for the choice of Python and certain libraries as a basis for the implementation of the toolbox can be found in Sect. 4. In Sect. 4.2 we explain how to apply the software for the different types of models and in Sect. 5 we demonstrate it for a specific example, the model of Smith et al. [18]. Finally in Sect. 6 we conclude and give an outlook.

2   The Application Domain

The PyNCT toolbox has been developed to investigate the response of stationary solutions of (large) dynamical systems to changes of control parameters. It can be used for any model consisting of a system of non-linear equations that are smooth and continuously differentiable.

In this paper we show the usefulness of our toolbox, using the auxin transport equations of a cell-based plant organ model as a demonstration vehicle. In Sect. 2.1 we present the mathematical description of a plant tissue of irregular cells. Based on this we describe a class of concentration-based auxin transport equations and look at the typical solutions that are of interest. PyNCT is able to generate solution branches automatically for any model that fits this framework. In Sect. 5 we demonstrate how to do so for a specific auxin transport model described by Smith et al. [18]. At the end of this section we discuss the current state-of the-art tools for this and compare them with our tool.

2.1 Auxin Transport Models

Network of Cells. In biology, cell tissues are represented by a graph with edges and vertices. The edges represent the cell walls of the plant organ and a cell is then a face in this graph. As a consequence a cell is a vertex in the weak dual graph G. See Fig. 1 for an example. This dual graph helps us to describe the tissue mathematically:

- The set of vertices V represents all cells in the tissue and we identify them with an index $i \in \{1, ..., n\}$.
- The set of edges E represents the connections between neighbouring cells. As a consequence the neighbouring cells of a cell i can be identified as all cells up to distance 1 from cell i. This subset of cells is denoted with $\mathcal{N}_i \subset V$.
- Every edge represents the connection between two neighbouring cells and thus we can uniquely associate the information about a cell wall with an edge. By labelling each edge with relevant information about the cell wall (for example the permeability of the cell wall), we get a weighted graph G.
- In every cell we can define various properties of the cell, such as the concentration of a specific hormone, protein,.... These are the variables of the models. We denote m as the number of the state variables per cell.

With the help of this representation of a tissue we can now describe easily how substances in cells are transported throughout the tissue with a system of equations.

The System of Equations. The toolbox contains software to easily calculate solutions for transport models that can be written as follows

$$\dot{\mathbf{y}}_i = \boldsymbol{\pi}(\mathbf{y}_i) - \boldsymbol{\delta}(\mathbf{y}_i) + D \sum_{j \in \mathcal{N}_i} (\mathbf{y}_j - \mathbf{y}_i) + T \sum_{j \in \mathcal{N}_i} (\boldsymbol{\nu}_{ji}(\mathbf{y}_1, ..., \mathbf{y}_n) - \boldsymbol{\nu}_{ij}(\mathbf{y}_1, ..., \mathbf{y}_n)). \quad (1)$$

The vector $\mathbf{y}_i$ contains the m time-dependent state variables in cell i. For instance, $\mathbf{y}_i$ may contain the auxin concentration ($m = 1$) or both auxin and PIN-FORMED1 concentrations ($m = 2$) in cell i. Further, the model consists of

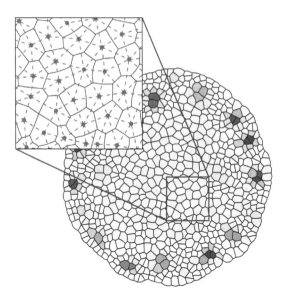

Fig. 1. The large picture shows a typical tissue of cells. The zoomed-in portion details the cell graph (full black) and its weak dual graph G of cellular connections (dashed red) (Color figure online).

π, δ, the production and decay functions, respectively, $\boldsymbol{D}$, a matrix with diffusion coefficients, T a matrix with the active transport parameters and $\boldsymbol{\nu}_{ij}$ the active transport functions. In our example, we assume active transport functions that can be expressed as

$$\left(\boldsymbol{\nu}_{ij}\right)_l \left(\mathbf{y}_1, ..., \mathbf{y}_n\right) = \psi_l(\mathbf{y}_i, \mathbf{y}_j)\frac{\varphi_l(\mathbf{y}_j)}{\sum_{k\in\mathcal{N}_i} \varphi_l(\mathbf{y}_k)}, \quad \text{for } l = 1, ..., m, \qquad (2)$$

where the functions ψ_l, φ_l depend on the model choices. Many concentration-based auxin transport models can be written in this form. More information about this class of models can be found in [7] and in Sect. 5 we demonstrate how the toolbox works for a specific example.

The Numerical Solutions. The models described above possess an inherent time-scale separation: the growth hormone dynamics involve short time scales (of the order of seconds) [4], while changes in cellular shapes and proliferation of new cells occur on much slower time scales (hours or days) [3]. In order to determine the distribution of auxin in the plant, it is sufficient to concentrate on the fast time scale of the hormone transport. Therefore we can assume a static cell structure and study the plant tissue as a dynamical system where we are only interested in the steady state solutions of the system and their dependence on the model parameters. The PyNCT toolbox is designed with sufficient functionality to calculate those steady state solutions immediately.

2.2   Current State-of-the-Art Tools

We can divide the current toolboxes that calculate these stationary solutions of dynamical systems in function of model parameters in two groups: tools based on dense linear algebra (limited to small tissues) and tools based on sparse linear algebra routines, which scale to large tissues.

The first group is the largest and well known tools like AUTO [6] and MAT-CONT [5] belong to this group. Also current new tools, especially developed for system biologist to investigate and analyse their new biological systems like Systems biology toolbox for Matlab [15] and Facile [17], heavily rely on dense linear algebra routines since they are all based on AUTO. The disadvantage of these tools is that the routines are not scalable to very large systems with many cells. The number of equations is typically limited to about 500.

An example of the second group is LOCA [14]. LOCA is developed around sparse linear algebra but it is designed for extremely large systems that need to be run on HPC infrastructure. LOCA is not easy to use and requires expert knowledge in C++ and HPC hardware.

Our toolbox is also designed to take advantage of sparse linear algebra but avoids C++ or HPC knowledge. The solutions of the typical large biological cell-based systems can be calculated very fast in contrast with existing system biology tools. Another big advantage of PyNCT compared to LOCA is the usability of the software. As explained in Sect. 4.2 in more detail, it is very easy to use the tool for a large class of transport models and even for many other models, only a routine with the equations must be provided.

3   Currently Available Functionalities

In this section we will explain briefly the main numerical algorithms implemented in PyNCT. Section 3.1 explains the numerical continuation methods. We also discuss the related functionalities available in PyNCT. Section 3.2 explains the principles and applications of bifurcation analysis.

If the reader is not interested in the mathematics under the hood of this toolbox and only wants to use it as a black box, he/she can skip Sect. 3.1 and jump to Sects. 4.2 and 5 where we explain how to use the toolbox.

3.1   Numerical Continuation Methods

The idea of continuation methods is to find a curve of approximate solutions $\mathbf{y}$ of a system of non-linear equations

$$F(\mathbf{y}, \boldsymbol{\lambda}) = \mathbf{0}, \tag{3}$$

as a function of the parameter vector $\boldsymbol{\lambda}$ with

$$F : \mathbb{R}^{v+w} \to \mathbb{R}^{v} : (\mathbf{y}, \boldsymbol{\lambda}) \mapsto F(\mathbf{y}, \boldsymbol{\lambda}). \tag{4}$$

Following the implicit function theorem we know that for a non-singular point $\left(\mathbf{y}^{(0)}, \boldsymbol{\lambda}^{(0)}\right)$ that satisfies $F\left(\mathbf{y}^{(0)}, \boldsymbol{\lambda}^{(0)}\right) = \mathbf{0}$, the solution set $F^{(-1)}\left(\mathbf{0}\right)$ can be locally parametrized about $\left(\mathbf{y}^{(0)}, \boldsymbol{\lambda}^{(0)}\right)$ with respect to a parameter of $\boldsymbol{\lambda}$. This means that the system of equations $F\left(\mathbf{y}, \boldsymbol{\lambda}\right) = 0$ defines an implicit curve $\mathbf{y}\left(\boldsymbol{\lambda}\left(s\right)\right)$ for any parametric curve $\boldsymbol{\lambda}\left(s\right) : \mathbb{R} \to \mathbb{R}^w$ in $\mathbb{R}^w$ [1]. To construct such a curve of subsequent solution points $\left(\mathbf{y}^{(i)}, \boldsymbol{\lambda}^{(i)}\right) = \left(\mathbf{y}^{(i)}, \boldsymbol{\lambda}^{(i)}\left(s\right)\right)$, continuation methods use a starting point $\left(\mathbf{y}^{(0)}, \boldsymbol{\lambda}^{(0)}\right)$, a solution of system (3), along with an initial continuation direction [12]. This starting point is typically a known trivial solution. An important family of continuation methods are predictor-corrector schemes. The idea of these algorithms is to predict a new solution point first. Then, in the corrector step, this predicted point is used as the initial guess for an iterative method that will converge to the solution up to a given tolerance. In our toolbox, the predictor step uses the secant method and a given step size to predict a guess for the next solution point on the curve. The corrector step improves the guess with Newton iterations.

Newton's Method. When applying the above continuation method, we improve the guess $\left(\tilde{\mathbf{y}}^{(i+1)}, \tilde{\boldsymbol{\lambda}}^{(i+1)}\right)$, found in the predictor step with Newton iterations [10]

$$\mathbf{y}^{(i+1)} = \tilde{\mathbf{y}}^{(i+1)} - \frac{F\left(\tilde{\mathbf{y}}^{(i+1)}, \tilde{\boldsymbol{\lambda}}^{(i+1)}\right)}{F'\left(\tilde{\mathbf{y}}^{(i+1)}, \tilde{\boldsymbol{\lambda}}^{(i+1)}\right)}, \tag{5}$$

until a sufficiently accurate new solution point $\left(\mathbf{y}^{(i+1)}, \boldsymbol{\lambda}^{(i+1)}\right)$ of F is reached. In every iteration step, the system

$$J\left(\mathbf{y}, \boldsymbol{\lambda}\right) \boldsymbol{x} = -F\left(\mathbf{y}, \boldsymbol{\lambda}\right) \tag{6}$$

is solved, with Jacobian matrix $J\left(\mathbf{y}, \boldsymbol{\lambda}\right)$ defined by

$$J\left(\mathbf{y}, \boldsymbol{\lambda}\right)_{ij} = \frac{\partial\left(F\right)_i}{\partial\left(\mathbf{y}\right)_j}\left(\mathbf{y}, \boldsymbol{\lambda}\right) \tag{7}$$

By default in PyNCT we can use a direct or an iterative solver for (6).

- *Direct sparse linear solver:* The direct linear solver from SciPy (`scipy.sparse.linalg.spsolve`) provides excellent performance for moderately sized systems. At the time of this writing, the solver is a wrapper around either SuperLU or UMFPack; both mature and widely used sparse direct solver libraries [8].
- *Iterative solver:* If the size of the system requires the use of an iterative solver, PyNCT enables the use of Generalized Minimal RESidual (GMRES), a Krylov based solver, implemented in SciPy. We chose this method because it is very robust and applicable on all types of linear problems. This is necessary because continuation methods calculate both the stable and unstable solutions. In many cases the latter degrades or even destroys convergence of most iterative solvers.

Although we only suggest these two linear solvers in PyNCT, SciPy provides a typical array of iterative solvers based on Generalized Minimal RESidual (GMRES), Conjugate Gradient (CG), and derived methods. All these methods can be used easily. More information about such Krylov subspace solvers can be found in [9] and up-to-date information about the linear solvers available in SciPy can be found in the SciPy documentation pages[1].

Jacobian. A continuation method requires the Jacobian matrix J for the calculations of the Newton corrections. The Jacobian of cell-based biological systems with local interactions is a very large sparse matrix. Therefore it is important to exploit our knowledge about the structure of the Jacobian. By ordering the variables in the right way it can be divided in different building blocks where every block represents the derivative of an equation of the model to a variable representing a substance in each cell. For instance, consider a system of m transport equations for every cell, with n the number of cells and m the number of unknowns (the different substances in a cell) as described in Eq. (1). The Jacobian then consists of m^2 blocks of size $n \times n$ if the vector of unknowns is grouped per substance type. The example available in PyNCT uses this ordering, but it is possible to order the variables in any way. All these blocks have a sparse structure because in every equation the changes over time only depend on the variables in the cell itself and the neighbours up to distance 2.

In the PyNCT toolbox it is possible to choose between using the exact Jacobian or an approximation:

- The Jacobian is calculated exactly by determining the derivatives of the system with the use of SymPy, a Python library for symbolic mathematics [19].
- The approximated Jacobian is calculated numerically by using finite differences. The jth column of the Jacobian matrix is found by a forward difference scheme

$$J\left(\mathbf{y}^{(i)}, \boldsymbol{\lambda}^{(i)}\right)_j = \frac{F\left(\mathbf{y}^{(i)} + \epsilon e_j, \boldsymbol{\lambda}^{(i)}\right) - F\left(\mathbf{y}^{(i)}, \boldsymbol{\lambda}^{(i)}\right)}{\epsilon}, \qquad (8)$$

where e_j is the jth unit vector and $\left(\mathbf{y}^{(i)}, \boldsymbol{\lambda}^{(i)}\right)$ is the ith calculated solution point on the branch as before.

We chose for forward finite differences because it is a very easy algorithm and do not need many calculations per iteration. For instance the value of $F\left(\mathbf{y}^{(i)}, \boldsymbol{\lambda}^{(i)}\right)$ is already calculated and saved. The default value for ϵ is $\epsilon = 10^{-10}$ but the user can customize it if desired.

Preconditioning. When using iterative methods to solve each Newton step, we can use a preconditioner to reduce the number of iterations or to guarantee convergence of the iterative method when working with complex systems [9].

---

[1] http://docs.scipy.org/doc/scipy/reference/sparse.linalg.html.

Instead of solving the original linear system $J(\mathbf{y}, \boldsymbol{\lambda})\, \boldsymbol{x} = -F(\mathbf{y}, \boldsymbol{\lambda})$ we solve the preconditioned system

$$P^{-1} J(\mathbf{y}, \boldsymbol{\lambda})\, \boldsymbol{x} = -P^{-1} F(\mathbf{y}, \boldsymbol{\lambda}), \qquad (9)$$

which is a better conditioned problem, leading to faster convergence of the Jacobian solve. By choosing the right preconditioner, preconditioned iterative solvers perform better then direct solvers. For problems where the diffusion between the cells dominates traditional preconditioners that approximately invert the Poisson equation such as incomplete factorizations or multigrid can be effective. However, when the active transport dominates different preconditioners need to be developed. This is still an open topic of research.

In PyNCT it is possible to use a preconditioner. Since a good preconditioner asks specific knowledge about the model, we did not provide any general preconditioners but it can be specified by the user. How this should be implemented is explained in Sect. 4.2

3.2 Bifurcation Analysis

The study of the relation between the stability of a solution and the parameters of the corresponding dynamical system is known as (local) bifurcation analysis [16]. Such an analysis identifies the stable and unstable solutions and the bifurcation points that mark the transitions between them. This is biologically relevant since it will allow us to predict the patterns that emerge in the time evolution as the parameters of the model are changed. A bifurcation point is a solution $\left(\mathbf{y}^{(i)}, \boldsymbol{\lambda}^{(i)}\right)$ of system (3) where the number of solutions changes when $\boldsymbol{\lambda}$ passes $\boldsymbol{\lambda}^{(i)}$. For a complete review of the different types of bifurcation points and their properties we refer to [16]. The analysis usually leads to a bifurcation diagram that highlights the connections between stable and unstable branches as the parameters change. It is useful to track all these solution branches that emerge, split or end in a bifurcation point which can be done with the help of numerical continuation methods explained in Sect. 3.1.

Our toolbox contains methods to calculate the stability of a solution point directly after each point or after calculating the whole solution branch. For the transport models, we chose to calculate the stability of the solutions as part of the post-processing since even without the stability information, the continuation data can be very useful. A great advantage of this choice is that the continuation data is much faster to compute because calculating the eigenvalues for every solution point on a branch is very time-consuming.

To calculate the eigenvalues of the Jacobian, we use the 'eig' routine in scipy.linalg based on dense linear algebra, although the Jacobian of transport models is a sparse matrix (see Sect. 3.1). Typically in transport models, around a bifurcation point, many eigenvalues cross the imaginary axis. As a consequence, the sparse routines of scipy.sparse.linalg for calculating eigenvalues fail to converge when searching for eigenvalues around zero. Although we are using dense linear algebra, calculating the stability for moderate system sizes can be performed in an acceptable time frame by parallelizing calculations with MPI (using mpi4py[2]).

Note that if interested in sparse routines, SciPy provides a sparse routine `scipy.sparse.linalg.eigs` that can be used easily in PyNCT.

After calculating the eigenvalues of the solution points, the bifurcation points must be indicated manually. It is then possible to start the continuation again from these bifurcation points in a new direction to find the branches that emerge. However, for now PyNCT does not contain methods for automatic branch switching.

4 Overview of Software Structure

4.1 Choice of Language and Libraries

Language. The PyNCT toolbox is implemented in Python. This choice is motivated by a number of factors:

- Python is a flexible language and is well-suited for rapid development. Adapting model code is straightforward and does not require an edit-compile-link cycle as does, for instance, C++.
- Python has a low entry barrier. It is easy to learn and to use and thus an ideal language for less technical users.
- Python has a large standard library with good documentation and a huge amount of contributed, community-maintained packages. PyNCT uses several existing libraries that include for example numerical methods so we don't have to 'reinvent the wheel'. More information about the packages included in PyNCT can be found below.
- Python is an open source programming language and also our software is freely available.

Numerical Libraries. The numerical part of our toolbox relies substantially on NumPy [2] and SciPy [8]. The former provides a foundation of linear algebra primitives in Python. The latter extends it by providing a huge variety of algorithms, solvers and support methods for "all things scientific" in Python. Both are high-quality, popular and well-documented open source libraries.

To enhance both speed and accuracy of the calculations we use symbolic expressions for the specification of the equations in the biological model and automatic differentiation to obtain the exact Jacobian expression. SymPy[19], a Python library for symbolic mathematics, is an excellent tool for these purposes in our case. The use of symbolic expressions, however, depends on the biological model under investigation and is not universally feasible for all applications

---

[2] https://pypi.python.org/pypi/mpi4py.

(Remark that as mentioned in Sect. 3.1 also the approximated Jacobian can be used if the user can't or don't want to use SymPy).

Other Libraries. The infrastructure for loading and storing virtual tissue representations and generating tissue geometries is provided by the Python Plant Tissue Simulation toolbox PyPTS [11]; an open source library. PyPTS uses a HDF5 based file format to store simulation results which makes pre/post-processing, visualisation and exchanging results with other tools such as VirtualLeaf [13] easy. It also provides an easy API for accessing and modifying tissue entities and attributes.

4.2   The Executable

When using the toolbox for a specific model, the system of equations must be specified.

For a specific class of transport models, the toolbox can be used by just providing the equations and parameters in configuration files (see Sect. 5 for an example).

For all other models a new class must be constructed. The class must contain an initialize method and a method that applies your system of equations. Additionally we also need a configuration file similar to the ones constructed for the transport models and explained in Sect. 5. It contains the parameter values of the model and the specifications of the numerical methods. At last an executable script, similar to the biology demo is necessary to start up the continuation. The PyNCT package already includes a basic template for this class, the executable and the configuration file which makes it very easy to start implementing your own model.

To extend this basic template, you can define an extra method that constructs the Jacobian in a given point. You can define an exact or an approximate Jacobian that differs from the standard approximation method described in Sect. 3.1. Then you can choose between the different Jacobian implementations to solve the Newton iterations. It is also possible to specify a preconditioner in this class to speed up the convergence to a solution point.

5   A Look at the Toolbox via an Example

In this section we show how easy it is to use the toolbox and find parts of the solution space of the model of Smith et al. [18]. The model satisfies Eqs. (1) and (2) and features 2 state variables per cell, namely the indole-3-acetic acid (IAA) concentration, $a_i(t)$, and the PIN-FORMED1 (PIN1) amount, $p_i(t)$. The model features IAA production, decay, active and passive transport terms, whereas for PIN1 only production and decay are included. This results in the following set of coupled non-linear ordinary differential equations (ODEs)

$$\frac{da_i}{dt} = \frac{\rho_{\text{IAA}}}{1 + \kappa_{\text{IAA}} a_i} - \mu_{\text{IAA}} a_i + \frac{D}{V_i} \sum_{j \in \mathcal{N}_i} l_{ij} (a_j - a_i)$$

$$+ \frac{T}{V_i} \sum_{j \in \mathcal{N}_i} \left[P_{ji}(\boldsymbol{a}, \boldsymbol{p}) \frac{a_j^2}{1 + \kappa_T a_i^2} - P_{ij}(\boldsymbol{a}, \boldsymbol{p}) \frac{a_i^2}{1 + \kappa_T a_j^2} \right], \tag{10}$$

$$\frac{dp_i}{dt} = \frac{\rho_{\text{PIN}_0} + \rho_{\text{PIN}} a_i}{1 + \kappa_{\text{PIN}} p_i} - \mu_{\text{PIN}} p_i, \tag{11}$$

for $i = 1, ..., n$ with n the number of cells. In this model D is a diffusion coefficient, V_i is the cellular volume, $l_{ij} = S_{ij}/(W_i + W_j)$ is the ratio between the contact area S_{ij} of the adjacent cells i and j, and the sum of the corresponding cellular wall thicknesses W_i and W_j. In addition, T is the active transport coefficient and P_{ij} is the number of PIN1 proteins on the cellular membrane of cell i facing cell j,

$$P_{ij}(\boldsymbol{a}, \boldsymbol{p}) = p_i \frac{l_{ij} \exp(c_1 a_j)}{\sum_{k \in \mathcal{N}_i} l_{ik} \exp(c_1 a_k)}. \tag{12}$$

More details on the model and the parameters can be found in [18].

The rest of the section is divided in three parts, the preparation, the actual calculations and the post-processing. In these sections we explain step by step how to find the steady state solutions starting from the above model.

5.1 Preparing for Continuation

Before we can calculate the solutions, we need to specify the model and choose from several solution methods implemented in PyNCT. Therefore we fill in a model file and a parameter file respectively.

The Model File. In the model file each part of the system (production, decay, diffusion, ...) is listed. For example for the model of Smith et al. this file becomes

```
1  {
2     "decayPIN": "muPIN*p",
3     "productionPIN": "(rhoPIN0 + rhoPIN * a) / (1.0 + kPIN * p)",
4     "decayIAA": "muIAA*a",
5     "productionIAA": "rhoIAA / (1.0 + kIAA * a)",
6     "passive_transport": "D * wall_length * (a_j - a_i)",
7     "phi": "wall_length*exp(c1*a_j)",
8     "psi": "p*a_i**2/(1.0+kT*a_j**2)"
9  }
```

The Parameter File. In the parameter file we specify all parameters that are necessary to perform the continuation. This includes the model parameters, information about the tissue, the solvers, the continuation and the saving process.

In this example, we consider a tissue with 742 irregular prismic cells that cover an almost-circular domain (geometry extracted from [13]) with free boundary conditions [7]. We choose as continuation parameter the model parameter T, and the trivial solution of this model in $T = 0$ as the starting point. We also specify a directory and file name to save the continuation data. A part of the parameter file reads

```
 1  {
 2    "input": "./location/of/cells_742.h5",
 3    "output": "./location/of/continuation.h5",
 4    "rhoIAA": 1.500,
 5    "D": 1.000,
 6
 7    ...
 8
 9    "T": 0.0,
10    "startpoint": "value",
11    "startpoint_a": "(-1.0 + sqrt(1.0 + 4.0*kIAA*rhoIAA/muIAA))/(2.0*kIAA)",
12    "startpoint_p": "(-1.0 + sqrt(1.0 + 4.0*kPIN*(rhoPIN0 + rhoPIN*a) /muPIN
           ))/(2.0*kPIN)",
13
14    ...
15  }
```

More information and examples of both the model file and the parameter file, can be found in the demos directory of the PyNCT toolbox.

5.2   Executing the Continuation

After specifying all parameters in the correct files, we can start the continuation by calling the 'doContinuation' method.

```
from pynct.biology import doContinuation
doContinuation.doContinuation('/location/of/parameterFile.json',
                              '/location/of/modelFile.json')
```

Every solution point is saved immediately after it is calculated in the specified output file. This method has the advantage that we can already start with the post processing before all points are calculated.

5.3   Post-processing

In the PyNCT biology demo, we already included two functionalities necessary for post-processing the calculated data. We can determine the stability properties of the solutions and we can visualize the solutions.

The Stability. The stability of the solutions on a continuation branch is determined with the function calculateEigenvaluesMpi in PyNCT.

```
from pynct.biology import calculateEigenvaluesMpi
calculateEigenvaluesMpi.main()
```

This function calculates the eigenvalues and saves them in a specified file. More information can be found in Sect. 3.2.

Plotting Tools. In order to process and interpret the calculated data in plant biology, it is very useful to visualize it. Although many tools already exist for plotting data, we added a number of basic functions in the PyNCT demo specifically aimed at visualizing continuation data from biological systems.

All plotting tools are implemented in the file 'plottools.py' which therefore needs to be imported. We can plot bifurcation diagrams with or without the stability of the calculated solutions and all the solution patterns. The functions work by just specifying the right data files. For example, the following code gives an interactive plot with the bifurcation diagram and a corresponding solution pattern.

```
from pynct.biology import plottools
plottools.bifDiagramInteractive('/location/of/continuation.h5')
```

We can change the highlighted solution point and thus the solution pattern interactively. Figure 2 displays such a plot where also the stability propertiees are shown.

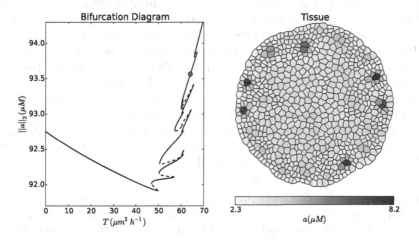

Fig. 2. Bifurcation diagram and corresponding solution pattern for the Smith et al. model for an almost-circular domain of 742 irregular cells (geometry taken from [13]). Left: An example of a bifurcation diagram that depicts the 2-norm of auxin concentration versus the continuation parameter T. Stable solutions are drawn with a full line and unstable solutions are dashed. Right: The solution pattern corresponding with the red dot on the left figure (Color figure online).

The bifurcation diagram shows the norm over all cells of IAA versus the continuation parameter T. As in the figure, we can highlight one solution point on the continuation branch. The distribution of IAA in the tissue in this solution point is then automatically displayed at the right in the figure. The darker the cell is coloured, the higher the concentration of the unknown (IAA).

More information about the plotting tools and how to use them can be found in the PyNCT biology demo.

6   Conclusion and Future Directions

We presented the Python Numerical Continuation Toolbox PyNCT, an open source library. The toolbox contains different state-of-the-art numerical algorithms for numerical continuation and is able to calculate the stability properties of the solutions.

The methods can be applied on coupled non-linear (smooth and continuously differentiable) equations and specifically on models describing the transport throughout a network of cells. For general models the system must be implemented in a new class but for a subset of concentration-based transport models (those models that satisfy Eqs. (1) and (2)) only a specification of the model parts in a configuration file is necessary. In the future we want to extend the class of models that can be analysed automatically.

The numerical methods implemented in PyNCT are based on sparse linear algebra since biological processes can be described often by just describing what happens in the direct neighbourhood. Therefore solutions can be calculated efficiently. Further there is no limitation on the number of unknowns or the size and shape of the tissue. These are the main advantages and differences of our toolbox in comparison with existing tools for system biologists.

PyNCT helps us to explore parts of the solution space. We can calculate a branch of solutions and determine the stability of each solution. Based on this information, we identify bifurcation points and calculate new branches that emerge. However, in the future we will include a method that detects the bifurcation points and performs automatic branch-switching.

Finally our toolbox allows quick experimentation and has a low entry barrier for less technical users. Biologists can now compare various transport models and explore different hypotheses very easily.

Acknowledgements. DD acknowledges financial support from the Department of Mathematics and Computer Science of the University of Antwerp. This work is part of the Geconcerteerde Onderzoeksactie (G.O.A.) research grant "A System Biology Approach of Leaf Morphogenesis" granted by the research council of the University of Antwerp. We acknowledge Giovanni Samaey for sharing basic version of a continuation code.

References

1. Allgower, E., Georg, K.: Numerical Path Following. Springer, Berlin (1994)
2. Ascher, D.: Numpy, Numerical Python (2001). http://www.numpy.org/. Accessed June 2015
3. Beemster, G., Baskin, T.: Analysis of cell division and elongation underlying the developmental acceleration of root growth in arabidopsis thaliana. Plant Physiol. **116**, 515–526 (1998)
4. Brunoud, G., Wells, D., Oliva, M., Larrieu, A., Mirabet, V., Burrow, A., Beeckman, T., Kepinski, S., Traas, J., Bennett, M., et al.: A novel sensor to map auxin response and distribution at high spatio-temporal resolution. Nature **482**, 103–106 (2012)

5. Dhooge, A., Govaerts, W., Kuznetsov, Y.A.: Matcont: a matlab package for numerical bifurcation analysis of odes. ACM Trans. Math. Softw. (TOMS) **29**(2), 141–164 (2003)

6. Doedel, E.J.: Auto: A program for the automatic bifurcation analysis of autonomous systems. Congr. Numer. **30**, 265–284 (1981)

7. Draelants, D., Avitabile, D., Vanroose, W.: Localized auxin peaks in concentration-based transport models of the shoot apical meristem. J. R. Soc. Interface 12(106) (2015). http://rsif.royalsocietypublishing.org/content/12/106/20141407.full

8. Jones, E., Oliphant, T., Peterson, P., et al.: SciPy: open source scientific tools for Python (2001). http://www.scipy.org/. Accessed June 2015

9. Kelley, C.T.: Iterative methods for linear and nonlinear equations. SIAM: Frontiers in Applied Mathematics 16 (1995)

10. Kelley, C.T.: Solving Nonlinear Equations with Newton's Method, vol. 1. Society for Industrial and Applied Mathematics, Philadelphia (2003)

11. Kłosiewicz, P.: Pypts, python plant tissue simulations (2015). https://pypi.python.org/pypi/PyPTS/. Accessed June 2015

12. Krauskopf, B., Osinga, H., Galán-Vioque, J.: Numerical Continuation methods for Dynamical Systems. Springer, Netherlands (2007)

13. Merks, R., Guravage, M., Inzé, D., Beemster, G.: Virtualleaf: an open-source framework for cell-based modeling of plant tissue growth and development. Plant Physiol. **155**(2), 656–666 (2011)

14. Salinger, A.G., Bou-Rabee, N.M., Pawlowski, R.P., Wilkes, E.D., Burroughs, E.A., Lehoucq, R.B., Romero, L.A.: Loca 1.0 library of continuation algorithms: theory and implementation manual. Sandia National Laboratories, Albuquerque, NM, Technical Report No. SAND2002-0396 (2002)

15. Schmidt, H., Jirstrand, M.: Systems biology toolbox for matlab: A computational platform for reasearch in systems biology. Bioinformatics Advance Access (2005)

16. Seydel, R.: Practical Bifurcation and Stability Analysis: From Equilibrium to Chaos, vol. 5. Springer, New York (1994)

17. Siso-Nadal, F., Ollivier, J.F., Swain, P.S.: Facile: a command-line network compiler for systems biology. BMC Syst. Biol. **1**(1), 36 (2007)

18. Smith, R., Guyomarc'h, S., Mandel, T., Reinhardt, D., Kuhlemeier, C., Prusinkiewicz, P.: A plausible model of phyllotaxis. PNAS **103**(5), 1301–1306 (2006)

19. SymPy Development Team: SymPy: Python library for symbolic mathematics (2014). http://www.sympy.org

Analysis of Cellular Proliferation and Survival Signaling by Using Two Ligand/Receptor Systems Modeled by Pathway Logic

Gustavo Santos-García[1]([✉]), Carolyn Talcott[2], and Javier De Las Rivas[3]

[1] Universidad de Salamanca, Salamanca, Spain
santos@usal.es
[2] Computer Science Laboratory, SRI International, 333 Ravenswood Ave,
Menlo Park, CA 94025, USA
clt@csl.sri.com
http://pl.csl.sri.com
[3] Cancer Research Center (CSIC/USAL) and IBSAL, Salamanca, Spain
jrivas@usal.es

Abstract. Systems biology attempts to understand biological systems by their structure, dynamics, and control methods. Hepatocyte growth factor (HGF) and interleukin 6 (IL6) are two proteins involved in cellular signaling that bind specific cell surface receptors (HGFR and IL6R, respectively) in order to induce cellular proliferation in different cell types or cell contexts. In both cases, the signaling is initiated by binding the ligand (HGF or IL6) to the membrane-bound receptors (HGFR or IL6R) so as to trigger two cellular signaling paths that have several common elements. In this paper we discuss the processes by which an initial cell leads to cellular proliferation and/or survival signaling by using one of these two ligand/receptor systems analyzed by "rewriting logic" methodology. Rewriting logic procedures are suitable computational tools that handle these dynamic systems, and they can be applied to the study of specific biological pathways behavior. Pathway Logic (PL) constitutes a rewriting logic formalism that provides a knowledge base and development environment to carry out model checking, searches, and executions of signaling systems. Moreover, Pathway Logic Assistant (PLA) is a tool that helps us visualize, analyze and understand graphically cellular elements and their relations. We compare the models of HGF/HGFR and IL6/IL6R signaling pathways in order to investigate the relation between these processes and the way in which they induce cellular proliferation. In conclusion, our results illustrate the use of a logical system that explores complex and dynamic cellular signaling processes.

Keywords: Signal transduction · Symbolic systems biology · Pathway logic · Rewriting logic · Maude

Pathway Logic development has been funded in part by NIH BISTI R21/R33 grant (GM068146-01), NIH/NCI P50 grant (CA112970-01), and NSF grant IIS-0513857. This work was partially supported by NSF grant IIS-0513857. Research was supported by Spanish projects Strongsoft TIN2012-39391-C04-04 and PI12/00624 (MINECO, Instituto de Salud Carlos III).

© Springer International Publishing Switzerland 2015
A. Abate and D. Šafránek (Eds.): HSB 2015, LNBI 9271, pp. 226–245, 2015.
DOI: 10.1007/978-3-319-26916-0_13

1   Modeling Signaling Pathways

The growth of genomic sequence information combined with technological advances in the analysis of global gene expression has revolutionized research in biology and biomedicine [13,39]. Investigation of mammalian signaling processes, the molecular pathways by which cells detect, convert, and internally transmit information from their environment to intracellular targets such as the genome, would greatly benefit from the availability of predictive models [9,17,26].

Various models for the computational analysis of cellular signaling networks have been proposed to simulate responses to specific stimuli [3,40]. The use of differential equations to represent changes in the concentrations from the input to the output is an adequate approach when for a given pathway or sub-pathway there is a large amount of quantitative information and a small number of reactions to be modeled [20,33]. However, in many cases complex cell signaling pathways have to be treated with other more qualitative modeling approaches, like logic modeling.

Symbolic models are based on formalisms that provide a language to represent the states of a system; mechanisms to model their changes (such as reactions); and tools for analysis based on computational or logical inference. A variety of formalisms have been used to develop symbolic models of biological systems, including Petri nets [16,19]; ambient/membrane calculi [29]; statecharts [10]; live sequence charts [30]; and rule-based systems [11,18].

Pathway Logic is a symbolic systems biology approach to modeling biological processes based on rewriting logic. It provides many benefits, including the ability to build and analyze models with multiple levels of detail, represent general rules, define new kinds of data and properties, and execute queries using logical inference. It allows us to develop abstract qualitative models (even quantitative and probabilistic models [1]) of metabolic and signaling processes that can be used as the basis for analysis by powerful tools, such as those developed in the formal methods community, to study a wide range of questions. For example, in [31] the use of Pathway Logic is described to model and analyze the dynamics in a well-known signaling transduction pathway, epidermal growth factor (EGF) pathway.

Rule-based modeling allows us to intuitively specify biological interactions while abstracting from the underlying combinatorial complexity. Other rule-based modeling formalisms similar to Pathway Logic are Kappa [8] and BioNet-Gen [4]. Kappa is a powerful tool in modeling biochemical systems, supporting efficient simulation and static analysis techniques.

The differences between Kappa and the BioNetGen Language are small both in syntax and in implementation. Such languages represent biological entities as agents. Agents are named sets of sites that can be used to hold state or bind and interact with other agents. Interactions are represented by rules in the form of precondition and effect, governed by an associated rate constant that determines how frequently the interaction occurs. The combination of different rule sets generates overall systems, thus allowing modular development of subsystems.

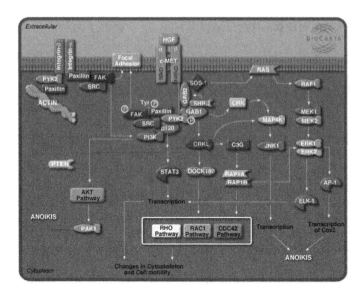

Fig. 1. Signaling of hepatocyte growth factor (HGF) inside the cell (cf. Biocarta pathway collections, http://www.biocarta.com)

In this way, rule-based approaches lighten the combinatorial explosion that results from molecular entities existing under multiple conditions.

Analogously to Pathway Logic, one benefit of Kappa is that Kappa tools use formal methods, such as causal summaries and reachability analysis, to aid information discovery in and debugging of large models. These techniques include the visual representations and the Kappa BioBrick Framework for modeling BioBrick parts.

Our approach focuses on the analysis and modeling of the biological processes by which an initial cellular system can lead to the activation of proliferation and/or survival signaling by using two ligand/receptor pathways.

The cells receive external signals by certain biomolecules (ligands) that are able to interact with certain receptors on the cellular surface producing some effects inside the cell. We select two ligand/receptor pathways that can trigger intra-cellular proliferation and survival signals through different molecular steps (i.e. through different reactions inside the cell). The ligand/receptor systems are HGF/HGFR (Fig. 1) and IL6/IL6R (Fig. 2). HGF is the protein known as hepatocyte growth factor and IL6 is interleukin 6. Each one of these ligand/receptor systems includes a pathway with multiple elements and reactions that are known and that have been modeled using Maude language by Pathway Logic.

In this paper, Sect. 2 contains a short introduction to rewriting logic and Maude. Section 3 gives an overview to Pathway Logic and Pathway Logic Assistant. In the following Sects. 4 and 5, we show the implementation of various rules and logical inferences in the signaling pathways. Conclusions are presented in Sect. 6.

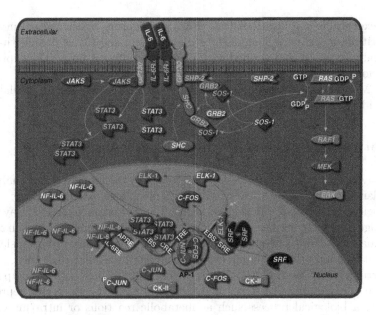

Fig. 2. Signaling pathway of interleukin 6 (IL6) inside the cell (cf. Biocarta pathway collections)

2   Rewriting Logic and Maude

Rewriting logic was first proposed by Meseguer in 1990 as a unifying framework for concurrency [21,22]. A large number of researchers have contributed to the development of several aspects of the logic and its applications in different areas of computer science [12,23].

The naturalness of rewriting logic for modeling and experimenting with mathematical and biological problems has been illustrated in a number of works [1,2,7]. Rewriting logic is a logic of concurrent change that can naturally deal with states and with highly nondeterministic concurrent computations. The basic idea is that we can model a cell as a concurrent system whose concurrent transitions are precisely its biochemical reactions. In this way we can develop symbolic models of biological systems which can be analyzed like any other rewrite theory.

A rewrite theory consists of a signature, which is taken to be an equational theory, and a set of labeled rewrite rules. The signature of a rewrite theory describes a particular structure for the states of a system so that its states can be distributed according to the laws of such a structure. The rewrite rules in the theory describe those elementary local transitions which are possible in the distributed state by concurrent local transformations.

Maude [7] is a high performance language and system supporting both equational and rewriting logic computation. Maude programs achieve a good agreement between mathematical and operational semantics. There are three different uses of Maude modules: (1) as programs that solve some applications; (2) as formal executable specifications that provide a rigorous mathematical model of an algorithm, a system, a language, or a formalism; and (3) as models that can be formally analyzed and verified with respect to different properties expressing various formal requirements.

The Maude system, its documentation, and related papers are available on the Maude web page at http://maude.csl.sri.com.

3   Pathway Logic and Pathway Logic Assistant

"Pathway Logic" (PL) [27, 36, 38] is an approach to the modeling and analysis of molecular and cellular processes based on rewriting logic. Pathway Logic models of biological processes are developed using Maude language. A Pathway Logic knowledge base includes data types representing cellular components such as proteins, small molecules, or complexes; compartments/locations; post-translational modifications and other dynamic events occurring in cellular reactions.

Rewrite rules describe the behavior of proteins and other components depending on modification states and biological contexts. Each rule represents a step in a biological process such as metabolic reactions or intra/inter cellular signaling reactions. A collection of such facts forms a formal knowledge base. A model is then a specification of an initial state (cell components and locations) interpreted in the context of a knowledge base. Such models are executable and can be understood as specifying possible ways in which a system can evolve. Logical inference and analysis techniques are used for simulation of possible ways in which a system could evolve, for the assemblage of pathways as answers to queries, and for the reasoning of the dynamic assembly of complexes, cascading transmission of signals, feedback-loops, cross talk between subsystems, and larger pathways. Logical and computational reflection can be used to transform and further analyze models.

Given an executable model such as the one described above, there are many kinds of computations that can be carried out, including: static analysis, forward simulation, forward search, backward search, explicit state model checking, and meta analysis.

Pathway Logic models are structured in four layers: sorts and operations, components, rules, and queries. The *sorts* and *operations* layer declares the main sorts and subsort relations, the logical analogue to ontology. The sorts of entities include `Chemical`, `Protein`, `Complex`, and `Location` (cellular compartments), and `Cell`. These are all subsorts of the `Soup` sort that represents unordered multisets of entities. The sort `Modification` is used to represent post-translational protein modifications (e.g., activation, binding, phosphorylating). Modifications are applied using the operator `[ - ]`. For example, the term `[Rac1 - GDP]` indicates that Ras-related C3 botulinum toxin substrate 1 (Rac1) is binding to guanosine diphosphate (GDP).

The *queries* layer specifies initial states or *dishes* to be studied and properties of interest. Initial states are in silico Petri dishes containing a cell and ligands of interest. An initial state is represented by a term of the form `PD(Soup)`,

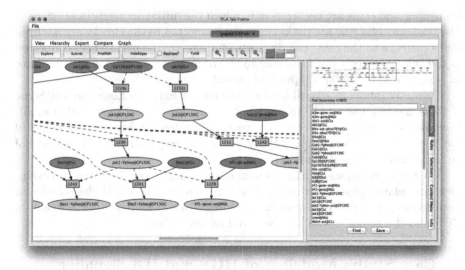

Fig. 3. A general view of a signaling pathway using Pathway Logic Assistant.

where Soup represents a soup of cell components and locations, ligands and other molecular components in the cell surroundings. Each location is represented by a term of the form {locName | components} where locName identifies the location (for example, CLm for cell membrane, CLc for cell cytoplasm, CLo for the outside of the cell membrane, CLi for the inside of the cell membrane, NUc for the nucleus, XOut for the outside of the cell, the medium or supernatant) while components stands for the mixture of proteins and other compounds in that location.

The *components* layer specifies particular entities (proteins, chemicals) and introduces additional sorts for grouping proteins in families. The *rules* layer contains rewrite rules specifying individual steps of a process. These rules correspond to reactions in traditional metabolic and interaction databases.

The Pathway Logic Assistant (PLA) [37] provides an interactive visual representation of Pathway Logic models and facilitates the following tasks: it displays the network of signaling reactions for a given dish; it formulates and submits queries to find and compare pathways; it visualizes gene expression data in the context of a network; or it computes and displays the downstream subnet of one or more proteins. Given an initial dish, the PLA selects the relevant rules from the rule set and represents the resulting reaction network as a Petri net. This provides a natural graphical representation that is similar to the hand drawn pictures used by biologists, as well as very efficient algorithms for answering queries.

Figure 3 shows a general view of Pathway Logic Assistant, which is a Java software that implements the Pathway Logic vision. It shows the Petri net representation of interleukin 6 signaling pathway using PLA. Rectangles are transitions and ovals are occurrences in which the initial occurrences are darker. The reactants of a rule are the occurrences connected to the rule by arrows from the occurrence to the rule. The products of a rule are the occurrences connected to the rule by arrows from the rule to the occurrence. Dashed arrows indicate an occurrence that is both input and output.

Currently there are several implementations of Pathway Logic models. Some of these models are: STM7 (a model of cellular response to external stimuli), Protease (a network model of gram+ bacterial proteases), Mycolate (a model of the Mycobacterial Mycolic Acid Biosynthesis Pathway), GlycoSTM (a model of glycosylation extending the KEGG pathways).

The Pathway Logic and PLA system, its documentation, a collection of examples, and related papers are available on http://pl.csl.sri.com. Models of cellular response to many different stimuli, including a much more complete model of HGF and IL6 signaling can also be found on our website.

4   Case Study: Modeling of HGF and IL6 Signaling Pathways (Dishes and Rewrite Rules)

In this section we define some rules of the Pathway Logic knowledge base. We will focus on the Pathway Logic models of response to HGF and IL6 stimulation. Hepatocyte growth factor (HGF) and interleukin 6 (IL6) signaling are important proteins involved in cellular signaling: HGF is a multifunctional growth factor which can induce cell dissociation, migration, protection from apoptosis, proliferation and differentiation; IL6 is a pleiotropic cytokine produced by various types of cells that can provoke a broad range of cellular and physiological responses, including the immune response, inflammation, hematopoiesis, cell growth, gene activation, proliferation and survival. In both cases, their signaling pathways include common reactions and circuits and, in fact, both can induce cellular proliferation activating proteins ERK and STAT inside the cells (see Figs. 1 and 2).

In our case study, a *dish* (called IL6Dish) with several locations is defined: the membrane (location tag CLm) contains IL6R; the inside of the membrane (location tag CLi) contains Rac1 binding to GDP; the cytoplasm (location tag CLc) contains Akt1, Erks, Gab1, Gab2, Hck, Jak1, Jak2, Mkk4, Pkcd, Shp2, Stat1, Stat3, Tyk2, Vav1, Vav2, and Vav3; and the nucleus (location tag NUc) contains A2m-gene, Foxo1, Irf1-gene, Lmo4, RankL-gene, and Socs3-gene. Moreover, there are other two locations: the outside (location tag XOut) contains the interleukin 6 (IL6) and the GP130C location contains the glycoprotein 130 (Gp130):

```
eq IL6Dish = PD( {XOut | IL6 }   {GP130C | Gp130 }
  {CLm | IL6R }   {CLi | [Rac1 - GDP] }
  {CLc | Akt1 Erks Gab1 Gab2 Hck Jak1 Jak2 Mkk4 Pkcd Shp2
      Stat1 Stat3 Tyk2 Vav1 Vav2 Vav3 }
  {NUc | A2m-gene Foxo1 Irf1-gene Lmo4 RankL-gene Socs3-
      gene }) .
```

In the same way, we define the `HgfDish` dish:

```
eq HgfDish = PD( {XOut | Hgf}  {HgfRC | HgfR}
  {CLm | empty}
  {CLi | [Cdc42 - GDP] [Hras - GDP] [Mras - GDP] [Rac1 -
    GDP] [Rala - GDP] [Rap1a - GDP] Src }
  {CLc | Akt1 Bad Cbl Crk CrkL Ctnnb1 Eif4ebp1 Erks Fak1
    Fak2 Gab1 Gab2 Grb2 [Gsk3s - act] Jnks Lkb1 Pak1
    Pi3k Plcg1 Pxn P38s Raf1 Rps6 Rsk1 Smad2 Smad3
    Stat3 S6k1}
  {NUc | Creb1 Elk1 Ets1 Fos-gene Foxo1 IL6-gene Mmp1-
    gene Mmp9-gene Myc-gene Pai1-gene} ) .
```

Note that in the two biological dishes we have defined the following common elements: `Foxo1` in the nucleus and `Akt1`, `Erks`, and `Stat3` in the cytoplasm.

Rewrite Rule 1334. One rule, as an example, that it is defined inside the HGF signaling pathway is rule 1334, directly sourced from MedLine database article *"Induction of epithelial tubules by growth factor HGF depends on the STAT pathway"* with ID 9440692 [5]. Boccaccio *et al.* determine that HGF stimulates recruitment of STAT3 to the receptor, tyrosine phosphorylation, nuclear translocation and binding to the specific promoter element SIE. Electroporation of a tyrosine-phosphorylated peptide, which interferes with both the association of STAT to the receptor and STAT dimerization, inhibits tubule formation in vitro without affecting either HGF-induced *scattering* or growth.

Our rule 1334 establishes: *When HGF (Hgf) binds to its receptor HGFR (HgfR), the Tyrosine (Y) kinase cytoplasmic domain of HGFR ([HgfR - Yphos]) phosphorylates STAT3 on tyrosine 705 ([Stat3 - phos(Y 705)]) in the presence of protein STAT3 (Stat3) in the cytoplasm (CLc).* In Maude syntax, this signaling process is described by the following rewrite rule:

```
rl [1334 . Stat3 . irt . Hgf] :
  {HgfRC | hgfrc ([HgfR - Yphos] : Hgf) }
  {CLc   | clc    Stat3                  }
  =>
  {HgfRC | hgfrc ([HgfR - Yphos] : Hgf) }
  {CLc   | clc    [Stat3 - phos(Y 705)] } .
```

Figure 4 shows this rule using PLA. Rectangles represent reaction rules. The label in a rectangle is its abbreviated identifier in the knowledge base. Solid arrows from an occurrence oval to a rule indicate that the occurrence is a reactant (rule input). Solid arrows from a rule to an occurrence oval indicate that the occurrence is a product (rule output). Dashed arrows from an occurrence oval to a rule indicate that the occurrence is a modifier/enzyme/control. It is necessary for the reaction to take place but is not changed by the reaction.

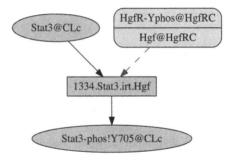

Fig. 4. Rule [1334.Stat3.irt.Hgf] using Pathway Logic Assistant.

Rewrite Rule 1282. According to [6] and [14], the rule 1282 describes that hepatocyte growth factor/scatter factor (HGF/SF) induces mitogenesis and cell dissociation upon binding to the protein-tyrosine kinase receptor encoded by the MET proto-oncogene (p190MET). The rule 1282 establishes: *When Hgf is outside the cell, in the presence of HgfR, then this receptor is phosphorylated and bound to Hgf.* In Maude syntax, this signaling process is described by the following rewrite rule (Fig. 5):

```
rl[1282.HgfR.irt.Hgf]:
  {XOut   | xout   Hgf                           }
  {HgfRC  | hgfrc  HgfR                          }
  =>
  {XOut   | xout                                 }
  {HgfRC  | hgfrc  ([HgfR - Yphos] : Hgf) } .
```

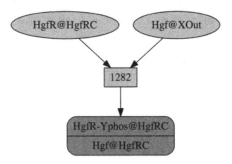

Fig. 5. Rule [1282.HgfR.irt.Hgf] using Pathway Logic Assistant.

Rewrite Rule 1237. Now we consider the binding of IL6 to the IL6R. According to [25] (*"Protein kinase C delta associates with the interleukin-6 receptor subunit glycoprotein (gp) 130 via Stat3 and enhances Stat3-gp130 interaction"*), the rule 1237 establishes the recruitment to Gp130 complex: *When IL6 binds to IL6R, and this receptor is also bound to GP130, in the presence of STAT3 in the cytoplasm (Clc), this protein (STAT3) is phosphorylated by the cytoplasmic domain of GP130 (GP130C).* In Maude syntax, this signaling process is described by the following rewrite rule (Fig. 6):

```
rl [1237.Stat3.to.Gp130C.irt.IL6]:
  {GP130C | gp130c (IL6 : IL6R : Gp130)                        }
  {CLc    | clc      Stat3                                     }
  =>
  {GP130C | gp130c (IL6 : IL6R : Gp130) [Stat3 - Yphos] }
  {CLc    | clc                                          } .
```

The interleukin (IL)-6-type cytokines play major roles in a variety of biological processes by signaling by means of a common receptor subunit—glycoprotein (gp) 130 [24].

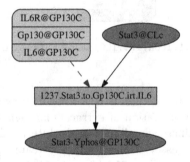

Fig. 6. Rule [1237.Stat3.to.Gp130C.irt.IL6] using Pathway Logic Assistant.

Rewrite Rule 1221. Interleukin 6 mediates pleiotropic functions in various types of cells through its specific receptor (IL6R). According to [15,35], the rule 1221 describes that an 80 kd single polypeptide chain (IL6R) is involved in IL6 binding and that IL6 triggers the association of this receptor with a non-ligand-binding membrane glycoprotein, gp130. The rule 1221 establishes: *When IL6 is outside the cell and the receptor IL6R is inside the cytoplasm, in the presence of GP130, the association of this receptor with a non-ligand-binding membrane glycoprotein occurs.* In Maude syntax, this signaling process is described by the following rewrite rule (Fig. 7):

```
rl[1221.IL6R.irt.IL6]:
  {XOut    | xout    IL6                        }
  {GP130C  | gp130c  Gp130                      }
  {CLm     | clm     IL6R                       }
  =>
  {XOut    | xout                               }
  {GP130C  | gp130c  (IL6 : IL6R : Gp130) }
  {CLm     | clm                               } .
```

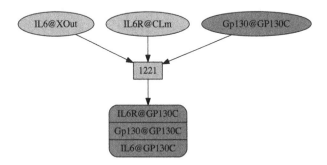

Fig. 7. Rule [1221.IL6R.irt.IL6] using Pathway Logic Assistant.

Rewrite Rule 1227. Interleukin-6 is a known growth and survival factor in multiple myeloma via activation of extracellular signal-regulated kinase and phosphatidylinositol 3-kinase signaling cascade. Interleukin-6 induces their tyrosine phosphorylation and association with downstream signaling molecules [28]. The rule 1227 establishes: *When IL6 binds to IL6R, and this receptor is also bound to GP130, in the presence of Hck in the cytoplasm (Clc), the protein Hck is activated ([Hck - act])*. In Maude syntax, this signaling process is described by the following rewrite rule (Fig. 8):

```
rl[1227.Hck.irt.IL6]:
  {GP130C | gp130c (IL6 : IL6R : Gp130) }
  {CLc    | clc    Hck                         }
  =>
  {GP130C | gp130c (IL6 : IL6R : Gp130) }
  {CLc    | clc    [Hck - act]               } .
```

Rewrite Rule 1224. According to [15,28,32], the rule 1224 describes phosphorylation on Tyrosine in the JAK/STAT signal transduction pathway in response to interleukin-6. The rule 1224 establishes: *When IL6 binds to IL6R, and this*

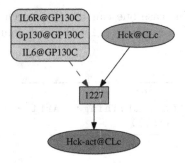

Fig. 8. Rule [1227.Hck.irt.IL6] using Pathway Logic Assistant.

receptor is also bound to GP130, in the presence of active HCK ([Hkc-act]) in the cytoplasm (Clc), the protein ERK is phosphorylated in TEY domain ([Erks -erksmods phos(TEY)]). In Maude syntax, this signaling process is described by the following rewrite rule (Fig. 9):

```
rl[1224.Erks.irt.IL6]:
   {GP130C | gp130c (IL6 : IL6R : Gp130)                    }
   {CLc    | clc [Hck - act] [Erks - erksmods]              }
   =>
   {GP130C | gp130c (IL6 : IL6R : Gp130)                    }
   {CLc | clc [Hck - act] [Erks - erksmods phos(TEY)] }  .
```

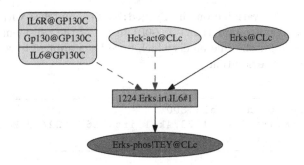

Fig. 9. Rule [1224.Erks.irt.IL6] using Pathway Logic Assistant.

5 Case Study: Understanding Dynamics on HGF and IL6 Signaling Pathways (Logical Inferences)

Our analysis begins with initial dish states IL6Dish and HgfDish defined in Sect. 3. Suppose we want to find out if there is a pathway leading to activation of Akt1 and Erks. In this case one can use the search command with a suitable search pattern and parameters ([1]: the first solution; =>+: at least one step):

```
Maude> search [1] IL6Dish =>+
  PD(soup:Soup {CLc | th:Things [Akt1 - act] [erks:ErkS -
      mod:ModSet act]}) .
```

The solution to this query given by Maude shows the matching in the search pattern PD(soup:Soup {CLc | th:Things [Akt1 - act] [erks:ErkS - mod:ModSet act]}):

```
Solution 1 (state 800)
states: 801   rewrites: 1832 in 465ms cpu (939ms real)
soup:Soup --> {GP130C | Gp130 : IL6 : IL6R}
  {XOut | empty}  {CLm | empty}
  {CLi | [Rac1 - GDP]}
  {NUc | A2m-gene Irf1-gene RankL-gene Socs3-gene Foxo1
      Lmo4}
th:Things --> Gab1 Gab2 Jak1 Jak2 Mkk4 Pkcd Shp2 Stat1
      Stat3 Tyk2 Vav1 Vav2 Vav3 [Hck - act]
erks:ErkS --> Erks
mod:ModSet --> phos(TEY)
```

We observe that the variable on the fly mod:ModSet matches with phos(TEY) in this solution. We also find [Rac1 - GDP] inside of the cell membrane. Then we can ask Maude for the rule labels which have been applied to reach the final state according to the solution:

```
Maude> show path labels 800 .
  1221.IL6R.irt.IL6   1223.Akt1.irt.IL6   1227.Hck.irt.IL6
  1224.Erks.irt.IL6   415c.Erks.act
```

Maude allows us to find all possible solutions and, in the case that the final state is not reachable from our initial dish, it indicates that there is no solution.

Now the matching for a new solution in our search is shown. In this solution, we also find [Rac1 - GDP] inside of the cell membrane and the protein ERK is also phosphorylated in TEY domain.

```
Maude> cont 1 .
Solution 2 (state 1435)
states: 1436  rewrites: 1892 in 73ms cpu (74ms real)
soup:Soup -->
  {CLm | empty}
  {CLi | [Rac1 - GDP]}
  {XOut | empty}
  {NUc | A2m-gene Foxo1 Irf1-gene Lmo4 RankL-gene Socs3-
    gene}
  {GP130C | Jak1 Gp130 : IL6 : IL6R}
th:Things --> Gab1 Gab2 Jak2 Mkk4 Pkcd Shp2 Stat1 Stat3
    Tyk2 Vav1 Vav2 Vav3 [Hck - act]
erks:ErkS --> Erks
mod:ModSet --> phos(TEY)
```

Then we can ask Maude for the rule labels which have been applied to reach the
final state according to the solution:

```
Maude> show path labels 1435 .
  1229c.Jak1.Gp130.complex   1221.IL6R.irt.IL6
  1223.Akt1.irt.IL6   1227.Hck.irt.IL6   1224.Erks.irt.IL6
  415c.Erks.act
```

Afterwards we consider the dish HgfDish and in order to find out the same
search with IL6Dish:

```
Maude> search [1] HgfDish =>+
  PD(soup:Soup {CLc | th:Things [Akt1 - act] [erks:ErkS -
    mod:ModSet act]}) .
```

The solution to this query given by Maude shows the matching in the same
search pattern with IL6:

```
Solution 1 (state 1469)
states: 1470  rewrites: 2618 in 508ms cpu (972ms real)
soup:Soup -->  {HgfRC | Hgf : [HgfR - Yphos]}
  {XOut | empty}  {CLm | empty}
  {CLi | Src [Cdc42 - GDP] [Hras - GDP] [Mras - GDP]    [
    Rac1 - GDP] [Rala - GDP] [Rap1a - GDP]}
  {NUc | Fos-gene IL6-gene Mmp1-gene Mmp9-gene Myc-gene
    Pai1-gene Creb1 Elk1 Ets1 Foxo1}
th:Things --> Bad Cbl Crk CrkL Ctnnb1 Eif4ebp1 Fak1 Fak2
    Gab1 Gab2 Grb2 Jnks Lkb1 P38s Pak1 Pi3k Plcg1 Raf1
    Rps6 Rsk1 S6k1 Smad2 Smad3 Stat3 [Gsk3s - act] [Pxn -
    Yphos]
erks:ErkS --> Erks
mod:ModSet --> phos(TEY)
```

We observe that the variable on the fly mod:ModSet matches with phos(TEY) in this solution. Moreover we find [Cdc42 - GDP], [Hras - GDP], [Mras - GDP], [Rac1 - GDP], [Rala - GDP], and [Rap1a - GDP] inside of the cell membrane. We also find, among other compounds, Fos-gene and IL6-gene in the nucleus. Hence we can ask Maude for the rule labels which have been applied to reach the final state according to the solution:

```
Maude> show path labels 1469 .
  1282.HgfR.irt.Hgf        1283.Akt1.irt.Hgf
  1292.Pxn.Yphos.irt.Hgf   1285.Erks.irt.Hgf
```

Now the matching for a new second solution in our search is shown. In this solution, we find: (1) Cdc42, Hras, Mras, Rac1, Rala, Rap1a are binding to guanosine diphosphate (GDP) inside of the cell membrane, as in the previous solution; (2) the glycogen synthase kinase-3 Gsk3s is activated in the cytoplasm; (3) Bad and Pxn are phosphorylated; and (4) the protein ERK is also phosphorylated in TEY domain.

```
Maude> cont 1 .
Solution 2 (state 10258)
states: 10259   rewrites: 26157 in 413ms cpu (414ms real)
soup:Soup -->
  {CLm | empty}
  {CLi | Src [Cdc42 - GDP] [Hras - GDP] [Mras - GDP] [
      Rac1 - GDP] [Rala - GDP] [Rap1a - GDP]}
  {XOut | empty}
  {NUc | Fos-gene IL6-gene Mmp1-gene Mmp9-gene Myc-gene
      Pai1-gene Creb1 Elk1 Ets1 Foxo1}
  {HgFRC | Hgf : [HgfR - Yphos]}
th:Things --> Cbl Crk CrkL Ctnnb1 Eif4ebp1 Fak1 Fak2 Gab1
    Gab2 Grb2 Jnks Lkb1 P38s Pak1 Pi3k Plcg1 Raf1 Rps6
    Rsk1
    S6k1 Smad2 Smad3 Stat3 [Bad - phos(S 75) phos(S 99)]
        [Gsk3s - act] [Pxn - Yphos]
erks:ErkS --> Erks
mod:ModSet --> phos(TEY)
```

Then we can ask Maude for the rule labels which have been applied to reach the final state according to the solution:

```
Maude> show path labels 10258 .
  1282.HgfR.irt.Hgf    1283.Akt1.irt.Hgf   1305.Bad.irt.Hgf
  1292.Pxn.Yphos.irt.Hgf   1285.Erks.irt.Hgf
```

In the same way that a search analysis was carried out with model checking, these queries could also be done using FindPath in Pathway Logic Assistant.

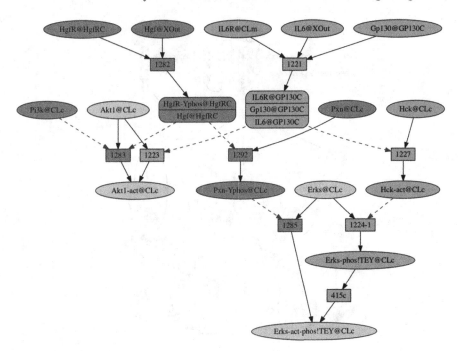

Fig. 10. A graphical comparison of the searches (activation of `Akt1` and `Erks` from `IL6Dish` and `HgfDish` dishes) using FindPath tool in Pathway Logic Assistant.

Because a network may contains more than one route to get to the goal (activation of `Akt1` and `Erks`), you will find one path (usually the shortest) by clicking on the FindPath button in the toolbar.

The advantage of PLA is that the resulting pathway can be shown as a nice graph. If a pathway exists its graph will be created and displayed. The set of initial occurrences of the pathway graph is the intersection of the occurrences of the pathway graph with the initial occurrences of the parent graph. The goal set of the pathway graph is the goal set of the query (Fig. 10). In this graph, IL6 rules and occurrences have a purple/darker color, and HGF ones have a blue-green/lighter color. The common part is peach colored. Dashed arrows from an occurrence oval to a rule indicate that the occurrence remains unchanged for the reaction.

The results provided by the two "searches" upon `IL6Dish` and `HgfDish` (i.e. upon IL6 and HGF signaling pathways) show a clear similarity that indicates the activation of the ERK proteins (phosphorylated in T and in Y) and therefore a common signal of activation of "proliferation" (as it is indicated in the theoretical and schematic map of pathways presented in Fig. 11). This activation goes through different ways for the case of IL6 versus HGF, since the resulting states end up with different `th:Things` and with different transcription activators in nucleus `NUc`.

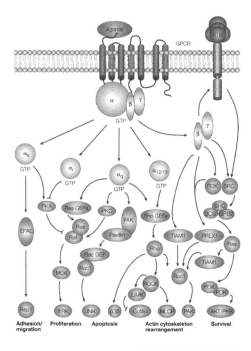

Fig. 11. Non-canonical G-protein-coupled-receptor signaling (cf. [34]).

6 Conclusions

Rewriting logic procedures are powerful symbolic methods that can be applied in order to understand the dynamics of complex biological systems. It provides many benefits, including the ability to build and analyze models with multiple levels of detail; to represent general rules; to define new kinds of data and properties; and to execute queries using logical inference.

In this work we show the application of a rewriting logic procedure based in Maude logic language to the dynamic modeling of biological signaling pathways. We are interested in formalizing models that molecular biologists can use to think about signaling pathways and their behavior, allowing them to computationally formulate questions about their outcomes and dynamics. In this way, as a case study, we compare the models of HGF/HGFR and IL6/IL6R signaling pathways investigating the relation and crosslinks between these processes and the way in which both can induce cellular "proliferation" and cellular "survival". In conclusion, our results provide a logical system that explores complex and dynamic cellular signaling processes.

Figure 11 shows a theoretical scheme taken from a cell biology review which presents the main molecular elements (i.e. proteins) involved in the transmission of signals from the cell membrane receptors (GPCR type) to the nucleus in order to trigger some specific biological processes: adhesion, proliferation, apoptosis, survival, etc. The figure is a schematic cartoon integrating several pathways to provide a general view of the main ways and interactions that occur in the cell to activate such specific processes. As it can be seen, this figure reflects well our results in the analyses of the logic and dynamics of HGF/HGFR and IL6/IL6R pathways, because our finding reveals the activation of ERK and AKT.

In this article, in order to model signal transduction processes, we also describe the use of Pathway Logic (PL) as a rewriting logic tool that is built using Maude, as well as the use of Pathway Logic Assistant (PLA) software to browse and analyze the logic models of many pathways already built. The models are derived from many experimental data and knowledge from the literature. Using these tools we achieve a very interesting comparison of two signaling pathways (HGF/HGFR and IL6/IL6R) finding their common gates and the key molecular elements that allow them to produce "proliferation" and "survival", despite the fact that most of the molecular elements and the ways of these two signaling processes are different.

References

1. Abate, A., Bai, Y., Sznajder, N., Talcott, C.L., Tiwari, A.: Quantitative and probabilistic modeling in pathway logic. In: Zhu, M.M., Zhang, Y., Arabnia, H.R., Deng, Y. (eds.) Proceedings of the 7th IEEE International Conference on Bioinformatics and Bioengineering, BIBE 2007, pp. 922–929. Harvard Medical School, Boston, MA, October 14–17, 2007. IEEE (2007)
2. Agha, G., Danvy, O., Meseguer, J. (eds.): Formal Modeling: Actors, Open Systems, Biological Systems - Essays Dedicated to Carolyn Talcott on the Occasion of her 70th Birthday. LNCS, vol. 7000. Springer, Heidelberg (2011)
3. Asthagiri, A.R., Lauffenburger, D.A.: A computational study of feedback effects on signal dynamics in a mitogen-activated protein kinase (MAPK) pathway model. Biotechnol. Progr. **17**(2), 227–239 (2001)
4. Blinov, M.L., Faeder, J.R., Goldstein, B., Hlavacek, W.S.: BioNetGen: software for rule-based modeling of signal transduction based on the interactions of molecular domains. Bioinformatics **20**(17), 3289–3291 (2004)
5. Boccaccio, C., Ando, M., Tamagnone, L., Bardelli, A., Michieli, P., Battistini, C., Comoglio, P.M.: Induction of epithelial tubules by growth factor HGF depends on the STAT pathway. Nature **391**(6664), 285–288 (1998)
6. Bottaro, D.P., Rubin, J.S., Faletto, D.L., Chan, A.M., Kmiecik, T.E., Vande Woude, G.F., Aaronson, S.A.: Identification of the hepatocyte growth factor receptor as the c-met proto-oncogene product. Science **251**(4995), 802–804 (1991)
7. Clavel, M., Durán, F., Eker, S., Lincoln, P., Martí-Oliet, N., Meseguer, J., Talcott, C. (eds.): All About Maude - A High-Performance Logical Framework, how to Specify, Program and Verify Systems in Rewriting Logic. LNCS, vol. 4350. Springer, Heidelberg (2007). doi:10.1007/978-3-540-71999-1
8. Danos, V., Laneve, C.: Formal molecular biology. Theor. Comput. Sci. **325**(1), 69–110 (2004). doi:10.1016/j.tcs.2004.03.065
9. Donaldson, R., Talcott, C.L., Knapp, M., Calder, M.: Understanding signalling networks as collections of signal transduction pathways. In: Quaglia, P. (ed.) Computational Methods in Systems Biology, CMSB, pp. 86–95. ACM (2010)

10. Efroni, S., Harel, D., Cohen, I.R.: Toward rigorous comprehension of biological complexity: modeling, execution, and visualization of thymic T-cell maturation. Genome Res. **13**(11), 2485–2497 (2003)
11. Faeder, J.R., Blinov, M.L., Hlavacek, W.S.: Rule-based modeling of biochemical systems with BioNetGen. Methods Mol. Biol. **500**, 113–167 (2009)
12. Fiadeiro, J.L. (ed.): WADT 1998. LNCS, vol. 1589. Springer, Heidelberg (1999)
13. Fisher, J., Henzinger, T.A.: Executable cell biology. Nat. Biotech. **25**(11), 1239–1249 (2007). doi:10.1038/nbt1356
14. Graziani, A., Gramaglia, D., dalla Zonca, P., Comoglio, P.M.: Hepatocyte growth factor/scatter factor stimulates the Ras-guanine nucleotide exchanger. J. Biol. Chem. **268**(13), 9165–9168 (1993)
15. Guschin, D., Rogers, N., Briscoe, J., Witthuhn, B., Watling, D., Horn, F., Pellegrini, S., Yasukawa, K., Heinrich, P., Stark, G.R.: A major role for the protein tyrosine kinase JAK1 in the JAK/STAT signal transduction pathway in response to interleukin-6. EMBO J. **14**(7), 1421–1429 (1995)
16. Hardy, S., Robillard, P.N.: Petri net-based method for the analysis of the dynamics of signal propagation in signaling pathways. Bioinformatics **24**(2), 209–217 (2008)
17. Heiser, L.M., Wang, N.J., Talcott, C.L., Laderoute, K.R., Knapp, M., Guan, Y., Hu, Z., Ziyad, S., Weber, B.L., Laquerre, S., Jackson, J.R., Wooster, R.F., Kuo, W.L., Gray, J.W., Spellman, P.T.: Integrated analysis of breast cancer cell lines reveals unique signaling pathways. Genome Biol. **10**(3), R31 (2009)
18. Hwang, W., Hwang, Y., Lee, S., Lee, D.: Rule-based multi-scale simulation for drug effect pathway analysis. BMC Med. Inform. Decis. Mak. **13**(Suppl 1), S4 (2013)
19. Li, C., Ge, Q.W., Nakata, M., Matsuno, H., Miyano, S.: Modelling and simulation of signal transductions in an apoptosis pathway by using timed petri nets. J. Biosci. **32**(1), 113–127 (2007)
20. Liu, X., Betterton, M.D., Saadatpour, A., Albert, R.: Methods in Molecular Biology, vol. 880, pp. 255–272. Humana Press, Clifton (2012)
21. Martí-Oliet, N., Ölveczky, P.C., Talcott, C. (eds.): Logic, Rewriting, and Concurrency- Essays dedicated to José Meseguer on the occasion of his 65th birthday. LNCS, vol. 9200. Springer, Heidelberg (2015). doi:10.1007/978-3-319-23165-5
22. Meseguer, J.: Conditional rewriting logic as a unified model of concurrency. Theor. Comput. Sci. **96**(1), 73–155 (1992)
23. Meseguer, J.: Twenty years of rewriting logic. J. Log. Algebr. Program **81**(7–8), 721–781 (2012). doi:10.1016/j.jlap.2012.06.003
24. Novotny-Diermayr, V., Lin, B., Gu, L., Cao, X.: Modulation of the interleukin-6 receptor subunit glycoprotein 130 complex and its signaling by LMO4 interaction. J. Biol. Chem. **280**(13), 12747–12757 (2005)
25. Novotny-Diermayr, V., Zhang, T., Gu, L., Cao, X.: Protein kinase C delta associates with the interleukin-6 receptor subunit glycoprotein (gp) 130 via Stat3 and enhances Stat3-gp130 interaction. J. Biol. Chem. **277**(51), 49134–49142 (2002)
26. Pais, R.S., Moreno-Barriuso, N., Hernandez-Porras, I., Lopez, I.P., De Las Rivas, J., Pichel, J.G.: Transcriptome analysis in prenatal IGF1-deficient mice identifies molecular pathways and target genes involved in distal lung differentiation. PLoS One **8**(12), e83028 (2013)
27. Panikkar, A., Knapp, M., Mi, H., Anderson, D., Kodukula, K., Galande, A.K., Talcott, C.L.: Applications of Pathway Logic modeling to target identification. In: Agha et al. (eds.): Formal Modeling: Actors, Open Systems, Biological Systems - Essays Dedicated to Carolyn Talcott on the Occasion of her 70th Birthday. LNCS, vol. 7000, pp. 434–445. Springer, Heidelberg (2011)

28. Podar, K., Mostoslavsky, G., Sattler, M., Tai, Y.T., Hayashi, T., Catley, L.P., Hideshima, T., Mulligan, R.C., Chauhan, D., Anderson, K.C.: Critical role for hematopoietic cell kinase (Hck)-mediated phosphorylation of Gab1 and Gab2 docking proteins in interleukin 6-induced proliferation and survival of multiple myeloma cells. J. Biol. Chem. **279**(20), 21658–21665 (2004)

29. Regev, A., Panina, E.M., Silverman, W., Cardelli, L., Shapiro, E.: BioAmbients: an abstraction for biological compartments. Theor. Comput. Sci. **325**(1), 141–167 (2004)

30. Sadot, A., Fisher, J., Barak, D., Admanit, Y., Stern, M.J., Hubbard, E.J.A., Harel, D.: Toward verified biological models. IEEE/ACM Trans. Comput. Biol. Bioinform. **5**(2), 223–234 (2008)

31. Santos-García, G., De Las Rivas, J., Talcott, C.L.: In: Saez-Rodriguez, J., Rocha, M.P., Fdez-Riverola, F., De Paz Santana, J.F. (eds.) A Logic Computational Framework to Query Dynamics on Complex Biological Pathways. AISC, vol. 294, pp. 207–214. Springer, Heidelberg (2014)

32. Schlessinger, J.: Cell signaling by receptor tyrosine kinases. Cell **103**(2), 211–225 (2000)

33. Smolen, P., Baxter, D.A., Byrne, J.H.: Mathematical modeling of gene networks. Neuron **26**(3), 567–580 (2000)

34. Sodhi, A., Montaner, S., Gutkind, J.S.: Viral hijacking of G-protein-coupled-receptor signalling networks. Nat. Rev. Mol. Cell Biol. **5**(12), 998–1012 (2004). doi:10.1038/nrm1529

35. Taga, T., Hibi, M., Hirata, Y., Yamasaki, K., Yasukawa, K., Matsuda, T., Hirano, T., Kishimoto, T.: Interleukin-6 triggers the association of its receptor with a possible signal transducer, gp130. Cell **58**(3), 573–581 (1995)

36. Talcott, C.: Pathway Logic. In: Bernardo, M., Degano, P., Zavattaro, G. (eds.) SFM 2008. LNCS, vol. 5016, pp. 21–53. Springer, Heidelberg (2008)

37. Talcott, C.L., Dill, D.L.: The pathway logic assistant. In: Plotkin, G. (ed.) Proceedings of the Third International Workshop on Computational Methods in Systems Biology, pp. 228–239 (2005)

38. Talcott, C.L., Eker, S., Knapp, M., Lincoln, P., Laderoute, K.: Pathway logic modeling of protein functional domains in signal transduction. In: Markstein, P., Xu, Y. (eds.) Proceedings of the 2nd IEEE Computer Society Bioinformatics Conference, CSB 2003, pp. 618–619, Stanford, CA, 11–14 August 2003. IEEE Computer Society (2003). doi:10.1109/CSB.2003.1227425

39. Vukmirovic, O.G., Tilghman, S.M.: Exploring genome space. Nature **405**(6788), 820–822 (2000). doi:10.1038/35015690

40. Weng, G., Bhalla, U.S., Iyengar, R.: Complexity in biological signaling systems. Science **284**(5411), 92–96 (1999)

Posters and Tool Demos

- *Md. Ariful Islam, Richard DeFrancisco, Chuchu Fan, Radu Grosu, Sayan Mitra and Scott Smolka.* Model Checking Tap Withdrawal in C. Elegans

- *Monika Varga, Aleš Prokop and Bela Csukas.* Unified dynamic modeling of conservation low and sign based, hybrid, multiscale biosystems

- *Hiroshi Yoshida.* A model towards multicell-turnover patterns using multi-variable polynomials - Polynomial Life

- *Matej Klement, David Šafránek, Jan Červený, Tadeáš Děd, Matej Troják, Luboš Brim and Stefan Müller.* E-cyanobacterium.org: A Web-based Platform for Systems Biology of Cyanobacteria

- *Delphine Draelants, Przemyslaw Klosiewicz, Jan Broeckhove and Wim Vanroose.* Numerical Continuation Toolbox in Python: PyNCT

- *Milan Češka, Caroline Schneider, Alessandro Abate, David Šafránek, Louis Mahadevan and Marta Kwiatkowska.* Stochastic Modelling of the Interface between Regulatory Enzymes and Transcriptional Initiation at Inducible Genes

- *Miriam García Soto.* Demo abstract of AVERIST: Algorithmic verifier of stability

- *Ratan Lal.* Bounded reachability analysis of parameterized linear hybrid systems

- *Katherine Casey and Jamie Vicary.* Graphical Logic for Biological Modelling

- *Daniel Figueiredo and Manuel A. Martins.* Two CS paradigms in biological boolean networks

- *Linar Mikeev and Verena Wolf.* SHAVE - Stochastic Hybrid Analysis of Markov Population Models

© Springer International Publishing Switzerland 2015
A. Abate and D. Šafránek (Eds.): HSB 2015, LNBI 9271, p. 247, 2015.
DOI: 10.1007/978-3-319-26916-0

Author Index

Angel, Peter 75

Bartocci, Ezio 156
Beica, Andreea 173
Bock, Christoph 141
Bortolussi, Luca 141, 156
Brim, Luboš 58
Broeckhove, Jan 211

Cinquemani, Eugenio 3

De Francisco, Richard 195
De Las Rivas, Javier 226
Demko, Martin 58
Draelants, Delphine 211

Fan, Chuchu 195
Fitime, Louis Fippo 75

Grosu, Radu 195
Guet, Calin C. 173
Guziolowski, Carito 75
Gyori, Benjamin M. 37, 96

Islam, Md. Ariful 195

Kłosiewicz, Przemysław 211
Krüger, Thilo 141
Kwiatkowska, Marta 119

Liu, Bing 96

Mereacre, Alexandru 119
Mikeev, Linar 141

Milios, Dimitrios 156
Mitra, Sayan 195

Nenzi, Laura 156

Paoletti, Nicola 119
Pastva, Samuel 58
Patanè, Andrea 119
Paul, Soumya 96
Petrov, Tatjana 173

Ramanathan, R. 37, 96
Roux, Olivier 75

Šafránek, David 58
Sanguinetti, Guido 156
Santos-García, Gustavo 226
Schuster, Christian 75
Siebert, Heike 20
Smolka, Scott A. 195
Streck, Adam 20

Talcott, Carolyn 226
Thiagarajan, P.S. 37, 96
Thobe, Kirsten 20

Vanroose, Wim 211

Wolf, Verena 141
Wong, Weng-Fai 37

Zhang, Yan 37
Zhou, Jun 37

Printed in the United States
By Bookmasters

Printed in the United States
By Bookmasters